AF761211

Contents

Preface

The human eye is a highly specialized organ, exquisitely sensitive and functionally complex. Its ability to transmit visual information makes it a vital window to the world. Yet this very complexity, comprising multiple anatomical and physiological barriers, presents formidable challenges in delivering drugs effectively to ocular tissues. For decades, the field of ophthalmic pharmacology has been constrained by limitations in traditional drug delivery systems, most notably, poor bioavailability, rapid drug clearance, non-specific targeting, and limited drug retention. These limitations have hindered the successful treatment of numerous eye disorders, including glaucoma, diabetic retinopathy, macular degeneration, and infectious and inflammatory diseases affecting both the anterior and posterior segments of the eye. In recent years, nanotechnology has emerged as a powerful and promising strategy to overcome these barriers. Nanotechnology-based systems offer a paradigm shift in ocular pharmacotherapy by enabling targeted, sustained, and patient-friendly drug delivery. From enhancing solubility and permeability to minimizing side effects and dosing frequency, nanomedicine has the potential to transform the landscape of ophthalmic care.

The book, *Advances in Nanotechnology for Ocular Drug Delivery: Innovations and Future Directions*, is designed to serve as a comprehensive and authoritative reference for researchers, pharmaceutical scientists, clinicians, postgraduate students, and industry professionals seeking a thorough understanding of this rapidly evolving field. This book is structured into 22 meticulously curated chapters, each focusing on a critical aspect of ocular drug delivery through nanotechnological innovations. Together, they span a wide range of systems and strategies, from polymeric nanoparticles and lipid-based nanocarriers to micelles, dendrimers, nanoemulsions, niosomes, and inorganic nanoparticles, shedding light on both their fundamental properties and their translational potential.

The opening chapters provide foundational insights into the structure and physiology of the eye, common ocular pathologies, and the pharmacokinetic hurdles that limit drug availability and efficacy. These introductory discussions lay the groundwork for understanding why conventional ophthalmic therapies are often inadequate and how nanocarriers have the potential to revolutionize treatment strategies. Subsequent chapters delve into specific nanocarrier systems and their application in ocular drug delivery. Polymeric nanoparticles, for example, offer advantages such as biocompatibility, tailored release kinetics, and surface modifiability, making them highly adaptable platforms for drug and gene delivery. Similarly, solid lipid nanoparticles (SLNs) and nanostructured lipid carriers (NLCs) offer favorable characteristics like enhanced encapsulation efficiency, stability, and controlled release, all essential features for improving therapeutic outcomes in ocular diseases. Special emphasis is placed on smart and stimuli-responsive nanocarriers, which represent the next frontier in personalized ophthalmic therapy. These systems respond to physiological or environmental cues, such as pH, temperature, or enzymes, to release drugs precisely where and when they are needed. Such intelligent delivery systems significantly reduce off-target effects and increase therapeutic precision, offering real promise for conditions that demand long-term, sustained treatment. In addition to material science and formulation design, the book covers advanced strategies such as ocular implants, iontophoresis, microneedles, and drug-eluting contact lenses. These approaches aim to enhance patient compliance and treatment convenience by minimizing the invasiveness and frequency of administration, which are often drawbacks of traditional intravitreal injections and eye drops. Notably, this book also dedicates several chapters to the application of nanotechnology in treating specific ocular diseases. Topics include the role of nanocarriers in glaucoma management, the use of biologics for posterior segment diseases, and the treatment of retinal disorders using nanostructured therapies. These disease-focused chapters provide a clinically relevant perspective and bridge the gap between laboratory innovation and real-world therapeutic needs. A distinctive feature of this book is its discussion of translational science and regulatory landscapes. While nanocarrier technologies offer immense promise, their successful clinical implementation demands rigorous evaluation in terms of safety, reproducibility, scale-up, and compliance with regulatory frameworks. The book explores these aspects in detail, offering guidance on the clinical translation and commercialization of ocular nanomedicine. Case studies of FDA-approved ocular nanotechnologies and those in clinical trials are presented to illustrate the path from benchtop discovery to bedside application. The final chapters of this book look toward the future, identifying key trends and potential breakthroughs on the horizon, including the integration of artificial intelligence in ocular formulation design, nanobiosensors for diagnostic applications, and the convergence of nanotechnology with gene editing tools like CRISPR-Cas9.

Through this book, we aim not only to disseminate knowledge but also to inspire continued innovation and research in ocular nanomedicine. We believe this book will be a valuable tool for both seasoned professionals and aspiring researchers in understanding and advancing the frontiers of this field. We express our heartfelt gratitude to all the chapter contributors who generously shared their expertise and time to enrich this book. Their scholarly contributions have given shape and substance to this ambitious undertaking. We also thank the editorial and production teams for their unwavering support throughout the development of this book. In closing, we hope this book will empower its readers to explore new ideas, foster interdisciplinary collaboration, and contribute meaningfully to the advancement of ocular health through nanotechnology.

Anjali Sharma
Ajmer Singh Grewal
Geeta Deswal
Devkant Sharma
Editors

About the Editors

Dr. Anjali Sharma is presently working as an associate professor (pharmaceutics) at Guru Gobind Singh College of Pharmacy, Yamuna Nagar, Haryana, India. She holds a B. Pharm. from Pt. B.D. Sharma University of Health Sciences, Rohtak, Haryana (India); an M. Pharm in Pharmaceutics from Uttarakhand Technical University, Dehradun, Uttrakhand, India (78.6%), and a Ph.D. from Shobhit University, Gangoh, Uttar Pradesh, India (2021). With over ten years of teaching experience, including five years at the postgraduate level, she has mentored 10 M. Pharm (02 ongoing) and 03 Ph.D. (ongoing) students. Her research focuses on nanotechnology-based drug delivery systems, enhancing drug efficacy and patient compliance. She has authored 50 research/review articles, 45 book chapters, 3 books, and holds 12 design patents and two copyrights. A committed educator and researcher, she is a Life Member of APTI and IPGA. Her excellence has been recognized through prestigious awards, including the Young Scientist Award (2021), Research Excellence Award (2023 & 2024), Global Faculty Award (2023), and Inspiring Academician Award (2024) by IPGA Uttarakhand State Branch. She has also been honored by the Bureau of Indian Standards (BIS) and other esteemed organizations for her contributions to science, education, and society.

Dr. Ajmer Singh Grewal, Professor (Pharmaceutical Chemistry), Guru Gobind Singh College of Pharmacy, Yamuna Nagar, Haryana (India), is having academic and research experience of 13+ years in medicinal chemistry. He completed his PhD in pharmaceutical sciences in the subject area of computational and medicinal chemistry from I.K. Gujral Punjab Technical University, Jalandhar, Punjab (India) in 2020. His present areas of research include structure-based drug design and medicinal chemistry of MMP2 and MMP9 inhibitors as anticancer agents, antibacterial drug resistance (NDM-1), GK activators as anti-diabetic agents, AR inhibitors for diabetic complications, PPAR agonists for metabolic syndrome, multi-potent anti-Alzheimer's agents, selective PDE4 & 7 inhibitors as anti-inflammatory agents and environmental chemistry. He owns to his credit 120 publications in national and international journals of repute, 25 book chapters, 20 books, 6 patents and 3 designs. He serves as a reviewer for reputed international journals published by Elsevier, Springer, Wiley, SAGE, MDPI, Bentham Science, and Frontiers. He is a life member of the Indian Society for Technical Education (ISTE). He has worked on various sophisticated analytical instruments including UV-Vis and FTIR spectrophotometer, flash chromatography, and HPLC; and also, on various molecular modeling softwares including ChemSketch, ChemDraw and Marvin Sketch; Open Babel and Frog2 for structure optimization; PyMOL for molecular graphics, Glide (Schrodinger), AutoDock Tools, AutoDock Vina, Discovery Studio (Dassault Systemes BIOVIA), Omega and Open Eye for *in silico* molecular docking studies.

Dr. Geeta Deswal is presently working as a professor (pharmacognosy) at Guru Gobind Singh College of Pharmacy, Yamuna Nagar, Haryana, India. She has obtained a PhD from M.M. University, Mullana, Haryana (India) in quality qudit of herbal formulations. She is an expert in bioactivity-guided extraction and isolation of phytoconstituents, standardization of drugs and herbal formulations, and phytochemical screening of medicinal plants and plant tissue culture. She is also working on the formulation and evaluation of herbal products as antimicrobial agents. She has worked on HPLC, UV spectrophotometer, Soxhlet Apparatus, Clevenger Apparatus, Sonicator, Rota evaporator, and Thin Layer and Column Chromatography. She has to her credit 50 publications in national and international journals of repute, 20 book chapters, and 15 books. She received the Young Achiever Award on Occasion of Pharma Vision 2K25 by Shree Balaji Books (New Delhi), and Innovational Ambassador by MHRD Innovational Cell, AICTE, New Delhi (India).

Dr. Devkant Sharma is currently an associate professor at Ch. Devi Lal College of Pharmacy, Jagadhri, Yamuna Nagar, Haryana (India). He holds a Bachelor of Pharmacy from Pt. B.D. Sharma University of Health Sciences, Rohtak, Haryana (India); a Master of Pharmacy (Pharmaceutics) from Punjab Technical University, Jalandhar, Punjab (India), and a Ph.D. in

Pharmaceutical Sciences from Shobhit University, Gangoh (Uttar Pradesh, India). With over 12 years of teaching experience, Dr. Sharma has developed deep expertise in lipid-based drug delivery systems, which form the core of his research interests. He is presently supervising one M. Pharm. and two Ph.D. scholars, contributing actively to academic mentorship and research development. Received the Pharma Excellence Award (2024) by IPGA Uttarakhand State Branch. Dr. Sharma has an impressive scholarly output, with 30 research and review articles, 40 book chapters, and three authored books to his credit. He also holds ten design patents and five copyrights, reflecting his commitment to innovation in pharmaceutical sciences.

List of Contributors

Abhinav Amar
Shaheed Baba Ajit Singh and Jujhar Singh Memorial College of Pharmacy, Bela, Ropar, 140111, Punjab, India

Abhishek Chandola
AVIPS Shobhit University, Gangoh, Saharanpur, Uttar Pradesh, 247341, India

Adnan Khan
Department of Microbiology, Faculty of Biological Sciences, Quaid-i-Azam University, Islamabad, Pakistan

Ajay Bilandi
Bhai Gurdas College of Pharmacy, Sangrur, Punjab, India

Akash Jain
M. M. College of Pharmacy, M. M. (Deemed to be University), Mullana, Ambala, Haryana, India

Ambika Nand Jha
School of Pharmacy, Sharda University, Greater Noida, Uttar Pradesh, India

Anjali Sharma
Guru Gobind Singh College of Pharmacy, Yamuna Nagar, Haryana, India

Anshuman Sinha
Department of Paediatrics, Ann & Robert H. Lurie Children's Hospital, Chicago, IL, USA; Feinberg School of Medicine, Northwestern University, Chicago, IL, USA

Arun Kumar Singh
Faculty of Pharmaceutical Sciences, Amrapali University, Haldwani, Uttarakhand, India

Ashwani K. Dhingra
Global Research Institute of Pharmacy, Radaur, Yamuna Nagar, Haryana, India

Athrv Arora
Chitkara College of Pharmacy, Chitkara University, Punjab, India

Avichal Kumar
Department of Pharmaceutics, KLE College of Pharmacy, Bengaluru, Karnataka, India

Balraj Saini
Chitkara College of Pharmacy, Chitkara University, Punjab, India

Bhawna Chopra
Guru Gobind Singh College of Pharmacy, Yamuna Nagar, Haryana, India

Bhawna Garg
Adesh Institute of Pharmacy and Biomedical Sciences, Adesh University, Bathinda, Punjab, India

Bhawna Sharma
Department of Pharmaceutics, Dr. K.N. MODI Institute of Pharmaceutical Education and Research, Modinagar, Ghaziabad, Uttar Pradesh, India

Brijesh Duvey
Ch. Devi Lal College of Pharmacy, Jagadhri, Haryana, India

Chaitanya Vinayak Narayan
I.T.S College of Pharmacy, Ghaziabad, Uttar Pradesh, India

Chirag Jain
Amity Institute of Pharmacy, Amity University, Noida, Uttar Pradesh, India

Darsh Gautam
Department of Pharmaceutics, Gautam College of Pharmacy, Hamirpur, Himachal Pradesh, India

Deepika Rani
AVIPS, Shobhit University, Gangoh, Saharanpur, Uttar Pradesh, 247341, India

Deepti Aggrawal
HR Institute of Pharmacy, HRIT University, 8th Km Stone, Delhi-Meerut Road, Ghaziabad, 201003, Uttar Pradesh, India

Devesh U. Kapoor
Dr. Dayaram Patel Pharmacy College, Bardoli, 394601, Gujarat, India

Drashti Dave
Department of Pharmaceutics, Sigma Institute of Pharmacy, Sigma University, Ajwa Nimeta Road, Bakrol, Vadodara, Gujarat, India

Gaurav Mude
K. C. Bajaj College of Pharmacy and Research, Jaripatka, Nagpur, Maharashtra, India, Institute of Pharmaceutical Sciences, SAGE University, Indore, Madhya Pradesh, India

Hemlata Kaurav
School of Pharmaceutical Sciences, Shoolini University of Biotechnology and Management Sciences, Bajhol, Solan, Himachal Pradesh, 173229, India

Himanshu Sharma
Chitkara College of Pharmacy, Chitkara University, Punjab, India

Humaira Nawaz
Department of Microbiology, Faculty of Biological Sciences, Quaid-i-Azam University, Islamabad, Pakistan

Indu Singh
Amity Institute of Pharmacy, Amity University, Noida, Uttar Pradesh, India

Iti Chauhan
Department of Pharmaceutics, I.T.S College of Pharmacy, Ghaziabad, Uttar Pradesh, India

Jasmine Chaudhary
M. M. College of Pharmacy, M. M. (Deemed to be University), Mullana, Ambala, Haryana, India

Ketan Shah
Department of Quality Assurance, Shree Naranjibhai Lalbhai Patel College of Pharmacy, Umrakh, 394345, Gujarat, India

Khurshid Iqbal
Molvi Ismail Mehta College of Pharmacy, Ambad, Dist. Jalna, 431204, Maharashtra, India

Kumar Guarve
Guru Gobind Singh College of Pharmacy, Yamuna Nagar, Haryana, India

Laxmi Gharti
School of Pharmaceutical Sciences, Shoolini University of Biotechnology and Management Sciences, Bajhol, Solan, Himachal Pradesh, 173229, India

Madhukar Garg
Chitkara College of Pharmacy, Chitkara University, Punjab, India

Manish Dhall
College of Pharmacy PGIMS (SDPGIPS). Pt B.D. Sharma University of Health Sciences, Rohtak, Haryana, India

Mansi
Himalayan Institute of Pharmacy, Kalaamb, Kala-Amb, Sirmour, Himachal Pradesh, India

Minky Mukhija
Ch. Devi Lal College of Pharmacy, Jagadhri, Yamuna Nagar, Haryana, India

Mohammad Tauqeer Sheikh
K. C. Bajaj College of Pharmacy and Research, Jaripatka, Nagpur, Maharashtra, India

Mohd Yasir
Department of Pharmacy, College of Health Sciences, Arsi University, Asella, Ethiopia

Mona Piplani
Maharaja Agrasen School of Pharmacy, Maharaja Agrasen University, Kalujhanda, Atal Shiksha Kunj, Dist. Solan, Baddi, Himachal Pradesh, 174103, India

Muskan
Guru Gobind Singh College of Pharmacy, Yamunanagar, Haryana, India

Nadia Ilyas
Department of Microbiology, Faculty of Biological Sciences, Quaid-i-Azam University, Islamabad, Pakistan

Navin Chandra Pant
Faculty of Pharmaceutical Sciences, Amrapali University, Haldwani, Uttarakhand, India

Himani Bajaj
AVIPS, Shobhit University, Gangoh, Saharanpur, Uttar Pradesh, India

Neelam Sharma
Amar Shaheed Baba Ajit Singh Jujhar Singh Memorial College of Pharmacy, Bela, Dist. Ropar, 140111, Punjab, India

Neelam Singh
HR Institute of Pharmacy, HRIT University, 8th Km Stone, Delhi-Meerut Road, Ghaziabad, 201003, Uttar Pradesh, India

Nikhil
M. M. College of Pharmacy, M. M. (Deemed to be University), Mullana, Ambala, Haryana, India

Nimbagal Raghavendra Naveen
Department of Pharmaceutics, Sri Adichunchanagiri College of Pharmacy, B. G. Nagara, Mandya, India

Nitasha Chauhan
Department of Pharmaceutical Sciences and Technology, Maharaja Ranjit Singh Punjab Technical University (MRSPTU), Bathinda, Punjab, India

Parul Gupta
Guru Gobind Singh College of Pharmacy, Yamunanagar, Haryana, India

Pooja Mittal
GITAM School of Pharmacy, GITAM (Deemed to be) University, Rudraram, Patancheru, Sangareddy District, Hyderabad, Telangana, India

Poonam Talwan
Department of Pharmaceutical Chemistry, Gautam College of Pharmacy, Hamirpur, Himachal Pradesh, India

Prabhjot Kaur
Guru Gobind Singh College of Pharmacy, Yamuna Nagar, Haryana, India

Prakash Goudanavar
Department of Pharmaceutics, Sri Adichunchanagiri College of Pharmacy, B. G. Nagara, Mandya, India

Prerna Mehta
Department of Biotechnology, GD Rungta College of Science & Technology, Bhilai- 490024, Chhattisgarh, India

Prerna Sharma
Shri Krishna Ayush University, Kurukshetra, Haryana, India

Punam Gaba
Amar Shaheed Baba Ajit Singh Jujhar Singh Memorial College of Pharmacy, Bela, Dist. Ropar, 140111, Punjab, India; Maharaja Agrasen School of Pharmacy, Maharaja Agrasen University, Kalujhanda, Atal Shiksha Kunj, Dist. Solan, Baddi, Himachal Pradesh, 174103, India

Rahul Kumar Sharma
Amar Shaheed Baba Ajit Singh Jujhar Singh Memorial College of Pharmacy, Bela, Dist. Ropar, 140111, Punjab, India

Ram Babu Sharma
Himalayan Institute of Pharmacy, Kalaamb, Kala-Amb, Sirmour, Himachal Pradesh, India

Ranjit Singh
AVIPS, Shobhit University, Gangoh, Saharanpur, Uttar Pradesh, 247341, India

Reenu Chauhan
HR Institute of Pharmacy, HRIT University, 8th Km Stone, Delhi-Meerut Road, Ghaziabad, 201003, Uttar Pradesh, India

Reeti Shah
Department of Pharmaceutical Sciences, University of Surrey, Guildford, Surrey GU2 7XH, England, United Kingdom

Ritu Chauhan
HR Institute of Pharmacy, HRIT University, 8th Km Stone, Delhi-Meerut Road, Ghaziabad, 201003, Uttar Pradesh, India

Riya Mehta
Department of Pharmaceutics, Shree Naranjibhai Lalbhai Patel College of Pharmacy, Umrakh, 394345, Gujarat, India

Rohit Pandey
Department of Pharmaceutics, Dr. K.N. MODI Institute of Pharmaceutical Education and Research, Modinagar, Ghaziabad, Uttar Pradesh, India

Saloni Sharma
Ch. Devi Lal College of Pharmacy, Jagadhri, Yamuna Nagar, Haryana, India

Samantha
Himalayan Institute of Pharmacy, Kalaamb, Kala-Amb, Sirmour, Himachal Pradesh, India

Samiullah Khan
Department of Microbiology, Faculty of Biological Sciences, Quaid-i-Azam University, Islamabad, Pakistan

Samrat Chauhan
Chitkara College of Pharmacy, Chitkara University, Punjab, India

Sanjay Kumar
Department of Pharmacognosy, Gautam College of Pharmacy, Hamirpur, Himachal Pradesh, India

Sanjay Nagdev
Shri. Prakashchand Jain College of Pharmacy and Research, Palaskheda, Jamner, Maharashtra, India

Santhosh Fattepur
School of Pharmacy, Management and Science University, Seksyen 13, Shah Alam, Selangor, Malaysia

Sapna Kumari
Shiva Institute of Pharmacy, Bilaspur, Himachal Pradesh, India

Shahzar Khan
Department of Microbiology, Faculty of Biological Sciences, Quaid-i-Azam University, Islamabad, Pakistan

Shailesh Sharma
Amar Shaheed Baba Ajit Singh Jujhar Singh Memorial College of Pharmacy, Bela, Dist. Ropar, 140111, Punjab, India

Shailesh Sharma
Amar Shaheed Baba Ajit Singh Jujhar Singh Memorial College of Pharmacy, Bela, Dist. Ropar, 140111, Punjab, India

Shantilal Singune
Institute of Pharmaceutical Sciences, SAGE University, Indore, Madhya Pradesh, India

Shikha Baghel Chauhan
Amity Institute of Pharmacy, Amity University, Noida, Uttar Pradesh, India

Shithin Ann Varghese
Department of Pharmaceutics, Sri Adichunchanagiri College of Pharmacy, B. G. Nagara, Mandya, India

Shivanki Joshi
Guru Gobind Singh College of Pharmacy, Yamuna Nagar, Haryana, India

Shruti
Amar Shaheed Baba Ajit Singh Jujhar Singh Memorial College of Pharmacy, Bela, Dist. Ropar, 140111, Punjab, India

Srikruthi Kunigal Sridhar
Department of Pharmaceutics, Sri Adichunchanagiri College of Pharmacy, B. G. Nagara, Mandya, India

Sruthi Laakshmi M.
Department of Pharmaceutics, SRM College of Pharmacy, SRMIST, Kattankulathur, Tamil Nadu, India

Swati
Himalayan Institute of Pharmacy, Kalaamb, Kala-Amb, Sirmour, Himachal Pradesh, India

Taufik Mulla
Institute of Pharmaceutical Sciences, Faculty of Pharmacy, Parul University, P.O. Limda, Waghodia, 391760, Dist. Vadodara, Gujarat, India

Ujjwal Nautiyal
Himalayan School of Pharmaceutical Sciences, Swami Rama Himalayan University, Swami Rama Nagar, Jollygrant, Uttarakhand, 248016, India

Umesh Telrandhe
Datta Meghe College of Pharmacy, DMIHER (DU), Sawangi, Wardha, Maharashtra, India

Vaishali Kamboj
M. M. College of Pharmacy, M. M. (Deemed to be University), Mullana, Ambala, Haryana, India

Vaseem Fateh
Burman Group of Institutions Roorkee, Haridwar, Uttarakhand, India

Vishnu Mittal
Adesh Institute of Pharmacy and Biomedical Sciences, Adesh University, Bathinda, Punjab, India

Yash Dudhwala
Department of Pharmaceutics, Shree Naranjibhai Lalbhai Patel College of Pharmacy, Umrakh, 394345, Gujarat, India

Yash Kalra
Amity Institute of Pharmacy, Amity University, Noida, Uttar Pradesh, India

Chapter 1

Introduction to Ocular Drug Delivery Systems

Vishnu Mittal[1]
Anjali Sharma[2]
Samrat Chauhan[3,*]
Saloni Sharma[4]
and Manish Dhall[5]
[1]Department of Pharmaceutics, Adesh Institute of Pharmacy and Biomedical Sciences, Adesh University, Bathinda, Punjab, India
[2]Guru Gobind Singh College of Pharmacy, Yamuna Nagar, Haryana, India
[3]Chitkara College of Pharmacy, Chitkara University, Punjab, India
[4]Ch. Devi Lal College of Pharmacy, Jagadhri, Yamuna Nagar, Haryana, India
[5]College of Pharmacy PGIMS (SDPGIPS). Pt. B.D. Sharma University of Health Sciences, Rohtak, Haryana, India

Abstract

An ocular drug delivery system is designed to administer medications to the eye for the treatment of various vision-related disorders. However, the unique anatomical and physiological barriers of the eye, both static and dynamic, pose significant challenges to effective drug delivery, limiting therapeutic efficacy and drug bioavailability. Traditional ocular dosage forms often fail to maintain sustained therapeutic levels due to rapid drug clearance and limited permeability. This chapter reviews recent advancements in ocular drug delivery, focusing on innovative strategies developed to overcome these challenges and enhance drug absorption. Techniques such as nanoparticle-based carriers, in situ gels, microemulsions, nanosuspensions, dendrimers, iontophoresis, and

* Corresponding Author's Email: samrat.chauhan11@gmail.com

In: Advances in Nanotechnology for Ocular Drug Delivery
Editors: Anjali Sharma, Ajmer Singh Grewal, Geeta Deswal et al.
ISBN: 979-8-89530-814-1
© 2025 Nova Science Publishers, Inc.

periocular injections have shown promise in increasing precorneal residence time, improving patient comfort, and achieving controlled drug release. These novel systems offer significant advantages over conventional approaches by optimizing pharmacokinetics and minimizing side effects. The chapter highlights both the potential and limitations of these emerging technologies while emphasizing the ongoing need for innovation in the field.

Keywords: controlled release, nanoparticles, ocular drug delivery, ophthalmic formulations, patient-friendly drug delivery, precorneal retention

Abbreviations

Apa-BSA-NPs: Apatinib-loaded Bovine Serum Albumin Nanoparticles
CMC: Critical Micelle Concentration
DES: Dry Eye Syndrome
EPOβ: Epoetin Beta
FA: Ferulic Acid
HA: Hyaluronic Acid
HuR: Human Antigen R
Mab: Monoclonal Antibody
miRNA: Micro RNA
MNPs: Magnetic Nanoparticles
MPDA: Mesoporous Polydopamine
mRNA: Messenger RNA
MSNs: Mesoporous Silica Nanoparticles
NPs: Nanoparticles
OCT: Octreotide
PEG: Polyethylene Glycol
siRNA: Small Interfering RNA
VEGF: Vascular Endothelial Growth Factor

Introduction

The eye is composed of several components that interact with one another in various ways. The anterior region & posterior region are two primary components of eye (Willoughby et al., 2010). About one-third of space is occupied by anterior portion of the eye, with the remaining portion being occupied by posterior section. The front portion contains the eye, cornea, watery fluid, pupil, ciliary body, and lens. The most popular non-invasive method of administering medication for conditions affecting the head is topical instillation. Nine out of ten ophthalmic formulae marketed are conventional dosage forms, including eye drops (Villate-Beitia et al., 2015). Patient consent and dosage simplicity may be the cause. However, direct drop delivery results in relatively limited ocular absorption. Eye turnover, nasolacrimal drainage, automatic blinking, & ocular static & dynamic barriers are only a few of physiological & physical limitations that hinder increased ocular medication absorption (Srivastava et al., 2023). As a result, the lower eye tissues get less than 5% of the directly delivered dosage (Kang-Mieler et al., 2014). The most popular and highly recommended method of medicine administration for treating conditions affecting the posterior eye is intravitreal injection. However, a number of adverse consequences, including endophthalmitis, bleeding, retinal detachment, and poor patient tolerance, are brought on by the need of frequent eye punctures with intravitreal injections (Lam et al., 2021, Wang et al., 2024). Static barriers, such as the choroid, retinal pigment epithelium (RPE), and sclera, as well as dynamic barriers, such as blood flow in the conjunctiva and choroid and lymphatic flow in the conjunctiva and episclera, are examples of ocular obstacles to transscleral drug transfer. To overcome the barriers to ocular drug delivery and improve ocular bioavailability, numerous conventional and novel drug delivery techniques have been developed, and in situ thermosensitive gels for the aforementioned ocular diseases. Ocular medication transport is crucial for treating various eye disorders since it minimizes systemic side effects while guaranteeing targeted and sustained drug release (Tsung et al., 2023). The eye's distinct chemistry and structure, which includes barriers like the ocular epithelium and blood-retinal barrier, significantly impairs effective medication availability. Advanced techniques for delivering eye medications, such as liposomes, in situ gels, and nanoparticles, enhance absorption, prolong drug retention, and increase patient compliance (Biswas et al., 2023). The administration of ocular medications is complicated by the complex structure and protective layers of the eye (Chandel & Kandav, 2024). Due to their short drug residence time, rapid tear change, and poor absorption, traditional dosage forms such eye drops must be used often. Systemic side effects and patient non-compliance further hinder effective eye therapy. To solve these issues and improve patient compliance and treatment effectiveness, better pharmaceutical delivery methods are required. Examples of nanocarriers that allow for controlled and targeted drug release include liposomes, nanoparticles, and micelles (Chamundeeswari et al., 2019).

Anatomical and Physiological Barriers in Ocular Drug Delivery

Tear Film Barrier

The tear film is the first line of protection against external agents because it rapidly eliminates foreign substances, including medications, via reflexive blinking and tear turnover (Sarkar et al., 2019). This barrier severely limits medication absorption since only a small percentage of the recommended dose stays in contact with the eye surface long enough to penetrate.

Corneal and Conjunctival Barriers

The ocular epithelium functions as a strong permeability barrier due to its lipophilia, preventing the entry of hydrophilic drugs. On the other hand, the stroma, which is mostly made up of hydrophilic materials, limits the spread of lipophilic drugs (Varela-Fernández et al., 2020).

Blood-Aqueous and Blood-Retinal Barriers

The BAB, which keeps hydrophilic drugs out of the watery humor, is produced by the ciliary body and ocular endothelial cells (Suri et al., 2020). Because BRB, which is made up of retinal pigment epithelium & retinal capillary endothelium, blocks systemic drugs from reaching the retina, treating problems of the posterior segment is particularly difficult. New delivery methods, including as intravitreal injections, plastic devices, and nanocarriers, are being developed to overcome these limiting obstacles.

Metabolic Barriers and Clearance Mechanisms

Numerous biochemical enzymes present in ocular tissues, such as esterases, oxidoreductases, and proteases, may quickly degrade drug molecules before they can reach their intended targets (Benedetti et al., 2009). Additionally, clearance processes including capillary and lacrimal drainage further limit the availability of medications, which lowers the effectiveness of treatment. Enzyme inhibitors, drug binding methods, and sustained-release delivery systems like hydrogels and biodegradable devices are some of the strategies being investigated to address these metabolic and clearance problems, ensure longer medication activity, and enhance treatment results (Kamaly et al., 2016).

Conventional Ocular Drug Delivery Systems

It is usually advised to provide drugs by topical drop instillation into the lower precorneal area when the patient is cooperative. Effective ocular pharmaceutical delivery using eye drops, however, requires a lengthy drug cornea contact time and significant corneal penetration. Numerous attempts have been made to increase precorneal dwell time and corneal penetration. Ocular permeability is increased by prodrugs, cyclodextrins, ion-pair generating agents, and iontophoresis (Willebrand & Ghuman, 2002).

Eye Drops and Solutions

Topical Eye Drops in a Solution or Liquid

Topical drops are the most practical, safe, quick-acting, patient-obedient, and uncomplicated way to provide ocular drugs. An eye drop solution offers a burst of drug absorption after contact drop instillation, but its concentration quickly drops (Gause et al., 2016). A first-order approximation to the rates of drug concentration declines could exist. To improve drug contact duration, penetration, and corneal absorption, several additives, including cyclodextrins, permeability enhancers, and viscosity enhancers, may be added to external eye drops. Viscosity boosters make the formulation more viscous, which improves absorption and the length of the precorneal stay after external drop administration. Penetration chemicals improve ocular transparency by changing the cornea's structure (Pawar et al., 2024). Bile salts, lubricants, surface active agents, and chelating agents were other substances that were studied as possible diffusion boosters. (Patel et al., 2013, Souto et al., 2019).

Ointments and Gels

Ophthalmic ointments are another kind of carrier system designed for direct treatment. A solid (paraffin) and a semisolid oil are combined to create ocular cream. Ointments improve ocular absorption and extend the drug's release (Pal Kaur & Kanwar, 2002). Even if VCM was more active, there was no suitable skin care product on the market. Better ocular tissue permeability of VCM was not expected in a normal eye, but there were few therapeutic advantages of VCM solution in the treatment of ocular disorders (Shifali et al., 2021).

Suspensions and Emulsions

Suspensions are another kind of non-invasive ocular chemical drop medication delivery method. Suspension is the process of dispersing finely split solid API in a liquid solvent with a good binding and spreading agent (Cholkar, 2015). Consequently, it is believed that the perfect particle size produces the best possible drug effect. Many fluid formulations are sold to treat bacterial eye disorders all over the world. The purpose of this formulation was to improve the solution formulation's tissue permeability, tear film quality, and tear film speed.

Emulsions

Using an emulsion-based preparation method has the potential to improve the solubility and absorption of drugs. For the administration of ocular medications, the o/w emulsion is often employed and recommended over the w/o method. Among the reasons are reduced pain and enhanced o/w emulsion ocular tolerance. AzaSite®, Refresh Endura® (a non-medicated ocular emulsion for eye moisture), and RestasiseTM are a few examples of recently introduced ocular emulsions in the United States. Emulsions have been shown in several studies to improve ocular bioavailability, uniform drug release, corneal penetration, and precorneal residence length (Yaqoob Khan et al., 2006).

Limitations of Conventional Systems

Despite a number of difficulties, traditional medication delivery methods for dosage form fitting are highly helpful and simple for people of all ages to utilize. When eye drops are applied to the conjunctiva's inferior fornix, only a little portion of the medication reaches the tissues of the eyeballs, and the majority is either absorbed systemically in the mouth and nasolacrimal duct or washed away with the lachrymal fluid. Low medication absorption from the eye's cell surface is another issue. One of the main causes of blindness is abnormalities with the posterior region of the eye (Bachu et al., 2018).

Role of Nanotechnology in Ocular Drug Delivery

Nanotechnology has transformed eye treatment by making it possible to develop nanoscale carriers that improve drug stability, solubility, and entry (Zhang et al., 2021). Nanocarriers such liposomes, polymeric nanoparticles, micelles, and dendrimers lessen exposure and harm overall by facilitating controlled and targeted drug release. The ability of these nanoparticles to penetrate ocular barriers allows for the efficient delivery of medications to inner eye tissues. Nanotechnology is a viable therapy option for illnesses affecting the anterior and posterior segments since surface alterations like PEGylation and ligand coupling both improve bioadhesion and prolong drug retention (Rai et al., 2016).

Enhancing Ocular Bioavailability

Improving ocular absorption continues to be a major priority because of enzymatic breakdown, frequent tear changes, and other physiological barriers (Gorantla et al., 2020). Techniques including prodrugs, mucoadhesive nanoparticles, permeability enhancers, and bioadhesive polymers have been developed to lengthen the period of pharmaceutical residence and improve eye absorption. Additionally, cutting-edge delivery techniques such in situ gels, ocular inserts, and sustained-release devices can maintain effective drug levels over extended periods of time, decreasing the need for frequent doses and improving patient adherence.

Controlled and Sustained Release Systems

The first controlled medication delivery device for ocular applications, Ocusert, was created in 1975. For a week, it was strong enough to release pilocarpine into the ocular conjunctival sac at a steady pace (Kesav et al., 2021). The process of delivering drugs to the tissues of the eyes in a controlled and prolonged way then started, and it remains a top goal for formulation scientists working on the creation of stronger drugs and biomolecules.

Nanocarrier-Based Approaches in Ocular Therapy

Nanoparticles (Polymeric and Lipid-Based)

Polymeric nanoparticles may be made from synthetic or naturally occurring polymers (Saini et al., 2024). Many retinal drugs are currently not working as intended because of the unique internal structure of the eye as well as the physical and chemical properties of the drugs. These drugs' toxicity was reduced, invasive procedures were avoided, and physiological control was achieved by encapsulating them in polymeric nanoparticles. Ganciclovir, voriconazole, latanoprost, cyclosporin, and dexamethasone are some of these drugs (Abdul-Al et al., 2023).

Liposomes and Niosomes

Liposomes

Phospholipid bilayers have caught the liquid core of liposomes, which are round lipid spheres. Because liposomes are disposable and harmless, they may be used to deliver medications to the eyes (Nimasha Rathnasinghe et al., 2024). Because liposomes can get up close and personal with eye tissues like the lens and conjunctiva, they improve medication absorption. They are distinctive and flexible in terms of eye care because of their facile surface alteration for targeting purposes and delayed release of bioactive at the delivery location. Additionally, liposomes shield the medication from the digestive enzymes found in tears and/or at membrane of eye. These have been investigated as potential delivery systems for hydrophilic and lipophilic medications, including ganciclovir, pilocarpine, ofloxacin, diclofenac sodium, and chloramphenicol (Wadhwa et al., 2009).

Niosomes

When nonionic surfactants self-assemble in a watery medium to generate nanocarriers, they form closed bilayer structures called niosomes, a kind of vesicular system that contains a nonionic surfactant (Patel et al., 2022). Because niosomal systems may increase the absorption and effectiveness of encapsulated medications, researchers have started employing them to treat serious inflammatory illnesses and disorders, including various malignancies.

Dendrimers and Micelles

Dendrimers

Dendrimers are spherical, negatively, positively, or neutrally branching nanostructured polymers. Their net charge comes from the functional groups at the ends of their stems (Bhatia, 2016). Side chain components called "dendrons" and a central unit called the "core," which is the main component, make up these molecules. Because dendrimers may precisely target inflammatory cells without compromising healthy tissue, they have been shown to be an effective medicine delivery method for the treatment of inflammatory eye disorders.

Micelles

Nanomicelles are the most widely used carrier techniques for introducing medications into clear water solutions. These nanomicelles are often made of amphiphilic chemicals. These substances might be surfactants or polymers (Tawfik et al., 2020). The development of nanomicellar-based technology for ocular drug administration is now receiving a lot of attention. The reasons may include their small size, hydrophilic nanomicellar corona that produces water solution, ease of production, and superior drug packing capabilities.

Nanoemulsions and Microemulsions

Nanoemulsions and microemulsions are potential vehicles for the delivery of medications into the eyes because of their ability to improve drug solubility, penetration, and absorption (Souto et al., 2022). These colloidal systems, which consist of water, oil, surfactants, and co-surfactants, allow drugs to cross ocular barriers. With droplet diameters usually between 20 and 200 nm, nanoemulsions provide prolonged drug release and better ocular retention, which may be advantageous for both hydrophilic and lipophilic drugs. Microemulsions, characterized by their very small droplet size (10–100 nm) and thermodynamic stability, further improve drug dispersion and absorption.

In Situ Gels and Hydrogels

"In-situ hydrogels" refers to polymeric liquids that undergo a sol-gel phase transition to form a flexible gel in response to external stimuli (Das et al., 2020). Gelation may be brought on by variations in temperature, pH, and ions as well as by UV light. The primary goal of research studies for ocular administration has been the development of thermosensitive gels that respond to temperature changes (Vashist et al., 2014). Among their primary benefits are longer drug stay durations at the application site, prolonged drug release at the intended site, and the ability to provide many drugs at once. Hydrogels' resistance to flushing and blinking eliminates a major challenge for traditional eye drops, which usually keep eye contact for a few minutes. Hydrogels may be applied externally, intracamerally, or intravitreal.

Implantable and Injectable Drug Delivery Systems

Ocular Inserts and Implants

Ocular implants provide several advantages over traditional treatment approaches and constitute a substantial advancement in ocular medication delivery (Pelusi et al., 2023). These are pharmaceutically designed to deliver medications to a particular spot in the eye in a predetermined amount of time, eliminating the need for repeated injections into the eye tissues, which may exacerbate inflammation and infection. Ocular implants give consistent beneficial drug levels at the site of action and localized distribution, which are the primary advantages over more conventional approaches of administering medications to the eye. Depending on the particular region of the eye, different implants are placed. For anterior segment abnormalities, a subconjunctival device is often used, but for posterior segment issues, intravitreal (IVI) and suprachoroidal therapies are used (Giri et al., 2024). Compared to conventional transscleral devices, they are closer to the site of action, which explains this. Drugs may directly reach the target tissues in the posterior segment when ocular devices are placed intravitreally. Although nonbiodegradable implants have the advantage of providing continuous, regulated medication release for extended periods of time (up too many years), they must be replaced or removed as soon as the drug is depleted.

Intravitreal and Periocular Injections

Through intravitreal and periocular injections, medications may now be administered directly to the back of the eye, avoiding the ocular barriers and systemic circulation. Conjunctivitis, diabetic retinopathy, and age-related macular degeneration are among the eye disorders that may be effectively treated with these focused approaches because they ensure high medication concentrations at the site of action. Intravitreal injections allow for precise targeting of retinal cells and prolonged medication action since the drugs are delivered directly into the vitreous fluid. Peripheral injections, including sub-Tenon, peribulbar, and retrobulbar routes, provide another means to deliver medicine near the eye without entering the globe. These injections improve drug absorption into the posterior region while reducing systemic exposure.

Iontophoresis and Microneedles

Ocular iontophoresis is becoming well-known for its non-invasive method of delivering medication to the eye's anterior and posterior regions. This technique employs migration and electro-osmosis to transfer ionized medications across membranes using low electrical currents. A ground electrode with the opposite charge is placed elsewhere on the body to complete the circuit, while an electrode with the same charge as the medicine is used for distribution. Compared to conventional ocular release techniques like topicals and needles, iontophoresis has a number of advantages.

Microneedle

A novel and somewhat intrusive technique for delivering medication to the posterior eye regions is the microneedle-based approach. Diabetic retinopathy, age-related macular degeneration, and posterior uveitis are examples of posterior eye conditions that may be effectively treated using this technique (Ahmed, 2021). Additionally, this technique could aid in overcoming the blood-reticular barrier and delivering therapeutic medication levels to the retina and choroid. In order to avoid harm to inner eye cells, microneedles are specially designed to penetrate the sclera by only hundreds of microns. These needles aid in inserting a medication or carrier system into the sclera or the "suprachoroidal space," which is the small area between the sclera and the choroid (SCS). Drug entrance into the choroid, neuronal retina, and deeper eye tissues may be facilitated by puncturing the sclera and inserting drug solution or carrier systems into the sclera or SCS. The use of drug-coated microneedle surfaces for retinal drug delivery was investigated. The function, scleral entrance, and intrascleral breakdown of the medication (sulforhodamine) coated on the microneedle surface were investigated using cadaver eyes. The findings demonstrated that the medication coated on the surface was rapidly dissolved in the scleral tissue, exhibiting elevated levels of scleral sulforhodamine inside the microneedle hole.

Clinical Applications and Therapeutic Efficacy

Treatment of Glaucoma

Primary open-angle glaucoma (POAG), angle-closure glaucoma (ACG), normal-tension glaucoma (NTG), and secondary glaucoma are the four kinds of glaucoma that may be distinguished by certain physical characteristics of the eye and underlying processes. POAG is the most prevalent kind of glaucoma; in this condition, the trabecular meshwork's decreased transparency weakens the aqueous humor flow, which raises intraocular pressure. The second most prevalent kind of glaucoma, known as ACG, is characterized by iris bulging, which reduces the eye's anterior chamber angle and blocks the passage of aqueous humor. The primary goals of treatment are to preserve the optic nerve and reduce intraocular pressure. External medications such beta-blockers, alpha-agonists, prostaglandin analogs, and carbonic anhydrase inhibitors are often used as first-line therapy (Chen & Zeng, 2024). For recurring occurrences of glaucoma, surgical procedures such as drainage device surgery and trabeculectomy provide up additional avenues for release. More recently, new treatment options with possibly reduced risk have been made available via less invasive techniques including microinvasive glaucoma surgery (MIGS).

Management of Retinal Disorders

Because it obstructs the retinal venous circulation, retinal vein occlusion (RVO), a dangerous vascular condition of the retina that affects over 16 million people in the US, is the second most common cause of vision loss. The degree and location of venous blockage distinguish the two main kinds of RVO, which are branch retinal vein occlusion (BRVO) and central retinal vein occlusion (CRVO). BRVO affects the lateral retinal vein, while CRVO affects the central retinal vein (Pulido et al., 2016).

Drug Delivery for Anterior and Posterior Segment Diseases

Many eye problems are chronic, necessitating long-term, consistent prescription therapies. Furthermore, eye shields lower the absorption of therapeutic drugs delivered locally to less than 5% (Molokhia et al., 2013). The limitations and shortcomings of traditional drug delivery techniques, such as inadequate drug absorption and poor drug penetration via ocular barriers, may be addressed by recent developments in nanotechnology. Nanocarriers have the potential to delay the effects of pharmaceuticals, shield them from ocular enzymes, and assist them in overcoming hurdles in the eyes via controlled and delayed drug release. This might greatly lower the frequency of doses and increase tissue levels of the medication for more restorative action. Biotech drugs and small chemicals may be delivered into the eyes using colloidal nanocarriers such nanoparticles, nanomicelles, nanowafers, and microneedles. The size range of the nanocarriers is 1-1000 nm. Nanoparticles greater than 10μm may produce abnormal body feelings and eye discomfort. Nanomedicine for ocular drug delivery may prove to be safer and more patient-compliant than traditional eye drops for anterior segment ocular diseases. All of the commercial eye devices that are used to treat retinal diseases are obtrusive. The intravitreal procedure is often used to transport biopharmaceutics to the back of the eye. Endophthalmitis, ocular bleeding, cataracts, glaucoma or elevated intraocular pressure, retinal detachment, and eye inflammation are among the conditions associated with this route (Chandel & Kandav, 2024).

Reducing Side Effects and Enhancing Drug Retention

One of the primary issues in ocular drug administration is minimizing systemic and local side effects while preserving prolonged pharmaceutical storage for continued therapeutic benefits. The risk of pain, toxicity, and patient non-compliance increases when systemic drugs and traditional eye drops need repeated dosing due to inadequate drug absorption. Advanced drug delivery methods are being developed to improve targeted medicine release and prolong ocular stay duration in order to overcome these problems (Morrison & Khutoryanskiy, 2014).

Regulatory Considerations and Safety Aspects

Legal requirements must be strictly followed while developing eye medication delivery systems to guarantee their quality, safety, and efficacy. Cleanliness, stability, biocompatibility, and medication release rates are among the stringent requirements required by regulatory bodies like the European Medicines Agency (EMA) and the Food and medication Administration (FDA) for the approval of novel eye products (SHARMA et al.). Through clinical and experimental research, safety tests look at systemic absorption, toxicity, and ocular irritation.

Future Perspectives and Challenges

Despite tremendous progress, there are still some obstacles to overcome in the administration of eye medications, including as poor absorption, quick drug clearance, and patient noncompliance (Agrahari et al., 2017). Future research aims to develop non-invasive drug delivery methods such gene therapy-based strategies, nanocarriers, and microneedles in order to improve treatment results. Personalized medicine, driven by biopharmaceuticals and AI-based drug optimization, is expected to change how eye care is provided.

Conclusion

Novel technologies targeted at improving medication retention, uptake, and patient adherence has significantly altered ocular drug administration. Overcoming ocular obstacles is still a difficult endeavor that requires ongoing innovation in sustained-release implants, in situ gels, and nanoparticles. Non-invasive techniques are being researched to increase patient comfort and safety, even though intravitreal and periocular injections provide customized delivery. With further research and regulatory advancements, ocular medication delivery is poised for innovations that will transform the way eye diseases are treated, increasing therapeutic efficacy and overall patient outcomes.

References

Abdul-Al, M., Saeinasab, M., Forouzideh, N., Keshel, S. H., Tuinea-Bobe, C., & Sefat, F. (2023). Encapsulation in the ocular system. In *Principles of Biomaterials Encapsulation* (pp. 323-357). Elsevier.

Agrahari, V., Agrahari, V., Mandal, A., Pal, D., & Mitra, A. K. (2017). How are we improving the delivery to back of the eye? Advances and challenges of novel therapeutic approaches. *Expert Opinion on Drug Delivery*, *14*(10), 1145-1162.

Ahmed, T. (2021). *Microneedle Mediated Approaches in Ocular Drug Delivery.* Brac University].

Al-Nimry, S., & Dayah, A. A. (2024). Implication of Microneedle for Ocular Delivery System. *Journal of Drug Delivery Science and Technology*, 105841.

Bachu, R. D., Chowdhury, P., Al-Saedi, Z. H., Karla, P. K., & Boddu, S. H. (2018). Ocular drug delivery barriers role of nanocarriers in the treatment of anterior segment ocular diseases. *Pharmaceutics*, *10*(1), 28.

Benedetti, M. S., Whomsley, R., Poggesi, I., Cawello, W., Mathy, F.-X., Delporte, M.-L., Papeleu, P., & Watelet, J.-B. (2009). Drug metabolism and pharmacokinetics. *Drug Metabolism Reviews*, *41*(3), 344-390.

Bhatia, S. (2016). Nanoparticles types, classification, characterization, fabrication methods and drug delivery applications. In: *Natural polymer drug delivery systems: nanoparticles, plants, and algae*, Springer, Cham., pp. 33-93.

Biswas, A., Choudhury, A. D., Bisen, A. C., Agrawal, S., Sanap, S. N., Verma, S. K., Mishra, A., Kumar, S., & Bhatta, R. S. (2023). Trends in formulation approaches for sustained drug delivery to the posterior segment of the eye. *AAPS PharmSciTech*, *24*(8), 217.

Chamundeeswari, M., Jeslin, J., & Verma, M. L. (2019). Nanocarriers for drug delivery applications. *Environmental Chemistry Letters*, *17*, 849-865.

Chandel, A., & Kandav, G. (2024). Insights into ocular therapeutics: a comprehensive review of anatomy, barriers, diseases and nanoscale formulations for targeted drug delivery. *Journal of Drug Delivery Science and Technology*, 105785.

Chen, Y.-F., & Zeng, W. (2024). Mechanism of selective laser trabeculoplasty: a systemic review. *International Journal of Ophthalmology*, *17*(5), 963.

Cholkar, K. (2015). *Topical clear aqueous Nanomicellar formulations for anterior and posterior ocular drug delivery*. University of Missouri-Kansas City.

Das, B., Chattopadhyay, D., & Rana, D. (2020). The gamut of perspectives, challenges, and recent trends for in situ hydrogels: A smart ophthalmic drug delivery vehicle. *Biomaterials science*, *8*(17), 4665-4691.

Gause, S., Hsu, K.-H., Shafor, C., Dixon, P., Powell, K. C., & Chauhan, A. (2016). Mechanistic modeling of ophthalmic drug delivery to the anterior chamber by eye drops and contact lenses. *Advances in Colloid and Interface Science*, *233*, 139-154.

Giri, B. R., Jakka, D., Sandoval, M. A., Kulkarni, V. R., & Bao, Q. (2024). Advancements in Ocular Therapy: A Review of Emerging Drug Delivery Approaches and Pharmaceutical Technologies. *Pharmaceutics*, *16*(10), 1325.

Gorantla, S., Rapalli, V. K., Waghule, T., Singh, P. P., Dubey, S. K., Saha, R. N., & Singhvi, G. (2020). Nanocarriers for ocular drug delivery: Current status and translational opportunity. *RSC advances*, *10*(46), 27835-27855.

Kamaly, N., Yameen, B., Wu, J., & Farokhzad, O. C. (2016). Degradable controlled-release polymers and polymeric nanoparticles: mechanisms of controlling drug release. *Chemical reviews*, *116*(4), 2602-2663.

Kang-Mieler, J. J., Osswald, C. R., & Mieler, W. F. (2014). Advances in ocular drug delivery: emphasis on the posterior segment. *Expert opinion on drug delivery*, *11*(10), 1647-1660.

Kesav, N. P., Young, C. E. C., Ertel, M. K., Seibold, L. K., & Kahook, M. Y. (2021). Sustained-release drug delivery systems for the treatment of glaucoma. *International Journal of Ophthalmology*, *14*(1), 148.

Lam, L. A., Mehta, S., Lad, E. M., Emerson, G. G., Jumper, J. M., Awh, C. C., & Services, T. F. o. I. I. S. (2021). Intravitreal injection therapy: current techniques and supplemental services. *Journal of VitreoRetinal Diseases*, *5*(5), 438-447.

Molokhia, S. A., Thomas, S. C., Garff, K. J., Mandell, K. J., & Wirostko, B. M. (2013). Anterior eye segment drug delivery systems: current treatments and future challenges. *Journal of Ocular Pharmacology and Therapeutics*, *29*(2), 92-105.

Morrison, P. W., & Khutoryanskiy, V. V. (2014). Advances in ophthalmic drug delivery. *Therapeutic delivery*, *5*(12), 1297-1315.

Murthy, R. Nano lipid Particles-A New Generation of Solid Lipid Drug Carriers in Drug Delivery. *Structure*, *2*(12), 15-17.

Nimasha Rathnasinghe, K., Rajapakshe, B., Awanthi De Silva, R., Nilmini, A., Tissera, N. D., Wijesena, R. N., & Priyadarshana, G. (2024). Current Trends on Unique Features and Role of Nanomaterials in Personal Care Products. *Cosmetics*, 11(5), 152.

Pal Kaur, I., & Kanwar, M. (2002). Ocular preparations: the formulation approach. *Drug development and industrial pharmacy*, *28*(5), 473-493.

Patel, A., Cholkar, K., Agrahari, V., & Mitra, A. K. (2013). Ocular drug delivery systems: An overview. *World journal of pharmacology*, *2*(2), 47.

Patel, P., Parashar, A. K., Kaurav, M., Yadav, K., Singh, D., Gupta, G., & Kurmi, B. D. (2022). Niosome: a vesicular drug delivery tool. In *Nanoparticles and Nanocarriers Based Pharmaceutical Preparations* (pp. 333-364). Bentham Science Publishers.

Pawar, P. K., Rathod, R. D., & Jagadale, S. R. (2024). A review on topical ophthalmic drug delivery system: Reference to viscosity enhancer. *Polymers in Medicine, 54*(1), 71-84.

Pelusi, L., Mandatori, D., Mastropasqua, L., Agnifili, L., Allegretti, M., Nubile, M., & Pandolfi, A. (2023). Innovation in the development of synthetic and natural ocular drug delivery systems for eye diseases treatment: focusing on drug-loaded ocular inserts, contacts, and intraocular lenses. *Pharmaceutics*, *15*(2), 625.

Pulido, J. S., Flaxel, C. J., Adelman, R. A., Hyman, L., Folk, J. C., & Olsen, T. W. (2016). Retinal vein occlusions preferred practice pattern® guidelines. *Ophthalmology*, *123*(1), P182-P208.

Rai, M., Ingle, A. P., Gaikwad, S., Padovani, F. H., & Alves, M. (2016). The role of nanotechnology in control of human diseases: perspectives in ocular surface diseases. *Critical Reviews in Biotechnology*, *36*(5), 777-787.

Saini, L., Dubey, A., Pal, R., Pandey, P., & Mandal, R. K. (2024). Synthetic and Natural Polymers Enhancing Drug Delivery and Their Treatment: A Comprehensive Review. *Journal of Drug Delivery & Therapeutics*, *14*(10).

Sarkar, P., Coffey, M., & Shawer, M. (2019). Development of Ophthalmic Formulations. In *Parenteral Medications, Fourth Edition* (pp. 249-275). CRC Press.

Sharma, D. K., Reddy, E. V., & Kumar, S. (2022). In-Depth Handbook of Pharmaceutical Studies. Addition Publisher.

Shifali, A., Kumar, P., & Pandit, V. (2021). Recent trends in Ocular Drug Delivery System: A Review. *Asian Journal of Research in Pharmaceutical Sciences*, 11(1), 71-80.

Souto, E. B., Cano, A., Martins-Gomes, C., Coutinho, T. E., Zielińska, A., & Silva, A. M. (2022). Microemulsions and nanoemulsions in skin drug delivery. *Bioengineering*, *9*(4), 158.

Souto, E. B., Dias-Ferreira, J., López-Machado, A., Ettcheto, M., Cano, A., Camins Espuny, A., Espina, M., Garcia, M. L., & Sánchez-López, E. (2019). Advanced formulation approaches for ocular drug delivery: state-of-the-art and recent patents. *Pharmaceutics*, *11*(9), 460.

Srivastava, V., Chary, P. S., Rajana, N., Pardhi, E. R., Singh, V., Khatri, D., Singh, S. B., & Mehra, N. K. (2023). Complex ophthalmic formulation technologies: advancement and future perspectives. *Journal of Drug Delivery Science and Technology*, *82*, 104317.

Suri, R., Beg, S., & Kohli, K. (2020). Target strategies for drug delivery bypassing ocular barriers. *Journal of Drug Delivery Science and Technology*, *55*, 101389.

Tawfik, S. M., Azizov, S., Elmasry, M. R., Sharipov, M., & Lee, Y.-I. (2020). Recent advances in nanomicelles delivery systems. *Nanomaterials*, *11*(1), 70.

Tsung, T.-H., Tsai, Y.-C., Lee, H.-P., Chen, Y.-H., & Lu, D.-W. (2023). Biodegradable polymer-based drug-delivery systems for ocular diseases. *International Journal of Molecular Sciences*, *24*(16), 12976.

Varela-Fernández, R., Díaz-Tomé, V., Luaces-Rodríguez, A., Conde-Penedo, A., García-Otero, X., Luzardo-Álvarez, A., Fernández-Ferreiro, A., & Otero-Espinar, F. J.

(2020). Drug delivery to the posterior segment of the eye: Biopharmaceutic and pharmacokinetic considerations. *Pharmaceutics*, *12*(3), 269.

Vashist, A., Vashist, A., Gupta, Y., & Ahmad, S. (2014). Recent advances in hydrogel based drug delivery systems for the human body. *Journal of Materials Chemistry B*, *2*(2), 147-166.

Villate-Beitia, I., Puras, G., Zarate, J., Agirre, M., Ojeda, E., & Pedraz, J. L. (2015). *First insights into non-invasive administration routes for non-viral gene therapy*. IntechOpen.

Wadhwa, S., Paliwal, R., Paliwal, S. R., & Vyas, S. (2009). Nanocarriers in ocular drug delivery: an update review. *Current pharmaceutical design*, *15*(23), 2724-2750.

Wang, T.-J., Rethi, L., Ku, M.-Y., Nguyen, H. T., & Chuang, A. E.-Y. (2024). A review on revolutionizing ophthalmic therapy: Unveiling the potential of chitosan, hyaluronic acid, cellulose, cyclodextrin, and poloxamer in eye disease treatments. *International journal of biological macromolecules*, *273*, 132700.

Weinreb, R. N., Aung, T., & Medeiros, F. A. (2014). The pathophysiology and treatment of glaucoma: a review. *Jama*, *311*(18), 1901-1911.

Willebrand, H., & Ghuman, B. S. (2002). *Free space optics: enabling optical connectivity in today's networks*. SAMS publishing.

Willoughby, C. E., Ponzin, D., Ferrari, S., Lobo, A., Landau, K., & Omidi, Y. (2010). Anatomy and physiology of the human eye: effects of mucopolysaccharidoses disease on structure and function–a review. *Clinical & Experimental Ophthalmology*, *38*, 2-11.

Yaqoob Khan, A., Talegaonkar, S., Iqbal, Z., Jalees Ahmed, F., & Krishan Khar, R. (2006). Multiple emulsions: an overview. *Current drug delivery*, *3*(4), 429-443.

Zhang, J., Jiao, J., Niu, M., Gao, X., Zhang, G., Yu, H., Yang, X., & Liu, L. (2021). Ten years of knowledge of nano-carrier based drug delivery systems in ophthalmology: current evidence, challenges, and future prospective. *International journal of nanomedicine*, 6497-6530.

Chapter 2

The Role of Nanotechnology in Enhancing Ocular Bioavailability

Punam Gaba[1,2,*]
Mona Piplani[1]
and Shailesh Sharma[2]
[1]Maharaja Agrasen School of Pharmacy, Maharaja Agrasen University, Himachal Pradesh, India
[2]Amar Shaheed Baba Ajit Singh Jujhar Singh Memorial College of Pharmacy, Punjab, India

Abstract

The eye is a highly complex organ with limited drug penetration and reduced therapeutic efficacy. Nanocarriers such as niosomes and liposomes, composed of phospholipid bilayers, are employed to encapsulate both hydrophilic and lipophilic drugs, thereby enhancing ocular bioavailability. These carriers improve drug solubility, corneal penetration, and enable sustained release, while minimizing systemic side effects. However, challenges such as stability issues and high production costs have prompted exploration of alternatives, including amphiphilic surfactant-based vesicles, which offer more cost-effective solutions. Recent innovations in ocular drug delivery have focused on surface-engineered nanocarriers and the integration of bioadhesive polymers to significantly enhance precorneal residence time and therapeutic efficacy. Despite notable progress, challenges such as cytotoxicity, scalability, and storage stability persist. Continued breakthroughs in nanocarrier optimization hold transformative potential

* Corresponding Author's Email: punam_gaba@yahoo.co.in

In: Advances in Nanotechnology for Ocular Drug Delivery
Editors: Anjali Sharma, Ajmer Singh Grewal, Geeta Deswal et al.
ISBN: 979-8-89530-814-1
© 2025 Nova Science Publishers, Inc.

for ocular therapeutics, paving the way for precisely targeted, sustained-release treatments that could redefine the management of vision-threatening conditions such as glaucoma and age-related macular degeneration.

Keywords: bioavailability, niosomes, liposomes, nanocarriers, phospholipids, biocompatible, solubility, penetration, stability, modification

Abbreviations

AMD: Age-Related Macular Degeneration
AuNPs: Gold Nanoparticles
BRB: Blood-Retinal Barrier
CNS: Central Nervous System
EMA: European Medicines Agency
IOP: Intraocular Pressure
ML: Machine Learning
MOFs: Metal-Organic Frameworks
NPs: Nanoparticles
PEG: Polyethylene Glycol
PLGA: Poly(lactic-co-glycolic acid)
PNPs: Polymeric Nanoparticles
RNA: Ribonucleic Acid
RNAi: RNA Interference
SLNs: Solid Lipid Nanoparticles

Introduction

Delivering medications to the eye presents unique challenges due to its natural protective barriers, such as the cornea and blood-retinal barrier, which often prevent drugs from effectively reaching their target (Gaudana et al., 2008). Traditional drug delivery methods like eye drops are easy to use but have poor bioavailability because most of the medication is washed away before absorption. For more serious posterior segment conditions, intravitreal injections or implants offer direct treatment but can be uncomfortable. Scientists are developing advanced approaches such as nanoparticles that bypass barriers, contact lenses that release drugs gradually, and gene therapies for inherited eye diseases (Thrimawithana et al., 2011). These innovations aim to enhance drug efficacy, reduce side effects, and improve quality of life in managing conditions like glaucoma and macular degeneration (Gote et al., 2019). The eye's complex anatomy, including structures like the sclera, conjunctiva, cornea, aqueous humor, iris, pupil, lens, retina, macula, choroid, and optic nerve, presents further challenges for treatment delivery. For example, the conjunctiva produces mucus and tears, the cornea regulates fluid and light, and the aqueous humor is involved in intraocular pressure control, which if dysregulated, leads to diseases like glaucoma. The macula and fovea support sharp central vision, while the optic nerve relays visual data to the brain, and its damage causes irreversible blindness. To navigate these complexities, various routes of ocular drug delivery have been developed, including topical (eye drops), subconjunctival, and intravitreal routes. Topical delivery is the most common but suffers from poor drug penetration and bioavailability (<5%) (Gaudana et al., 2010). Subconjunctival injections bypass barriers and allow deeper drug delivery (Kim et al., 2014), while intravitreal injections directly treat retinal conditions but carry risks like endophthalmitis (Purvin et al., 2004). In this context, nanotechnology has emerged as a promising solution. Defined as the manipulation of materials between 1 and 100 nanometers, nanotechnology enables drug carriers to overcome ocular barriers via enhanced surface area-to-volume ratios and quantum effects (Sanders et al., 2018; Roco et al., 2020). Key concepts include nanomaterials (e.g., nanoparticles, nanotubes), quantum effects, and self-assembly (Sanders et al., 2018). Applications span medicine, electronics, consumer products, and environmental sustainability (Farokhzad et al., 2009). Despite these advances, ocular drug delivery still faces several challenges: (1) anatomical barriers, such as the lipophilic corneal epithelium that limits hydrophilic drug penetration (Gaudana et al., 2008); (2) low bioavailability due to tear washout and blinking, which leads to higher prescribed doses and potential systemic absorption (Patel et al., 2013); (3) solubility and stability issues, where many drugs either do not dissolve in ocular fluids or degrade too quickly to be effective (Patel et al., 2013); (4) patient adherence concerns, as complex dosing regimens or fear of injections reduce compliance (Urtti et al., 2006); and (5) safety and toxicity risks, where prolonged use of formulations may lead to irritation, allergic responses, or off-target effects (Urtti et al., 2006). Therefore, developing efficient, safe, and patient-friendly ocular treatments requires a multidisciplinary approach combining anatomical insight, advanced drug delivery technologies, and patient-centric design.

Nanotechnology-Based Approaches for Ocular Drug Delivery

Nanotechnology offers promising solutions to several challenges in ocular drug delivery, including enhanced targeting accuracy, extended drug release, and improved bioavailability. Below are key nanotechnology-based strategies currently being explored for ocular applications.

Nanoparticles (NPs)

Nanoparticles, such as liposomes, polymeric nanoparticles, and SLNs, can encapsulate both hydrophilic and lipophilic drugs. Their primary advantages include controlled drug release and enhanced permeability across ocular barriers (Diebold et al., 2010).

Liposomes

Composed of phospholipid bilayers, liposomes can encapsulate drugs to improve their retention time, ocular penetration, and enable sustained drug release.

Polymeric Nanoparticles (PNPs)

Formulated using biocompatible polymers like PLGA, polymeric nanoparticles offer targeted drug delivery, protection of the drug from degradation, and controlled release over extended periods.

Nanomicelles

These are self-assembling colloidal structures formed by amphiphilic surfactants. They are generally more stable and soluble than other forms, and are capable of encapsulating both hydrophilic and hydrophobic drugs, thereby enhancing ocular drug delivery (Gote et al., 2019).

Nanosuspensions

Nanosuspensions, or colloidal dispersions, are especially beneficial for drugs with poor water solubility. They increase drug bioavailability and penetration into ocular tissues when administered topically or via injection (Patravale et al., 2004).

Dendrimers

Dendrimers are highly branched, tree-like macromolecules that can carry multiple drug molecules on their surface. They offer advantages such as improved drug solubility, stability, and ocular penetration, along with controlled and sustained drug release (Iezzi et al., 2004).

Nanoemulsions

These are oil-in-water or water-in-oil emulsions with droplet sizes typically between 20–200 nm, stabilized by surfactants. Nanoemulsions can enhance the solubility and bioavailability of both hydrophilic and lipophilic drugs.

Nanofibers and Nanostructured Hydrogels

Nanofibers and hydrogels are engineered to allow sustained, localized drug release, improving drug retention time and patient adherence. These are often used in ocular inserts or patches (Li et al., 2010).

Nanocarriers for Targeted Delivery

Advanced nanocarriers like polymeric micelles and nanoliposomes can be engineered to target specific ocular tissues such as the retina. Functionalization with ligands (e.g., small molecules, peptides, or monoclonal antibodies) enables receptor-specific drug delivery and enhances therapeutic efficacy (Kompella et al., 2013).

Nanocrystals for Improved Drug Delivery

Nanocrystals are employed to improve the solubility, dissolution rate, and ocular penetration of poorly water-soluble drugs. They support controlled drug release and address formulation limitations found in conventional dosage forms.

Safety, Toxicity, and Regulatory Considerations in Ocular Drug Delivery

Ocular drug delivery systems, mainly those integrating advanced nanotechnologies, present exciting and innovative treatment options. However, it is crucial to conduct thorough assessments of their safety and potential toxicity. Additionally, adherence to regulatory standards is vital to ensure not only the effectiveness of these treatments but also to diminish any adverse effects that could influence patient health and well-being.

Safety Concerns

The eye's sensitivity makes safety a critical factor in drug delivery. Potential risks include:

- Irritation & Inflammation: Nanoparticles or synthetic carriers may trigger immune responses, causing redness, discomfort, or chronic inflammation.
- Ocular Toxicity: Nanoparticle accumulation in ocular tissues (cornea, retina) can lead to long-term damage, influenced by size, charge, and material composition.
- Corneal Penetration: While enhancing drug absorption, deep penetration may cause unintended drug release, harming delicate structures like the retina or optic nerve (Aicha et al., 2006).

Nanoparticle Toxicity Mechanisms

- Size-Dependent Effects: Smaller nanoparticles penetrate deeper but may induce oxidative stress; larger ones may cause surface irritation.
- Surface Charge: Cationic nanoparticles interact more with cell membranes, increasing toxicity risks.
- Material Composition: Metal-based nanoparticles (e.g., gold, silver) can release toxic ions, damaging retinal cells.
- Retinal Toxicity: Accidental exposure may lead to photoreceptor degeneration and vision loss (Hoshyar et al., 2016).

Regulatory Considerations

Regulatory agencies (FDA, EMA) require rigorous testing for ocular nanomedicines:

- Approval Process: Nanoparticles are classified as novel entities, requiring preclinical (animal) and clinical (human) trials.
- Biocompatibility: Materials must be biodegradable, non-immunogenic, and avoid long-term ocular accumulation.
- Testing Standards: *In vivo* studies assess toxicity, pharmacokinetics, and efficacy.
- Clinical trials validate safety and therapeutic performance.
- Nanoparticle-Specific Guidelines: FDA regulations mandate detailed characterization (size, charge, stability) and toxicity profiling.

While advanced ocular drug delivery systems improve treatment outcomes, their development must prioritize safety, minimize toxicity, and comply with strict regulatory standards. Continuous research and innovation are essential to balance therapeutic benefits with patient safety.

Risk Management Strategies

Several techniques are used to reduce the dangers related to ocular drug administration, particularly in systems based on nanotechnology.

Surface Modification of Nanoparticles

Reducing toxicity can be achieved by altering the surface characteristics of nanoparticles. Nanoparticle stability, immunogenicity, and cytotoxicity can all be decreased by coating them with biocompatible polymers like PEG (Suk et al., 2016).

Controlled Release Mechanisms

This allows us to reduce systemic exposure as well as the toxicity for ocular medication administration. Reducing the negative effects on the ocular tissues as well as increasing the therapeutic efficacy through controlled release mechanism via nanoparticles is highly anticipated (De Campos et al., 2009).

Types of Nanoparticles and Their Applications in Ocular Drug Delivery

Nanoparticles have attracted substantial attention in ocular medication delivery due to their potential to penetrate barriers such as the corneal epithelium and improve drug solubility, stability, and bioavailability. These nanoparticles are engineered at the nanoscale (usually 1-1000 nm) to improve the pharmacokinetic properties of eye medicines. Numerous kinds of nanoparticles have been created and investigated for their use in the delivery of drugs into the eyes. Table 1 discusses some popular nanoparticle varieties and their uses in ocular medication delivery. These are listed below, along with pertinent references.

Table 1. Nanoparticles and their applications in ocular drug delivery

Types of Nanoparticles	Applications in Ocular Drug Delivery	Ref.
Polymeric Nanoparticles (PNPs)	In particular, polymeric nanoparticles help to treat glaucoma, cataracts, and dry eye disease because of their prolonged release, ability to improve the retention of medications on the ocular surface, and ability to transfer medications across ocular barriers.	(Diebold et al., 2010)
Liposomes	Liposomes are frequently utilized to deliver medications that target the retina, antibiotics, and anti-inflammatory medicines. They can be utilized for illnesses of the anterior and posterior segments and increase the ocular bioavailability of poorly soluble medications.	(Bochot et al., 2012)

Types of Nanoparticles	Applications in Ocular Drug Delivery	Ref.
Solid Lipid Nanoparticles (SLNs)	SLNs are used for the sustained release of anti-glaucoma drugs, corticosteroids for inflammation, and antioxidants. They enhance residence time on the ocular surface, making them ideal for prolonged drug delivery.	(Sutter et al., 2007)
Nanomicelles	Nanomicelles are used to deliver antibiotics, anti-inflammatory drugs, and anti-glaucoma medications. They are used to enhance the controlled release of the drug, provided that they also expand ocular bioavailability and drug solubility.	(Gote et al., 2019)
Polymeric Micelles	Polymeric micelles are used in ocular drug delivery for diseases like macular degeneration, retinal diseases, and intraocular inflammation. They augment drug retention and enable targeted delivery to ocular tissues.	(Kompella et al., 2013)
Gold Nanoparticles (AuNPs)	Gold nanoparticles are the first line of defense for targeted drug delivery in treating glaucoma, uveitis, and retinal disorders. Their exterior can be altered to enhance cellular absorption and improve therapeutic outcomes.	(Jeong et al., 2009)
Silica Nanoparticles	Silica nanoparticles are used for carrying ocular medications across the ocular barriers for conditions like diabetic retinopathy and AMD. Their porous structure allows for sustained release and improved drug retention.	(Howell et al., 2015)
Dendrimers	Dendrimers are being studied for ocular drug delivery in conditions like glaucoma and retinal disorders. Their structure allows for controlled release and optimal drug loading, boosting their ability to target specific ocular tissues and deliver medications efficiently.	(Iezzi et al., 2004)
Nanocrystals	Nanocrystals are used to distribute poorly water-soluble drugs, like corticosteroids and anti-inflammatory agents, to the eye. They augment dissolution rate, improve ocular infiltration, and enable continuous release.	(Kompella et al., 2013)
Chitosan Nanoparticles	Antibiotics, antifungal agents, and anti-glaucoma medications are delivered via chitosan nanoparticles, whose mucoadhesive qualities improve drug retention on the surface of the eye.	(De et al., 2004)

Future Directions and Emerging Trends in Ocular Drug Delivery

With an emphasis on conquering problems brought on by the unique anatomical and physiological barriers of the eye, the field of ocular medication administration is evolving quickly. Recent developments in drug delivery technologies are intended to enhance the bioavailability, targeting, and effectiveness of therapies for a range of eye conditions, including cataracts, AMD, glaucoma, and diabetic retinopathy. Key prospects and new developments in ocular medication administration are covered in Table 2.

Table 2. Future directions and emerging trends in ocular drug delivery

Emerging Trends in Ocular Drug Delivery	Future Directions	Ref.
Nanoparticles and Nanocarriers	To minimize adverse effects, nanoparticles that are particularly made to target ocular tissues, such as the retina, and steer clear of systemic distribution will be essential. Stimuli-responsive nanoparticles, which release medications in reaction to environmental variables (such as pH or temperature), may also provide more individualized therapy choices.	(Patravale et al., 2004)
Smart Nanocarriers	Smart drug delivery devices that can be regulated externally, such as light-triggered release, can improve treatment outcomes for disorders like retinal degeneration by enabling on-demand medication release.	(Salem et al., 2009)
Gene Therapy	Gene expression in ocular tissues will be sustained with the development of more effective gene delivery methods, such as viral and non-viral vectors. One particularly interesting strategy for resolving the issues with ocular gene therapy, such as immunological response and poor gene stability, is gene delivery via nanoparticles.	(Han et al., 2012)
RNA Interference (RNAi)	Therapeutic silence of genes implicated in ocular illnesses is made possible by the development of RNAi delivery technologies that efficiently target retinal cells using nanoparticles or lipid-based formulations.	(Aharonovich et al., 2019
Injectable Depot Systems	One of the main priorities will be the creation of biodegradable injectable depots that release medications gradually and with few side effects. These preparations offer a more pleasing and easier treatment choice, which can enhance patient results.	(Zhu et al., 2020)

Emerging Trends in Ocular Drug Delivery	Future Directions	Ref.
Microneedles	Microneedle-based ocular delivery systems can provide controlled and continued medication release in the posterior part of the eye, potentially substituting traditional injections with a non-invasive alternative.	(Sedhom et al., 2021)
Personalized Medicine	The creation of customized treatments that are suited to each patient's unique requirements based on their genetic composition, the course of their disease, and their ocular features will be made easier by the incorporation of artificial intelligence and machine learning into ocular drug delivery systems.	(Wang et al., 2021)
Nanostructured Hydrogels	Non-invasive medication delivery methods for the anterior part of the eye will be made possible by the advancement of smart hydrogels that can react to ocular circumstances.	(Sun et al., 2023)

Conclusion

Nanotechnology has reformed ocular drug delivery by accounting for key barriers of ocular anatomy, such as poor bioavailability, rapid clearance, and limited corneal penetration associated with conventional treatments. To enhance drug solubility, prolong precorneal retention, and enable targeted delivery to both anterior and posterior eye segments, we have developed advanced nanocarriers including liposomes, niosomes, polymeric nanoparticles, and dendrimers. These novelties improve therapeutic outcomes for many diseases like glaucoma, age-related macular degeneration, and diabetic retinopathy while minimizing systemic side effects. Despite these developments, barriers remain, including nanoparticle stability, potential cytotoxicity, and large-scale manufacturing constraints. Future research needs to focus on improving biocompatible materials, refining stimuli-responsive nanocarriers (e.g., pH or temperature-sensitive systems), and mixing gene therapy with nanotechnology for sustained treatment of hereditary retinal illnesses. Additionally, supervisory frameworks must evolve to ensure the wellbeing and efficacy of nano-based ocular formulations, requiring severe preclinical and clinical evaluations. The union of nanotechnology with artificial intelligence and personalized medicine holds potential for next-generation therapies. To transform ocular drug delivery, offering precise, patient-specific treatment innovations like microneedles, smart hydrogels, and RNAi-loaded nanoparticles are needed to advance. As these technologies mature, they may redefine standards of care, providing more effective, less aggressive solutions for vision-threatening diseases.

References

Aharonovich, I. (2019). Clearly identical photons. *Nature Nanotechnology*, 14(6), 502–503.

Aicha L., Vissvesvaran, G. P., Kuo Lee Lieu, Murata, K., McGinn, T., Manmoto, C. C., Sinclair, J., I. Karliga, David, Fawzi, A., & Kubota, R. (2006). Protective Effect of Crocin against Blue Light– and White Light Mediated Photoreceptor Cell Death in Bovine and Primate Retinal Primary Cell Culture. *Investigative Ophthalmology & Visual Science*, 47(7), 3156–3156.

Bochot, A., & Fattal, E. (2012). Liposomes for Intravitreal Drug delivery: a State of the Art. *Journal of Controlled Release*, 161(2), 628–634.

de Campos, A. M., Diebold, Y., Carvalho, E. L. S., Sánchez, A., & José Alonso, M. (2004). Chitosan nanoparticles as new ocular drug delivery systems: In vitro stability, in vivo fate, and cellular toxicity. *Pharmaceutical Research*, 21(5), 803–810.

Diebold, Y., & Calonge, M. (2010). Applications of Nanoparticles in Ophthalmology. *Progress in Retinal and Eye Research*, 29(6), 596–609.

Farokhzad, O. C., & Langer, R. (2009). Impact of Nanotechnology on Drug Delivery. *ACS Nano*, 3(1), 16–20.

Gaudana, R., Ananthula, H. K., Parenky, A., & Mitra, A. K. (2010). Ocular Drug Delivery. *The AAPS Journal*, 12(3), 348–360.

Gaudana, R., Jwala, J., Boddu, S. H. S., & Mitra, A. K. (2008). Recent Perspectives in Ocular Drug Delivery. *Pharmaceutical Research*, 26(5), 1197–1216.

Gote, V., Sikder, S., Sicotte, J., & Pal, D. (2019). Ocular Drug Delivery: Present Innovations and Future Challenges. *Journal of Pharmacology and Experimental Therapeutics*, 370(3), 602–624.

Han, Z., Conley, S. M., Makkia, R. S., Cooper, M. J., &Naash, M. I. (2012). DNA nanoparticle-mediated ABCA4 delivery rescues stargardt dystrophy in mice. *Journal of Clinical Investigation*, 122(9), 3221–3226.

Hoshyar, N., Gray, S., Han, H., & Bao, G. (2016). The Effect of Nanoparticle Size on in Vivo Pharmacokinetics and Cellular Interaction. *Nanomedicine*, 11(6), 673–692.

Howell, D. W., Tsai, S., Churion, K., Patterson, J., Abbey, C., Atkinson, J. T., Porterpan, D., You, Y., Meissner, K. E., Bayless, K. J., &Bondos, S. E. (2015). Identification of multiple dityrosine bonds in materials composed of the drosophila protein ultrabithorax. *Advanced Functional Materials*, 25(37), 5988–5998.

Iezzi, R., Guru, B. R., Glybina, I. V., Mishra, M. K., Kennedy, A., & Kannan, R. M. (2012). Dendrimer-based Targeted Intravitreal Therapy for Sustained Attenuation of Neuroinflammation in Retinal Degeneration. *Biomaterials*, 33(3), 979–988.

Jeong Hun Kim, Kim, J.-S., Kim, K.-W., Myung Kyum Kim, & Young Suk Yu. (2009). Intravenously administered gold nanoparticles pass through the blood-retinal barrier depending on the particle size, and induce no retinal toxicity. *IOP Publishing*, 20(50), 505101–505101.

Kompella, U. B., Amrite, A. C., Pacha Ravi, R., & Durazo, S. A. (2013). Nanomedicines for Back of the Eye Drug delivery, Gene delivery, and Imaging. *Progress in Retinal and Eye Research*, 36, 172–198.

Patel, A. (2013). Ocular Drug Delivery systems: an Overview. *World Journal of Pharmacology*, 2(2), 47.

Patravale, V. B., Date, A. A., & Kulkarni, R. M. (2004). Nanosuspensions: a Promising Drug Delivery Strategy. *Journal of Pharmacy and Pharmacology*, 56(7), 827–840.

Purvin, V., & Maturi, R. (2004). Sprint Car Visual Loss. *Survey of Ophthalmology*, 49(1), 90–95.

Roco, M. C., Hersam, M. C., & Mirkin, C. A. (2011). Nanotechnology Research Directions for Societal Needs in 2020. In Springer eBooks. Springer Nature.

Salem, B., Zhang, Q., Chen, J., Cobley, C. M., Zhang, Q., Rycenga, M., Xie, J., In San Kim, Kwang Yong Song, Andrea Wershof Schwartz, Wang, L. V., & Xia, Y. (2009). Gold nanocages covered by smart polymers for controlled release with near-infrared light. *Nature Materials*, 8(12), 935–939.

Sanders, W. C. (2018). Basic Principles of Nanotechnology. Taylor and Francis, a CRC title, part of the Taylor and Francis imprint, a member of the Taylor and Francis Group, the academic division of T&F Informa, plc.

Sedhom, R., McShea, M. J., Cohen, A. B., Webster, J. A., & Mathews, S. C. (2021). Mobile app validation: A digital health scorecard approach. *Npj Digital Medicine*, 4(1),111.

Suk, J. S., Xu, Q., Kim, N., Hanes, J., & Ensign, L. M. (2016). PEGylation as a strategy for improving nanoparticle-based drug and gene delivery. *Advanced Drug Delivery Reviews*, 99, 28–51.

Sun, H., Saurav Maji, Chandrakasan, A. P., & Marelli, B. (2023). Integrating biopolymer design with physical unclonable functions for anticounterfeiting and product traceability in agriculture. *Science Advances*, 9(12).

Sutter, S. C., M. Pilar Buera, & Elizalde, B. E. (2007). β-Carotene Encapsulation in a Mannitol Matrix as Affected by Divalent Cations and Phosphate Anion. *International Journal of Pharmaceutics*, 332(1-2), 45–54.

Thrimawithana, T. R., Young, S., Bunt, C. R., Green, C., & Alany, R. G. (2011). Drug Delivery to the Posterior Segment of the Eye. *Drug Discovery Today*, 16(5-6), 270–277.

Urtti, A. (2006). Challenges and Obstacles of Ocular Pharmacokinetics and Drug Delivery. *Advanced Drug Delivery Reviews*, 58(11), 1131–1135.

Wang, B., Díaz-Payno, P. J., Browe, D. C., Freeman, F. E., Nulty, J., Burdis, R., & Kelly, D. J. (2021). Affinity-bound growth factor within sulfated interpenetrating network bioinks for bioprinting cartilaginous tissues. *Acta Biomaterialia*, 128, 130–142.

Yoo Min Kim, Chiang, B., Wu, X., &Prausnitz, M. R. (2014). Ocular Delivery of Macromolecules. *Journal of Controlled Release*, 190, 172–181.

Zhu, Y., Jiao, X., Chen, W., Bai, S., Zheng, T., He, C., Guo, Z., Jiang, M., Du, G., & Sun, X. (2020). Albumin-biomineralized nanoparticles to synergize phototherapy and immunotherapy against melanoma. *Journal of Controlled Release*, 322, 300–311.

Chapter 3

Nanostructured Lipid Carriers for Ocular Drug Delivery

Shikha Baghel Chauhan[1,*]
Yash Kalra[1]
Chirag Jain[1]
Indu Singh[1]
and Anshuman Sinha[2,3]

[1]Amity Institute of Pharmacy, Amity University, Noida, Uttar Pradesh, India
[2]Department of Paediatrics, Ann & Robert H. Lurie Children's Hospital, Chicago, IL, USA
[3]Feinberg School of Medicine, Northwestern University, Chicago, IL, USA

Abstract

Nanostructured lipid carriers (NLCs) have emerged as a promising platform for ocular drug delivery, effectively addressing challenges such as low bioavailability, rapid drug elimination, and limited corneal permeability. This chapter explores the design, formulation, and therapeutic potential of NLCs in enhancing ocular drug delivery. Composed of a blend of solid and liquid lipids, NLCs form an imperfect crystalline matrix that allows for higher drug loading and controlled release. Their nanoscale dimensions facilitate extended retention in the precorneal area, thereby improving drug absorption and therapeutic efficacy. The chapter discusses critical formulation parameters, including lipid selection, surfactants, and preparation techniques such as high-pressure homogenization and ultrasonication. It also provides a detailed

---

[*] Corresponding Author's Email: shikha.pharma@gmail.com

In: Advances in Nanotechnology for Ocular Drug Delivery
Editors: Anjali Sharma, Ajmer Singh Grewal, Geeta Deswal et al.
ISBN: 979-8-89530-814-1

© 2025 Nova Science Publishers, Inc.

examination of ocular drug delivery mechanisms, including transcorneal and transconjunctival pathways. Surface modifications, such as the incorporation of mucoadhesive polymers and targeting ligands, are reviewed for their role in enhancing bioavailability. Furthermore, the chapter evaluates preclinical and clinical developments, alongside safety considerations, to assess the translational potential of NLCs. As a significant advancement in ocular therapeutics, NLCs offer prolonged and targeted drug release, reduced dosing frequency, and improved patient compliance. Future directions for optimizing NLC-based ocular formulations for commercial application are also highlighted.

Keywords: ocular drug delivery, nanostructured lipid carriers, bioavailability, mucoadhesion, penetration enhancement, nanoparticle

Abbreviations

AMD: Age-related Macular Degeneration
APIs: Active Pharmaceutical Ingredients
CNLCs: Cationic Nanostructured Lipid Carriers
DNA: Deoxyribonucleic Acid
IOP: Intraocular Pressure
NLCs: Nanostructured Lipid Carriers
RNA: Ribonucleic Acid
siRNA: Small Interfering RNA
SLNs: Solid Lipid Nanoparticles

Introduction

Ocular drug delivery remains a significant challenge due to the eye's complex anatomical and physiological barriers. Protective mechanisms such as the corneal epithelium, tear film, rapid drainage, and blood-ocular barriers hinder effective drug absorption and reduce therapeutic efficacy. Conventional ophthalmic formulations like eye drops, ointments, and suspensions often suffer from low retention, fast precorneal clearance, and poor penetration into deeper ocular tissues. These limitations necessitate frequent dosing, which can lead to poor patient compliance and inconsistent treatment outcomes. Nanostructured lipid carriers (NLCs) have emerged as a promising alternative to traditional delivery systems by enhancing drug permeability, prolonging ocular residence time, and improving bioavailability. Their lipophilic nature facilitates better penetration through the corneal epithelium and allows for sustained, controlled release of therapeutic agents (Youssef et al., 2022). Additionally, NLCs help overcome challenges such as efflux pump activity and tear turnover, which typically hinder drug absorption and retention. NLC-based formulations have shown success across various ocular conditions. In glaucoma, they enable long-term intraocular pressure reduction with drugs like timolol or latanoprost, minimizing systemic side effects. For dry eye syndrome, NLCs enhance cyclosporine retention, stabilizing the tear film and reducing irritation. In conditions like diabetic retinopathy and age-related macular degeneration, they effectively deliver antioxidants and anti-VEGF agents, reducing the need for invasive treatments (Chakole et al., 2021). The effective delivery of drugs to ocular tissues is significantly challenged by various physiological barriers and dynamic clearance mechanisms. These include poor corneal permeability, rapid tear turnover, and limited drug residence time on the ocular surface. The cornea's multi-layered structure forms a dual-permeability barrier that restricts drug penetration (Singh et al., 2024). A major obstacle is anterior abolition, where processes like blinking, tear turnover, and nasolacrimal drainage rapidly remove drugs from the precorneal space, reducing their therapeutic efficacy. Blinking alone can shorten the contact time of topically applied drugs, while tear turnover constantly flushes them away, limiting absorption (Lv et al., 2024). The corneal epithelium acts as a selective barrier, further restricting drug entry into deeper ocular tissues such as the anterior chamber and posterior segment. Additionally, systemic absorption via conjunctival blood vessels can prematurely remove drugs from target sites, lowering their local concentrations and therapeutic effectiveness (Chen et al., 2021). Blood-ocular barriers, including the blood-aqueous and blood-retinal barriers, impede drug distribution to intraocular compartments, making the treatment of retinal diseases particularly difficult. Conventional topical formulations rarely achieve therapeutic levels in the posterior segment, often requiring invasive methods like intravitreal injections. Enzymatic degradation and efflux transporters further diminish drug effectiveness. These limitations highlight the urgent need for innovative drug delivery systems that can improve ocular bioavailability while minimizing both systemic and local side effects (Rajan et al., 2024). NLCs have gained significant attention in ophthalmic drug delivery due to their ability to overcome limitations associated with conventional formulations. These carriers are composed of a blend of solid and liquid lipids that form a stable nanostructure with enhanced drug-loading capacity, improved bioavailability, and controlled drug release properties (Singh et al., 2024). Their lipid-based composition promotes better adhesion to the ocular surface, resulting in prolonged retention time and increased corneal permeability. The small particle size of NLCs allows for deeper penetration into ocular tissues, making them effective for delivering drugs to both anterior and posterior segments of the eye (Wathoni et al., 2024). One of the key benefits of NLCs is their ability to sustain drug release, thereby reducing the frequency of administration and improving patient compliance. Unlike traditional eye drops that are quickly washed away, NLCs form a mucoadhesive film on the ocular surface, allowing for gradual drug diffusion and extended therapeutic action. Surface modifications, such as PEGylation, ligand functionalization, or incorporation of bioadhesive polymers, can further enhance tissue specificity and targeting (Borges et al., 2021). Additionally, NLCs present a biocompatible and biodegradable alternative to polymeric nanoparticles, reducing the risk of irritation and cytotoxicity. The nanostructure of NLCs is characterized by a disordered crystalline matrix that prevents drug expulsion and improves entrapment efficiency. This structure is influenced by the ratio of solid to liquid lipids and the inclusion of surfactants or stabilizers. Their physicochemical properties, such as particle size, surface charge, and zeta potential, are critical to their performance, with neutral or negative charges being preferable for minimizing mucosal repulsion and enhancing corneal interaction (Shekhar et al., 2021; Hassan et al., 2022). NLCs can encapsulate a wide range of therapeutic agents, including hydrophilic and lipophilic drugs, proteins, peptides, and nucleic acids, offering versatility in ocular applications (Putchakayala et al., 2025). Clinically, NLCs have demonstrated promising outcomes in the treatment of glaucoma, dry eye syndrome, ocular infections, and retinal diseases. For instance, NLCs loaded with antiglaucoma drugs such as timolol or latanoprost can maintain intraocular pressure with fewer systemic side effects. In dry eye syndrome, cyclosporine-loaded NLCs enhance ocular surface retention and provide better tear film stability. Looking forward, future research should aim at optimizing formulations through targeted delivery, stimuli-responsive systems, and hybrid nanocarrier technologies to maximize their therapeutic potential in ocular medicine (Rajoriya et al., 2024).

Mechanisms of Ocular Drug Delivery Using Nanostructured Lipid Carriers

NLCs have emerged as a transformative innovation in ocular drug delivery, providing an advanced platform that significantly enhances therapeutic efficacy through multiple mechanisms. These lipid-based nanosystems possess unique physicochemical characteristics that improve drug bioavailability, prolong ocular retention time, and facilitate targeted delivery to specific ocular tissues. The eye's natural defense mechanisms, such as the corneal epithelium, rapid tear turnover, and active efflux pumps, pose significant barriers to drug absorption. NLCs, composed of a combination of solid and liquid lipids, overcome these challenges by enhancing drug encapsulation efficiency and promoting corneal penetration. They facilitate drug transport through various mechanisms, including paracellular and transcellular diffusion, endocytosis, and bioadhesion, which collectively extend the drug's residence time at the site of action. The mucoadhesive nature of NLCs helps maintain prolonged precorneal presence, thereby reducing the need for frequent dosing. Additionally, their controlled release profile not only minimizes systemic and local side effects but also boosts therapeutic outcomes. These advantages make NLCs highly suitable for treating a range of ocular disorders such as glaucoma, dry eye syndrome, and age-related macular degeneration (Baig et al., 2024). The various mechanisms involved in NLC-mediated drug delivery are illustrated in Figure 1.

Mucosal Adhesion and Extended Retention Time

One of the major limitations of conventional eye drops is rapid drug clearance due to blinking and tear turnover, which significantly reduces ocular bioavailability. NLCs address this issue by exhibiting mucoadhesive properties that enable strong interaction with the mucin layer of the corneal and conjunctival surfaces. The incorporation of polymers or surfactants in NLC formulations enhances this adhesion, effectively prolonging the drug's residence time in the precorneal area. This improved retention facilitates sustained drug absorption, reducing the frequency of administration and improving patient compliance. Additionally, NLCs can be engineered to form a gel-like structure upon contact with the ocular surface, further enhancing their ability to remain in place. This feature is particularly beneficial for the management of chronic ocular conditions such as dry eye syndrome and glaucoma, where long-term treatment is essential (Adwan et al., 2024).

Improved Penetration of the Cornea and Conjunctiva

The corneal epithelium, due to its lipophilic nature and tight junctions, serves as a major barrier to drug permeation. In contrast, while the conjunctival epithelium is more permeable, it contributes to rapid drug clearance due to tear drainage. NLCs enhance drug transport across these barriers through their nanoscale particle size and lipid-based composition, enabling both transcorneal and transconjunctival absorption. The lipid components of NLCs mimic the natural lipid layer of the tear film, promoting stronger adhesion to ocular tissues and facilitating drug diffusion across epithelial membranes. Additionally, surface modifications such as PEGylation or ligand functionalization can further augment receptor-mediated endocytosis, thereby enhancing corneal penetration. These properties allow NLCs to efficiently deliver drugs to the anterior segment of the eye, making them particularly beneficial for treating conditions such as glaucoma, conjunctivitis, and uveitis (Chandel et al., 2024).

Endocytosis Facilitates Intracellular Drug Delivery and Overcoming Efflux Transporters and Ocular Barriers

Endocytosis, the process by which cells ingest nanoparticles and allow them to reach deeper ocular tissues, is critical in medication delivery. This method prevents premature drug clearance and bypasses efflux transporters such as P-glycoprotein (P-gp) and multidrug resistance-associated proteins (MRPs), which pump medicines out of ocular cells, lowering treatment effectiveness. The tiny particle size and lipid-based composition of NLCs allow for attachment to corneal and conjunctival epithelial cells, boosting cellular absorption via multiple endocytotic routes (Varela-Fernández et al., 2022). Drugs are actively pumped out of corneal and retinal cells by efflux transporters such P-gp and MRPs, which restrict their intracellular accumulation. NLCs' lipid-based nature allows them to evade these efflux processes by imitating endogenous lipids and enabling passive diffusion across cell membranes. Encapsulating hydrophobic medicines inside the lipid matrix protects them from direct identification by efflux proteins, hence increasing intracellular retention (Younis et al., 2021). Surface changes to NLCs can reduce efflux transporter function, which improves drug accumulation in target ocular tissues. NLCs also cross important ocular barriers, including the corneal epithelium, conjunctival epithelium, blood-aqueous barrier (BAB), and blood-retinal barrier (BRB). Mucoadhesive NLCs can improve penetration by enabling transcellular transport via endocytosis, whereas receptor-mediated endocytosis allows them to cross the blood-aqueous and blood-retinal barriers and transfer drugs into aqueous fluid and retinal tissues (Adwan et al., 2024; Varela-Fernández et al., 2022).

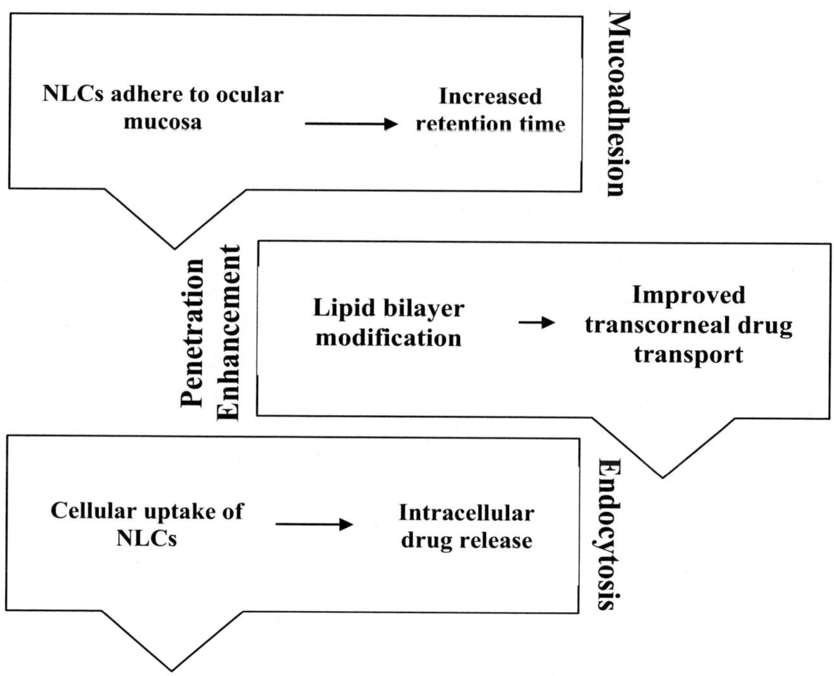

Figure 1. Mechanisms of Ocular Drug Delivery Using NLCs.

Strategies for Formulating Ocular Nanostructured Lipid Carriers

To create NLCs for ocular medication delivery, lipid selection, drug loading, production processes, and surface changes must be optimized in a systematic manner. Because the eye contains several protective barriers, including the corneal epithelium and tear film, NLC formulations must be formulated to promote drug retention, permeability, and prolonged drug release while maintaining biocompatibility and minimum discomfort. To improve therapeutic efficacy, ocular NLC formulation strategies focus on meticulous lipid component selection, drug loading and release kinetics optimization, precision production procedures, and targeted delivery systems (Onugwu et al., 2024). The lipid components used in nanoparticles (NLCs) have a substantial impact on their stability, drug encapsulation effectiveness, and ocular tolerance. NLCs are made up of a combination of solid and liquid lipids, with solid lipids providing structural stability and sustained release qualities, and liquid lipids enhancing drug solubility and preventing crystallisation. Glyceryl monostearate, Compritol 888 ATO, stearic acid, and cetyl palmitate are examples of common solid lipids, whereas liquid lipids such as oleic acid, medium-chain triglycerides, and isopropyl myristate improve medication dispersion and permeability. The lipid content also influences the viscosity and mucoadhesive characteristics of NLCs, both of which are required for long-term medication retention on the ocular surface. The drug loading is determined by the medicinal agent's solubility in both solid and liquid lipids, as well as the emulsification efficiency of the NLC preparation. The drug release mechanism is based on diffusion and lipid matrix disintegration, which may be controlled by selecting the right lipids and stabilizers (Cimino et al., 2024). NLCs that are pH-sensitive and enzyme-responsive have been created to produce drug release in the ocular environment under stimulus control.

High-Pressure Homogenization

Given its effectiveness in lowering particle size and improving drug encapsulation, high-pressure homogenization is a popular approach for producing NLCs. This method includes treating a lipid mixture to high pressures and temperatures, causing lipid droplets to break down into nanoparticles. Solid lipids are first melted, then combined with liquid lipids and active pharmaceutical ingredients (APIs) to create a homogenous lipid blend. To stabilize the emulsion, the mixture is disseminated in an aqueous phase containing surfactants. The lipid droplets are shrunk to nanoscale sizes, generally 100-500 nm. Because organic solvents are not used, high-pressure homogenization is scalable for industrial applications and provides safety and environmental viability. However, it also has drawbacks, such as heat-induced breakdown of thermosensitive medicinal molecules. To reduce these dangers and increase NLC encapsulation efficiency and therapeutic efficacy, process variables such as homogenization pressure, temperature, and cycle number must be optimized (Alatawi et al., 2024).

Solvent Evaporation

The solvent evaporation technique is another well-established method for NLC fabrication, particularly suitable for drugs that are sensitive to heat. This method involves dissolving the lipids and drug molecules in an organic solvent, such as dichloromethane or chloroform, to create a homogenous lipid-drug solution. This solution is then emulsified into an aqueous phase containing surfactants, resulting in the formation of a lipid emulsion. Following emulsification, the organic solvent is evaporated under reduced pressure, leading to the solidification of lipid nanoparticles and the formation of NLCs. A key advantage of the solvent evaporation technique is its ability to maintain the stability of heat-sensitive drugs, as the process does not involve exposure to high temperatures. However, the use of organic solvents raises concerns regarding toxicity, environmental impact, and regulatory compliance. Therefore, meticulous solvent selection, proper handling, and post-processing techniques, such as ultrafiltration or dialysis, are essential to ensure the complete removal of residual solvents from the final formulation. Despite these challenges, the solvent evaporation method remains a widely used approach due to its simplicity and effectiveness in producing NLCs with controlled particle size and drug loading capacity (Fathi et al., 2024).

Ultrasonication

Ultrasonication is a versatile technique employed either as a standalone method or in combination with other approaches, such as high-pressure homogenization or solvent evaporation. This method utilizes ultrasonic energy to facilitate the reduction of particle size and enhance drug dispersion within the lipid matrix. During ultrasonication, high-frequency sound waves generate cavitation bubbles in the liquid medium. The collapse of these bubbles produces localized high temperatures and pressures, which aid in breaking down larger lipid particles into nanoscale dimensions (Stefanov et al., 2023). The ultrasonication process offers several advantages, including its relatively simple operation and the ability to produce nanoparticles without requiring complex equipment. However, certain limitations must be addressed, such as the potential contamination from sonicator probes if not properly maintained. Additionally, excessive exposure to ultrasonic energy can lead to thermal degradation of heat-sensitive compounds, necessitating careful control of sonication time and intensity (Duong et al., 2020). Despite these challenges, ultrasonication remains a valuable technique for achieving fine-tuned nanoparticle size and improved drug encapsulation in NLC formulations.

Microemulsion Technique

The microemulsion technique is a straightforward and efficient method for NLC preparation that does not necessitate high-energy input or specialized equipment. In this approach, molten lipids are blended with an aqueous phase containing surfactants or cosurfactants to form a thermodynamically stable microemulsion. The microemulsion is then cooled or subjected to further processing to facilitate lipid solidification, resulting in the formation of NLCs. One of the major benefits of this technique is its ability to provide precise control over particle size by adjusting the droplet size of the microemulsion before solidification. Additionally, microemulsions can encapsulate both hydrophilic and lipophilic drugs, making them highly versatile in drug delivery applications. Moreover, the thermodynamic stability of microemulsions ensures long-term stability of the prepared NLCs. However, the technique may require the use of relatively high surfactant concentrations, which could pose challenges related to cytotoxicity and biocompatibility in certain formulations (Fathi-Karkan et al., 2024).

Surface Alterations and Targeting Methods

Surface changes are important for improving the bioavailability, retention, and targeting ability of ocular NLCs. Drug stability, clearance, and site-specific delivery can all benefit from modifications like PEGylation and ligand conjugation. PEGylation is the process of attaching polyethylene glycol (PEG) chains to the surface of NLCs to provide a hydrophilic barrier that resists opsonization and premature removal from the ocular surface. PEGylated NLCs have a longer residence duration in the tear film and better mucoadhesion, which ensures long-term medication absorption. PEGylation also lowers aggregation and improves colloidal stability, making it a viable long-term ocular medication delivery method. Ligand-based surface functionalization improves NLC targeting by directing them to specific ocular structures or receptors. Antibodies, peptides, aptamers, and folic acid derivatives can be attached to the surface of NLCs to promote receptor-mediated endocytosis. Transferrin-functionalized NLCs, for example, have been designed to boost medication penetration into retinal tissues, whilst hyaluronic acid-coated NLCs increase corneal absorption. By assuring accurate medication localization at the site of action, such targeted delivery methods increase therapeutic efficacy while reducing systemic adverse effects (Giri et al., 2024).

Applications of Nanostructured Lipid Carriers in Ocular Drug Delivery

Table 1. Applications of NLCs in ocular disease management

Applications	Description and Mechanism	NLC Advantages
Managing glaucoma	NLCs increase corneal permeability and reduce systemic adverse effects, hence improving the sustained release and absorption of anti-glaucoma medicines (e.g., Timolol, Brimonidine).	Prolonged medication retention at the ocular location, improved corneal penetration due to lipophilic composition, and reduced dose frequency all promote patient compliance (Shree et al., 2024).
Infections and inflammation of the eyes (conjunctivitis, uveitis)	NLCs encapsulate antibiotics (e.g., Ciprofloxacin, Moxifloxacin) and anti-inflammatory drugs (e.g., Dexamethasone, Cyclosporine A) for controlled drug release and inflammation reduction.	Increased medication solubility and retention Delivery to affected tissues is targeted. Reduced medication toxicity as compared to traditional formulations (Wu et al., 2024).
Retinal disorders (diabetic retinopathy and age-related macular degeneration - AMD)	NLCs enable the intravitreal and periocular administration of medications such as Bevacizumab, Ranibizumab, and Dexamethasone, assuring long-term therapeutic benefits.	Sustained drug release lowers the need for repeated injections, shields pharmaceuticals from enzymatic degradation, and enhances the bioavailability of hydrophobic medications (Wu et al., 2024).
Antiviral and antifungal agents were delivered	NLCs improve the ocular residence duration of antiviral medications (acyclovir, ganciclovir) and antifungal treatments (natamycin, amphotericin B).	Improved medication penetration into ocular tissues, Effective against bacterial, fungal, and viral eye infections. Side effects are less severe than with regular eye drops (Mahapatra et al., 2025).

Table 1. (Continued)

Applications	Description and Mechanism	NLC Advantages
Gene and protein delivery for ocular regeneration	NLCs provide for the prolonged release of growth factors, nucleic acids, and peptides for retinal regeneration and corneal healing.	Protects genetic materials from damage Allows for the regulated release of therapeutic proteins, perhaps leading to non-invasive gene therapy procedures (Mahapatra et al., 2025).
Dry eye syndrome	Cyclosporine A and hyaluronic acid are delivered via NLCs, providing long-lasting lubrication and anti-inflammatory benefits.	Increases medication retention in the tear film, relieves pain, and increases hydration (Shree et al., 2024).
Corneal wound healing	NLCs supply growth factors, peptides, and antibiotics to help regenerate corneal tissue and prevent infections.	Increased bioavailability at the corneal surface, promoting quicker healing and tissue regeneration (Shree et al., 2024).

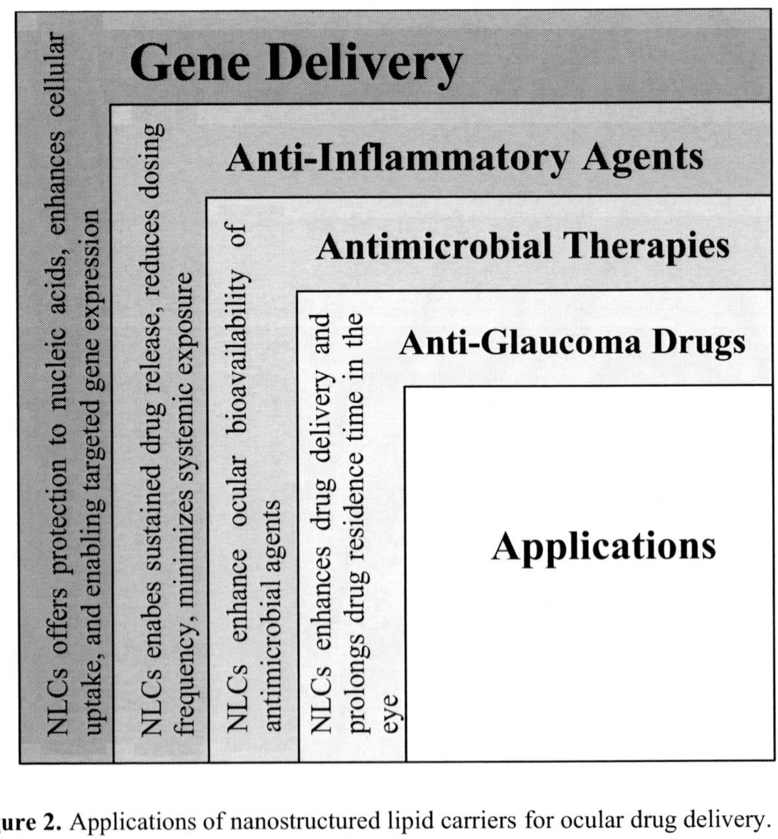

Figure 2. Applications of nanostructured lipid carriers for ocular drug delivery.

NLCs represent a major advancement in ocular drug delivery, offering an innovative platform to improve the therapeutic performance of a wide range of pharmaceutical compounds. These nanocarriers are distinguished by their excellent biocompatibility, the ability to encapsulate both hydrophilic and lipophilic drugs, and their capacity to provide controlled and sustained release. These features make NLCs particularly well-suited for direct ocular application, effectively overcoming the challenges associated with conventional formulations such as poor retention, rapid clearance, and limited drug penetration. As highlighted by Shree et al. (2024), Table 1 presents the diverse therapeutic applications of NLCs in ophthalmic drug delivery, while Figure 2 provides a visual overview of their multifaceted roles in enhancing ocular pharmacotherapy.

Challenges and Limitations

Given the delicate and sensitive nature of the eye, the safety of NLCs in ocular medication administration is critical. Ocular biocompatibility studies examine the interactions of NLCs with ocular tissues such as the cornea, conjunctiva, sclera, and retina to ensure that they do not cause irritation, inflammation, or structural damage when used. Corneal epithelial cell lines are frequently subjected to in vitro experiments such as the MTT cytotoxicity test, lactate dehydrogenase release assay, and reactive oxygen species production investigations. In vivo biocompatibility investigations involve administering NLC formulations to animal models, followed by clinical assessments utilizing slit-lamp biomicroscopy and histological examination to detect inflammatory reactions (Kailasam et al., 2024). Despite the apparent benefits of NLCs in ocular medication administration, possible dangers and long-term consequences must be extensively studied before broad implementation. The nanoscale size of these carriers enables for deep penetration into ocular tissues, but also raises concerns about aggregation, long-term retention, and possible ocular physiologic consequences. Long-term studies are required to determine the fate of NLCs after delivery, including biodegradability, systemic absorption, and clearance mechanisms. Chronic exposure to NLC-based formulations must be evaluated in terms of retinal toxicity, intraocular pressure changes, and possible immunogenicity (Wu et al., 2024). NLC-based ocular preparations must be rigorously evaluated for safety, effectiveness, and quality before they can be approved by regulators. Before giving marketing clearance, regulatory organizations such as the United States Food and Drug

Administration (FDA), the European Medicines Agency (EMA), and India's Central Drugs Standard Control Organisation (CDSCO) require comprehensive preclinical and clinical evidence (Arabpour et al., 2024).

Future Directions

NLCs have a potential future in ocular medication administration, with advances aimed at increasing bioavailability, targeted distribution, and patient compliance. Stimuli-responsive NLCs can release medications in reaction to environmental stimuli, which improves therapeutic effects while lowering dose frequency. Incorporating NLCs with sophisticated biomaterials and surface-modification methods might improve ocular retention and targeted administration, especially in glaucoma, diabetic retinopathy, and age-related macular degeneration. Surface-modified NLCs may give more effective transport to the posterior portion of the eye, a problem in ophthalmic medication development. Another developing field in ocular therapies is NLC-mediated gene and protein delivery. Nucleic acid encapsulation and stabilization inside NLCs has the potential to transform therapy for hereditary and degenerative retinal disorders (Hatem et al., 2024). Future research should look on hybrid NLCs that carry small-molecule medicines and genetic materials at the same time, potentially providing a combination therapy approach for illnesses that require both pharmacological and genetic intervention. Nanotechnology combined with artificial intelligence (AI) and machine learning can optimize NLC formulation design, anticipate drug release patterns, and improve preclinical and clinical screening processes. However, regulatory progress and clinical translation are critical for the general use of NLC-based ocular therapeutics. To commercialize NLC-based ophthalmic formulations, interdisciplinary collaboration among pharmaceutical scientists, ophthalmologists, and regulatory bodies is required (Iqbal et al., 2024).

Conclusion

NLCs provide a promising development in ocular medication delivery, addressing the constraints of traditional eye formulations. Their unusual structure, which combines solid and liquid lipids, increases medication solubility, retention time, and ocular permeability. NLCs provide for regulated and prolonged medication release, lowering dose frequency and systemic adverse effects. Surface alterations can enable for targeted and mucoadhesive administration, which increases therapeutic effectiveness. Preclinical and clinical investigations show that they have the potential to treat glaucoma, infections, and retinal diseases. However, issues like as large-scale manufacturing, long-term stability, and regulatory barriers persist. To modernize ophthalmic therapies and enhance patient compliance, future research should focus on refining formulation tactics and combining smart drug delivery technology.

References

Adwan, S., Qasmieh, M., Al-Akayleh, F., & Agha, A. S. (2024). Recent Advances in Ocular Drug Delivery: Insights into Lyotropic Liquid Crystals. *Pharmaceuticals*, *17*(10), 1315.

Alatawi, H. M., Alhwiti, S. S., Alsharif, K. A., Albalawi, S. S., Abusaleh, S. M., Sror, G. K., & Qushawy, M. (2024). Nanostructured Lipid Carriers (NLCs) as Effective Drug Delivery Systems: Methods of Preparation and their Therapeutic Applications. *Recent Patents on Nanotechnology*, *18*(2), 179-189.

Arabpour, Z., Salehi, M., An, S., Moghtader, A., Anwar, K. N., Baharnoori, S. M., Shah, R. J., Abedi, F., & Djalilian, A. R. (2024). Exploring Hydrogel Nanoparticle Systems for Enhanced Ocular Drug Delivery. *Gels*, 10(9), 589.

Baig, M. S., Karade, S. K., Ahmad, A., Khan, M. A., Haque, A., Webster, T. J., Faiyazuddin, M., & Al-Qahtani, N. H. (2024). Lipid-based nanoparticles: innovations in ocular drug delivery. *Frontiers in Molecular Biosciences*, *11*, 1421959.

Borges, G. S. M., Prazeres, P. H. D. M., De Souza, Â. M., Yoshida, M. I., Vilela, J. M. C., Silva, A. T. M. E., Oliveira, M. S., Gomes, D. A., Andrade, M. S., De Souza-Fagundes, E. M., & Ferreira, L. A. M. (2021). Nanostructured lipid carriers as a novel tool to deliver sclareol: physicochemical characterisation and evaluation in human cancer cell lines. *Brazilian Journal of Pharmaceutical Sciences*, *57*, e18497.

Chakole, C. M., & Chauhan, M. K. (2021). Research progress of nanostructured lipid carriers in ocular drug delivery. *Drug Delivery Letters*, *11*(3), 203-219.

Chandel, A., & Kandav, G. (2024). Insights into ocular therapeutics: a comprehensive review of anatomy, barriers, diseases and nanoscale formulations for targeted drug delivery. *Journal of Drug Delivery Science and Technology*, 97, 105785.

Cheng, K., Hsieh, C., Nepali, K., & Liou, J. (2020). Ocular Disease Therapeutics: Design and delivery of drugs for diseases of the eye. *Journal of Medicinal Chemistry*, *63*(19), 10533-10593.

Cimino, C., Zingale, E., Bonaccorso, A., Musumeci, T., Carbone, C., & Pignatello, R. (2024). From preformulative design to in vivo tests: a complex path of requisites and studies for nanoparticle ocular application. Part 1: design, characterization, and preliminary in vitro studies. *Molecular Pharmaceutics*, *21*(12), 6034-6061.

Fathi, F., Machado, T. O. X., de A C Kodel, H., Portugal, I., Ferreira, I. O., Zielinska, A., Oliveira, M. B. P. P., & Souto, E. B. (2024). Solid lipid nanoparticles (SLN) and nanostructured lipid carriers (NLC) for the delivery of bioactives sourced from plants: part I–composition and production methods. *Expert Opinion on Drug Delivery*, *21*(10), 1479-1490.

Fathi-Karkan, S., Amiri Ramsheh, N., Arkaban, H., Narooie-Noori, F., Sargazi, S., Mirinejad, S., Roostaee, M., Sargazi, S., Barani, M., Malahat Shadman, S., Althomali, R. H., & Rahman, M. M. (2024). Nanosuspensions in ophthalmology: Overcoming challenges and enhancing drug delivery for eye diseases. *International Journal of Pharmaceutics*, *658*, 124226.

Giri, B. R., Jakka, D., Sandoval, M. A., Kulkarni, V. R., & Bao, Q. (2024). Advancements in ocular therapy: a review of emerging drug delivery approaches and pharmaceutical technologies. *Pharmaceutics*, *16*(10), 1325.

Hassan, D. H., Shohdy, J. N., El-Nabarawi, M., El-Setouhy, D. A., & Abdellatif, M. M. (2022). Nanostructured lipid carriers for transdermal drug delivery. *International Journal of Applied Pharmaceutics*, 88–93.

Hatem, S., Mohammed, D., & Ezzat, N. (2024). Nanotechnology-based strategies overcoming the challenges of retinoblastoma: a comprehensive overview and future perspectives. *Future Journal of Pharmaceutical Sciences*, *10*(1), 14.

Iqbal, H., Razzaq, A., Zhou, D., Lou, J., Xiao, R., Lin, F., & Liang, Y. (2024). Nanomedicine in glaucoma treatment; Current challenges and future perspectives. *Materials Today Bio*, 28, 101229.

Kailasam, V., Sai Veda Koduganti, S., Dasgupta, O., Garg, P., & Nirmal, J. (2024). Ocular delivery of Amphotericin B: Current challenges and future perspectives. *Expert Opinion on Drug Delivery*, *21*(12), 1793-1805.

Lv, Z., Li, S., Zeng, G., Yao, K., & Han, H. (2024). Recent progress of nanomedicine in managing dry eye disease. *Advances in Ophthalmology Practice and Research*, 4(1), 23-31.

Mahapatra, M., Mohapatra, S., & Jena, G. K. (2025). Nanostructured Lipid Carriers: A Revolutionary Approach in Antifungal Drug Delivery. *Biomedical Materials & Devices*, 2025, 1-17.

Onugwu, A. L., Emencheta, S. C., Sabastine, R. N., Onugwu, S. O., Echezona, A. C., Nwagwu, C. S., Chime, S. A., Ezike, T. C., Eze, O. P., Onyeyilim, E. L., Attama, A. A., & Prajapati, B. G. (2024). Review of nanoformulations for treating ocular fungal infections. *Journal of Drug Delivery Science and Technology*, 101, 106219.

Putchakayala, N. S. D., Morusu, K., Ramisetty, S., & Ravoru, N. (2025). Nanostructured Lipid Carriers: A Novel Platform in the Formulation of Targeted Drug Delivery Systems. *Current Nanomedicine*, *15*(1), 26-49.

Rajan, P. B., Koilpillai, J., & Narayanasamy, D. (2024). Advancing Ocular Medication Delivery with Nano-Engineered Solutions: A Comprehensive Review of Innovations, Obstacles, and Clinical Impact. *Cureus*, *16*(8), e66476.

Rajoriya, V., Gupta, R., Vengurlekar, S., & Singh, U. S. (2024). Nanostructured lipid carriers (NLCs): A promising candidate for lung cancer targeting. *International Journal of Pharmaceutics*, 655, 123986.

Shekhar, S., Sharma, S., Kumar, A., Taneja, A., & Sharma, B. (2021). The framework of nanopesticides: a paradigm in biodiversity. *Materials Advances*, *2*(20), 6569-6588.

Shree, D., Patra, C. N., & Sahoo, B. M. (2024). Applications of nanotechnology-mediated herbal nanosystems for ophthalmic drug. *Pharmaceutical Nanotechnology*, *12*(3), 229-250.

Singh, A. K., Kumar, M., & Upadhyay, P. K. (2024). Nanostructured Lipid Carriers Mediated Drug Delivery to Posterior Segment of Eye and their In-vivo Successes. *Current Pharmaceutical Biotechnology*, *25*(6), 713-723.

Singh, C. P., Rai, P. K., Kumar, M., Tiwari, V., Tiwari, A., Sharma, A., & Sharma, K. (2024). Emphasis on Nanostructured Lipid Carriers in the Ocular Delivery of Antibiotics. *Pharmaceutical Nanotechnology*, *12*(2), 126-142.

Stefanov, S., Gugleva, V., & Andonova, V. (2023). Technological strategies for the preparation of lipid nanoparticles: an updated review. *Pharmacia*, *70*(3), 449–463.

Varela-Fernández, R., García-Otero, X., Díaz-Tomé, V., Regueiro, U., López-López, M., González-Barcia, M., Lema, M. I., & Otero-Espinar, F. J. (2022). Lactoferrin-loaded nanostructured lipid carriers (NLCs) as a new formulation for optimized ocular drug delivery. *European Journal of Pharmaceutics and Biopharmaceutics*, *172*, 144-156.

Wathoni, N., Suhandi, C., Elamin, K., Lesmana, R., Hasan, N., Mohammed, A., El-Rayyes, A., & Wilar, G. (2024). Advancements and challenges of nanostructured lipid carriers for wound healing applications. *International Journal of Nanomedicine*, *Volume 19*, 8091–8113.

Wu, K. Y., Gao, A., Giunta, M., & Tran, S. D. (2024). What's new in ocular drug delivery: advances in suprachoroidal injection since 2023. *Pharmaceuticals*, *17*(8), 1007.

Wu, Y., Li, X., Fu, X., Huang, X., Zhang, S., Zhao, N., Ma, X., Saiding, Q., Yang, M., Tao, W., Zhou, X., & Huang, J. (2024). Innovative Nanotechnology in Drug Delivery Systems for Advanced Treatment of Posterior Segment Ocular Diseases. *Advanced Science*, 11(32), 2403399.

Younis, M. A., Tawfeek, H. M., Abdellatif, A. A., Abdel-Aleem, J. A., & Harashima, H. (2021). Clinical translation of nanomedicines: Challenges, opportunities, and keys. *Advanced Drug Delivery Reviews*, *181*, 114083.

Youssef, A. A. A., Dudhipala, N., & Majumdar, S. (2022). Dual drug loaded lipid nanocarrier formulations for topical ocular applications. *International Journal of Nanomedicine*, 17, 2283-2299.

Chapter 4

Solid Lipid Nanoparticles for Ocular Drug Delivery

Bhawna Sharma[1]
Iti Chauhan[2,*]
Rohit Pandey[1]
and Mohd Yasir[3]
[1]Department of Pharmaceutics, Dr. K.N. MODI Institute of Pharmaceutical Education and Research, Modinagar, Ghaziabad, Uttar Pradesh, India
[2]Department of Pharmaceutics, I.T.S College of Pharmacy, Ghaziabad, Uttar Pradesh, India
[3]Department of Pharmacy, College of Health Sciences, Arsi University, Asella, Ethiopia

Abstract

Ocular drug delivery presents substantial challenges due to the eye's complex anatomy and robust protective barriers. Conventional treatment approaches, such as eye drops, often fail to achieve therapeutic drug concentrations at the target site because of rapid tear drainage, limited precorneal residence time, and poor bioavailability. Additionally, the physicochemical properties of many drugs further compromise their efficacy when administered through traditional routes. These limitations underscore the urgent need for innovative drug delivery strategies to improve therapeutic outcomes in ocular diseases. Solid lipid nanoparticles (SLNs) have emerged as a promising and versatile platform to address these challenges. SLNs offer several notable advantages, including enhanced drug permeation across ocular barriers and sustained drug release, thereby improving bioavailability and reducing dosing

* Corresponding Author's Email: iti.pharma@gmail.com

In: Advances in Nanotechnology for Ocular Drug Delivery
Editors: Anjali Sharma, Ajmer Singh Grewal, Geeta Deswal et al.
ISBN: 979-8-89530-814-1
© 2025 Nova Science Publishers, Inc.

frequency. As lipid-based nanocarriers, SLNs also protect therapeutic agents from enzymatic degradation within the tear film and enable targeted delivery to specific ocular tissues. This chapter examines the design and development of SLN-based formulations for the treatment of various ocular disorders. It highlights the capacity of SLNs to overcome ocular barriers, enhance therapeutic efficacy, and improve patient compliance. In addition, future perspectives and the challenges associated with translating SLN-based drug delivery systems from laboratory research to clinical application are discussed.

Keywords: Anterior and posterior eye disorders, controlled release systems, lipid-based drug delivery, nanocarriers, ocular bioavailability, ocular drug delivery, ocular nanotechnology, solid lipid nanoparticles (SLNs), sustained drug release, tear film barriers

Abbreviations

API: Active Pharmaceutical Ingredient
CAM: Chorioallantoic Membrane
DLS: Dynamic Light Scattering
GRAS: Generally Recognized As Safe
HET: Hen's Egg Test
HET-CAM: Hen's Egg Test on Chorioallantoic Membrane
HL: Hydration Level
HLB: Hydrophilic-Lipophilic Balance
HPLC: High Performance Liquid Chromatography
IOP: Intraocular Pressure
mOsm/L: Milliosmoles per Liter
o/w: Oil-in-Water
PDI: Polydispersity Index
PEG: Polyethylene Glycol
PIT: Phase Inversion Temperature
PVA: Polyvinyl Alcohol
QbD: Quality by Design
RI: Refractive Index
S-HCE: SkinEthic Human Corneal Epithelium
SIRC: Statens Serum Institut Rabbit Corneal
SLNs: Solid Lipid Nanoparticles
TGA: Thermogravimetric Analysis

USP: United States Pharmacopeia
UV: Ultraviolet
VEGF: Vascular Endothelial Growth Factor
W/O: Water-in-Oil

Introduction

Ocular drug delivery poses significant challenges due to the eye's complex structure and robust protective mechanisms. Despite the convenience of traditional eye drops, their therapeutic efficacy is often limited by rapid drainage, short residence time, and poor bioavailability. The physicochemical properties of many drug molecules further hinder their effectiveness when delivered via conventional routes. Recent advancements in ocular drug delivery systems aim to overcome these limitations and enhance therapeutic outcomes. Researchers have explored numerous innovative approaches such as nano-formulations, in situ gels, implants, contact lenses, and microneedles, all designed to improve drug penetration, prolong residence time, and sustain drug release in both the anterior and posterior segments of the eye (Pawar et al., 2018). The anterior segment comprises the ciliary body, aqueous humor, cornea, conjunctiva, iris, and lens, while the posterior segment includes the sclera, choroid, retina, and vitreous body (Ita, 2022; Vaneev et al., 2021). The cornea, located at the forefront of the iris and pupil, is densely innervated and maintains a stable curvature throughout life, with a central optic zone that provides the majority of its refractive function (Schermer et al., 1989; DelMonte et al., 2011). The conjunctiva, a thin, transparent mucous membrane, secretes fluid, mucus, and electrolytes and is exposed to the air along with the tear film covering the cornea (Xu et al., 2022). The pupil, a central circular aperture within the iris, regulates the amount of light entering the eye and is controlled by a light-dependent reflex mechanism (Szabadi, 2018). The iris divides the anterior ocular chamber into two parts and is suspended in aqueous humor, which flows from the posterior to the anterior chamber through the pupil and drains via the Canal of Schlemm (Jakubiak et al., 2019). The ciliary body, located near the ora serrata, produces aqueous humor and is connected to the lens via suspensory ligaments (Achouri et al., 2012). The aqueous humor, an alkaline fluid rich in sodium and chloride ions, plays a vital role in nutrient transport and intraocular pressure regulation (Ameeduzzafar et al., 2016). The lens, situated behind the pupil and in front of the vitreous chamber, functions to refract and focus light onto the retina

(Ruan et al., 2020). In the posterior segment, the sclera serves as a protective outer layer and anchorage point for the extraocular muscles (Malhotra et al., 2011). The choroid, a vascular layer beneath the sclera, supplies oxygen and nutrients to the retina but also acts as a barrier to drug diffusion due to the blood–retinal barrier (Del Amo et al., 2017). The retina lines the inner surface of the eye and interfaces with both the choroid and the vitreous body (Naylor et al., 2019). The vitreous humor, located behind the lens, is a transparent, jelly-like substance that maintains intraocular pressure and remains constant throughout life (Mishra et al., 2024). The optic nerve transmits visual signals from the retina to the brain (Lou et al., 2021). Physiologically, the eye remains hydrated and functional through various mechanisms. The aqueous humor provides oxygen and nutrients to the cornea, lens, and vitreous body, while the cornea also absorbs oxygen directly from the atmosphere. Insufficient oxygen leads to lactic acid accumulation, resulting in corneal edema and visual impairment (Balyen, 2022). The sclera maintains intraocular pressure and structural integrity. The lacrimal system keeps the eye clean and moist through continuous blinking and tear secretion, distributing fluid evenly across the eye's surface. Tear fluid contains antimicrobial agents like lysozyme and is replenished rapidly to prevent dryness and infection (Tóth-Molnár, 2020). The ciliary body contributes to aqueous humor production, which helps regulate intraocular pressure. Imbalances in this system may lead to conditions like glaucoma (Ficarrotta, 2020). Ocular drug delivery is further limited by various barriers. The cul-de-sac, a recess in the eyelids, can hold only about 30 μL of fluid, which is reduced during blinking or inflammation, affecting drug concentration and efficacy (Singh et al., 2020). Drug loss from lacrimal fluid due to drainage, lacrimation, and protein binding further reduces effectiveness (Lee et al., 1979). The cornea is divided into the epithelium, stroma, and endothelium, each selectively permeable to different drug types; hydrophilic drugs are blocked by the epithelium, while lipophilic drugs face resistance in the stroma. The endothelium allows selective entry of macromolecules while maintaining corneal clarity (Ahmed et al., 2022; Singh et al., 2020). The degree of ionization, hydrophobicity, and molecular weight of the drug significantly influence corneal permeability, making trans-corneal diffusion a rate-limiting step. Blood-ocular barriers, including the blood-aqueous barrier (BAB) and blood-retinal barrier (BRB), also limit systemic drug penetration. While small, lipophilic molecules can cross the BAB, they are cleared quickly. The BRB, composed of retinal pigment epithelial and endothelial cells, restricts access to the retina, preventing harmful substances from entering (Singh et al., 2020). Conventional ocular drug delivery systems face considerable limitations in maintaining therapeutic concentrations at the target site for sufficient durations. Frequent dosing is often required due to the rapid clearance of drugs from the ocular surface, but this can lead to ocular surface irritation and toxicity. The unique anatomy, physiology, and barrier functions of the eye contribute to low drug absorption, with less than 1% of the administered dose typically reaching intraocular tissues. Achieving therapeutic efficacy requires a careful balance between hydrophilicity, lipophilicity, and prolonged ocular contact time (Anand et al., 2020).

Solid Lipid Nanoparticles: Formulation and Characterization

SLNs were first developed in the mid-1990s and have since been recognized as promising alternatives to traditional drug delivery systems such as liposomes, emulsions, micelles, and polymeric nanoparticles due to their numerous advantages (Mehnert et al., 2012). These nanocarriers are typically composed of "*Generally Recognized as Safe* (GRAS)" materials, including physiologically biocompatible and biodegradable lipids, making them suitable and safe for pharmaceutical applications. One of the key benefits of SLNs is their ability to protect encapsulated drugs within their solid lipid matrix, thereby enhancing drug stability (Qushawy et al., 2019). SLNs also offer higher entrapment efficiency compared to liposomes for both hydrophilic and hydrophobic drugs. Furthermore, the lipid composition of SLNs can be modified to control drug release profiles, allowing for either immediate or sustained delivery. To enhance targeting and further improve stability, surface modifications can be made to SLNs, facilitating the delivery of therapeutic agents to specific organs or tissues (Baek et al., 2017).

Types of SLN Models

The incorporation of drugs into SLNs typically follows three main models, which are influenced by factors such as the drug's solubility, the composition of the lipid matrix, the manufacturing temperature, and the chosen production method. These models include the homogeneous matrix model, the drug-enriched shell model, and the drug-enriched core model (Ganesan & Narayanasamy, 2017; Shirodkar et al., 2019). Each model determines the spatial distribution of the drug within the nanoparticle, thereby impacting the drug release profile, stability, and overall therapeutic efficacy (Figure 1).

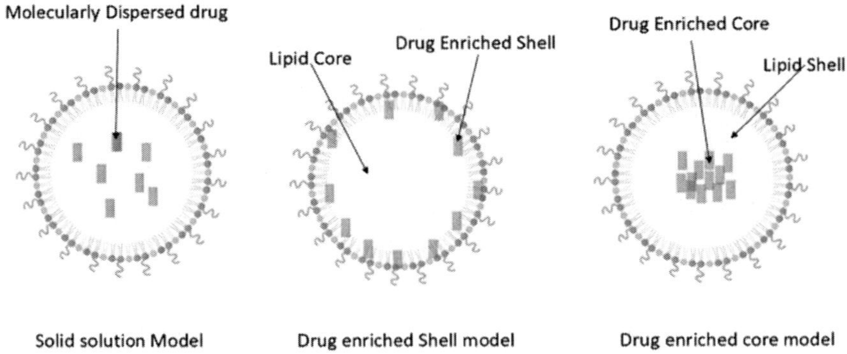

Figure 1. Type of SLN models.

SLN Type 1: Homogenous Matrix (Solid Solution) Model

In this model, the drug is uniformly dispersed within the lipid matrix in a molecularly dissolved or amorphous form. The homogeneous matrix structure arises when the active pharmaceutical ingredient is evenly distributed throughout the lipid core as molecular clusters or in an amorphous state. This type is typically prepared using high-pressure homogenization or cold homogenization techniques conducted at temperatures above the lipid's melting point.

SLN Type 2: Drug-Enriched Shell Model

This model is produced through a heat-induced homogenization process. As the lipid reaches its recrystallization temperature, a solid lipid core begins to form. During this stage, the drug tends to remain concentrated in the surrounding liquid shell of the nanoparticle. Upon cooling, the outer shell solidifies, effectively trapping the drug within this layer.

SLN Type 3: Drug-Enriched Core Model

In this model, drug precipitation occurs before the lipid solidifies. When the drug concentration within the lipid matrix approaches its saturation solubility limit, the cooling of the nanoemulsion results in the drug becoming supersaturated and precipitating. Lipid recrystallization then occurs around the drug-enriched core, forming a structure that resembles a membrane. This model is particularly suitable for drugs that require a sustained or extended-release profile over time.

Preparation Methods for Fabrication SLNs

The various methods employed for the preparation of SLNs are summarized in Table 1.

Table 1. Preparation methods for SLNs

Method	Instrumentation	Procedure	Reference
Hot homogenization	• High pressure homogenizer • Extruder	The lipid is heated beyond its melting point while being subjected to constant agitation, at which point the drug is amalgamated with a heated surfactant solution. Following this, the dispersion is homogenized using a high-shear mixing or extrusion apparatus at temperatures sustained above the melting point of the lipid, subsequently undergoing high-pressure homogenization in a thermally controlled environment to yield a pre-emulsion. The homogenization procedures are reiterated until the desired particle size is attained. The dispersion is then cooled to ambient temperature to facilitate the formation of a colloidal solution of lipid nanoparticles.	Dhanya et al., 2023
Cold homogenization	• Ball mill, • High pressure homogenizer • Bath sonicator	A molten lipid dissolves the drug. Liquid nitrogen or dry ice are used to solidify this drug-lipid combination. The solidified bulk is ground into powder with particle sizes as small as 100 μm. To create pre-suspension, the resulting powder is dissolved in a cold aqueous surfactant solution. Lastly, to create colloidal suspensions of lipid LNPs, the pre-suspension is put through several cycles of HPH.	Hernández-Esquivel et al., 2022

Table 1. (Continued)

Method	Instrumentation	Procedure	Reference
Melt emulsification/ microemulsion	• Ultra-Turrax® T25 • Ultrasonicator (probe sonicator)	This approach does not use an organic solvent. The medication is dissolved into the lipid when it melts over its melting point temperature. To generate an o/w microemulsion, this molten lipid phase is introduced into a heated aqueous surfactant solution while being vigorously stirred. To create SLN, it is ultimately dissolved in cold water with the use of high-speed agitation and ultrasonication.	Chantaburanan et al., 2023
Solvent emulsification, diffusion	• High speed stirrer • Ultra Turrax® • Ultra-homogenizer	The cosolvent was used to dissolve the drug and lipid mixture. After high-speed stirring and ultrasonication, the resulting solution is gradually added to an aqueous surfactant solution. This is when the cosolvents start to permeate into the aqueous medium, which causes lipids to precipitate and SLN to be produced.	Kumar et al., 2022
Phase inversion temperature (PIT)	• Magnetic stirrer • Mechanical stirrer	The phase inversion temperature (PIT) method facilitates the transition from a w/o to an o/w emulsion by rapidly cooling a mixture of surfactants to room temperature following an elevation in temperature above the phase transition temperature of 10°C. A nanoemulsion is created by altering the phase composition with an HLB converting agent. Mechanical and magnetic stirrers are employed to continually stir the mixes during the heating stage.	Spadaro et al., 2021
Coacervation	• Magnetic stirrer	During the coacervation procedure, a surfactant solution is elevated above the Krafft Point while being continuously agitated. Alkaline salts of fatty acids precipitate until the lipids are completely separated when the pH is reduced by the incremental addition of acidifying agents (coacervating solutions) to the mixture of molten lipids and polymer (PVA) solution.	Malekpour-Galogahi et al., 2018

Method	Instrumentation	Procedure	Reference
		Subsequently, the mixture undergoes cooling while maintaining constant stirring.	
Supercritical fluid	• Supercritical fluid assembly	A supercritical fluid is used to remove the organic solvent once the o/w emulsion has been created. Supercritical CO_2 is delivered from the bottom in a countercurrent fashion, while the emulsion is injected top-down within an extraction column. Because the solvent is quickly and completely removed, lipids precipitate.	Saad et al., 2020

Characterization of SLNs

Table 2. Characterization of SLNs

Parameter	Method/Technique	Significance	Reference
Particle size and polydispersity index (PDI)	Photon correlation spectroscopy, Dynamic Light Scattering (DLS), Electron Microscopy, Atomic force microscopy	Affects drug loading, release kinetics, and cellular uptake. PDI indicates uniformity; lower PDI values suggest better quality and stability	Mukherjee et al., 2009
Zeta potential or electrokinetic potential	Laser Doppler Electrophoresis	Determines stability by predicting particle aggregation. Higher zeta values, (positive or negative), indicate higher dispersion stability. In ocular drug delivery, corneal epithelial cells have a negative charge. Cationic SLNs can enhance drug residence time and penetration by interacting ionically with the cornea, potentially modifying its morphology.	Bhattacharjee, 2016
Encapsulation efficiency and drug loading	Ultracentrifugation, Ultrafiltration, Spectrophotometry, Gravimetric Analysis, filtration using a sephadex column, Microdialysis	Measures drug retention in SLN formulation. Affects therapeutic efficacy and drug release profile.	Yang & Juhu, 2002; Jain et al, 2005; Heaiti et al.,1998

Table 2. (Continued)

Parameter	Method/Technique	Significance	Reference
Particle morphology	Transmission Electron Microscopy, Scanning Electron Microscopy, Atomic force microscopy	Provides insights into surface characteristics and integrity.	Severs, 2007; Manjunath et al., 2005
Crystallinity	Differential Scanning Calorimetry, X-ray Diffraction (XRD), Wide angle X-ray diffraction	Investigation of polymorphic modifications, melting, crystallization behaviour, solid-to-solid transition temperatures of lipids, and solid fat content of the excipient	Kakkar et al., 2014
Thermal stability	Thermogravimetric Analysis (TGA), Differential Scanning Calorimetry	Ensures SLNs remain stable during storage and processing.	Seyfoddin et al., 2010
Drug release profile	Dialysis Method, UV-Visible Spectroscopy, HPLC	Essential for determining controlled or sustained release behavior.	Seyfoddin et al., 2010
Surface hydrophobicity	Contact Angle Measurement	Affects cellular uptake and biocompatibility, Influences adhesion to ocular tissues and drug penetration	Seyfoddin et al., 2010
Stability	Zeta Potential, Particle Size Analysis, Stability studies	Ensures long-term performance and effectiveness of the formulation. Determination of shelf life	Seyfoddin et al., 2010
In-vitro drug release	Franz Diffusion Cell, Dialysis Method	Predicts therapeutic potential and release dynamics.	Pandey et al., 2021
In vivo biocompatibility	Animal Studies	Confirms suitability for therapeutic use.	Seyfoddin et al., 2010

The characterization of SLNs (as outlined in Table 2) is primarily conducted to ensure the quality, stability, and performance of the formulated nanoparticles. Due to the inherent complexity of the system and the extremely small particle size, characterizing SLNs presents a significant challenge for researchers. However, a thorough evaluation of certain key parameters is essential, as these characteristics directly influence the stability of the formulation and the release kinetics of the encapsulated drug. Accurate characterization helps in predicting the behavior of SLNs in biological environments and in optimizing their therapeutic efficacy.

In Vitro and *In Vivo* Studies for Ocular Delivery

pH and Osmolarity

The pH of ocular drug formulations is a critical parameter influencing patient comfort and drug bioavailability. The normal pH of human tears is approximately 7.2, and significant deviations from this value can irritate, increase lacrimation, and premature drug elimination, leading to reduced bioavailability. For ocular SLNs, an optimum pH of 7.2 ± 0.2 is recommended to ensure maximum comfort and minimal irritation. The pH of the formulation can be determined using a pH meter (Yadav et al., 2020; Larkins et al., 2024). The osmolarity of human tears typically ranges between 290-310 mOsm/L. However, when a solution is instilled into thc cyc, evaporation can increase its osmolarity. If a hypertonic solution is administered, the evaporation-induced rise in osmolarity can further exacerbate tear film instability. This can lead to ocular surface damage and discomfort. Hyperosmotic drops can cause dehydration of corneal and conjunctival epithelial cells, leading to irritation, inflammation, and potential cytotoxicity. Hypotonic solutions can help in rehydrating the ocular surface and maintaining epithelial integrity, making them more suitable for ocular drug delivery. The osmolarity of the developed SLN formulation can be determined using an osmometer, which measures the osmolar concentration of solutes in the formulation. The results are expressed as osmolality (in mmol/kg) or osmolarity (in mmol/L) (Yadav et al., 2020; Larkins et al., 2024).

Refractive Index (RI)

The refractive index of SLNs can be determined using Abbe's refractometer. For minimum interference with vision, a RI value close to that of ocular tissues (1.34-1.41) is considered ideal. To minimize visual disturbance, topically applied eyedrops should be optically transparent and should ideally have a refractive index like tear fluid (Yadav et al., 2020; Agarwal et al., 2021).

Viscosity

The viscosity of developed SLN dispersion can be determined using a Brookfield Rheometer or Ubbelohde viscometer at 25 ± 0.1 °C. For ocular formulations, an optimal viscosity of ≤ 20 cP is recommended. Higher values can lead to reflex tearing and increased blinking, as the eye attempts to restore the natural lacrimal fluid viscosity (1.05-5.97 cP). The optimized viscosity ensures that the formulation does not cause excessive blinking or reflex tearing while enhancing its retention on the ocular surface for a longer duration. To further improve ocular drug delivery, viscosity modifiers can be incorporated to enhance corneal retention, increase the residence time of the formulation in contact with the cornea, and reduce lacrimal drainage. This helps improve ocular bioavailability and ensures sustained drug absorption for effective therapeutic outcomes (Vicente-Pascual et al., 2020; Yadav et al., 2020).

Adhesion Test: Flow Rates

An adhesion test is conducted to evaluate the flow behavior of the formulations, where a higher flow rate indicates lower adhesiveness. An in vitro model-based method can be employed to simulate the flow rate of ocular formulations on the corneal surface. This method utilizes a gel substrate with a surface tension-like tear fluid (28 dyne/cm) to mimic physiological conditions. A gel formulation (5% hydroxyethyl cellulose + 1% decylpolyglucoside + 3% hyaluronic Acid + water) is prepared to resemble properties of tear film. The prepared gel is evenly spread onto a glass support and allowed to set until it achieves the desired hardness. The glass support is inclined at an 18° angle to simulate the natural flow dynamics on the ocular surface. A fixed volume (e.g., 100 μL) of the test formulation is placed at the top of the inclined gel surface. The time required for the sample to travel a specified distance is recorded. The flow rate is expressed as distance covered per unit time (cm/s) (Vicente-Pascual et al., 2020).

In Vitro Drug Release Study

The in vitro drug release from ocular SLNs occurs primarily through diffusion and/or erosion of the lipid matrix. Various diffusion-based methods are used to simulate ocular physiological conditions and evaluate drug release profiles. The most employed techniques include the dialysis bag diffusion method, Franz diffusion cell, USP dissolution apparatus. The cumulative drug release (%) is plotted against time to assess release kinetics. Mathematical models such as zero-order, first-order, Higuchi, Hixon-Crowell, or Korsmeyer-Peppas equations help determine the predominant release mechanism. The manufacturing technique, drug solubility in the lipid matrix, drug/lipid interactions, type and concentration of surfactant, concentration of lipid, lipid matrix composition, and particle size are the key variables affecting drug release from SLNs. SLNs typically exhibit a biphasic release profile, characterized by an initial burst release followed by sustained drug release. The initial burst effect occurs due to the rapid diffusion of surface-adsorbed drug molecules, while the entrapped drug is released gradually through diffusion and lipid matrix degradation. This controlled and sustained release mechanism enhances ocular bioavailability and prolongs drug retention at the target site (Sawant & Dodiya, 2008).

Corneal Permeation

The nanosystem's ability to pass through the cornea is investigated using a variety of in vitro, ex vivo, and in vivo assays. Although rodents were typically used in in vivo models, epithelial cell layer cultures, rebuilt corneas, or excised corneas were used in in vitro and ex vivo models (Ahmed et al., 2022; Singh et al., 2020). For the evaluation of corneal permeability of drugs, rabbit, porcine, and bovine corneas have been investigated with rabbit being the most commonly used animal model due to extensive ocular pharmacokinetic data availability. in vitro cell culture models can be classified into three different groups:

1) Primary cell culture- Rabbit corneal epithelial or conjunctival cells
2) Immortalized cell lines- Human corneal epithelial cells, Transformed human corneal epithelial cells, Clonetics human corneal epithelium, SV40-immortalized human corneal epithelial cell line, Statens Serum Institut rabbit corneal (SIRC) cells, SkinEthic human corneal epithelium (S-HCE)
3) Reconstructed tissue cultures- bovine corneal tissue construct, Human corneal construct, multi-layered human and porcine corneal tissue construct

Corneal permeation studies are generally performed by preparing the tissue in vitro or ex vivo and placing it in a diffusion cell assembly. A variety of permeation chambers are available, including horizontal perfusion cells, modified Erlenmeyer flask diffusion cells, polycarbonate corneal perfusion chambers, modified Franz diffusion cells, Ussing chamber, modified ussing chambers, Valia-Chien diffusion cell and Franz-type diffusion cells. To assess the nanosystem's permeation potential, many permeation parameters are evaluated. The permeability coefficient, enhancement ratio (ER), average flux (J max), and the amount of medication penetrated per unit area (μg/cm2) are examples of permeability metrics (Sanatana et al., 2023; Agarwal & Rupenthal, 2016).

In Vivo Ocular Irritation Test

The Draize test, developed in 1944, is an acute ocular toxicity assay designed to assess the irritation potential of materials that may come in accidental contact with human eyes. In the standard Draize method 10 μL of a test liquid (or 10 mg of a test solid) is instilled into the lower conjunctival cul-de-sac of the test subject's eye, followed by a saline rinse. Ocular reactions are systematically evaluated based on parameters such as:

1) Corneal opacity and area of corneal involvement
2) Conjunctival hyperemia (redness) and chemosis (swelling)
3) Ocular discharges
4) Iris abnormalities and light reflex responses

Observations are recorded at predefined time points: 1 hour, 24 hours, 48 hours, 72 hours, 7 days, and 21 days post-administration. The test is typically conducted in rabbits, given their large eyes, well-described anatomy, ease of handling, cost-effectiveness, and availability (Walig et al.,2017). To enhance precision and reliability, additional modern assessment techniques are incorporated:

1) Slit-lamp biomicroscopy for detailed evaluation of corneal lesions
2) Fluorescein staining to detect micro-ulcerations and epithelial damage
3) Optical or ultrasonic pachymetry to measure corneal thickening and assess edema

Hen's Egg Test or Hühner-Embryonen-Test (HET)

The Hen's Egg Test on the Chorioallantoic Membrane (HET-CAM) was introduced by Luepke in 1985 as an alternative to the Draize rabbit eye test. This method evaluates the toxicity and safety of various substances, including chemicals, pharmaceuticals, and cosmetics. HET has gained prominence as a humane alternative to traditional animal testing. HET provides critical insights into embryo lethality, embryonic growth and development, teratogenicity and systemic toxicity, including immunopathological aspects and metabolic pathways. It is the only test that closely mimics the conjunctivae of the human eye, enabling the semi-quantitative assessment of a chemical's irritant potential on the CAM. The irritation effects on the membrane are observed immediately after applying the pure or diluted chemical. This method is valuable for evaluating potential ocular discomfort caused by test nanosystems due to the structural and vascular similarities between human mucosal tissue and CAM (Lupke, 1986; Viera et al., 2022). The irritant response is assessed based on the appearance of CAM blood vessel reactions such as clotting, bleeding, and hyperemia, which are used to determine an irritation score. The test relies on observing these effects within five minutes of applying the test substance to the CAM of a fertilized chicken egg that has been incubated for ten days. Freshly fertilized hen eggs are incubated under controlled conditions (37 ± 0.5°C and 67 ± 5% relative humidity) for ten days, with rotation every 12 hours (Lupke, 1986; Viera et al., 2022).

In Vivo Release Studies and Toxicological Tests

Animal studies are typically essential for examining the ocular bioavailability and release characteristics of drugs formulated within diverse drug delivery systems. While larger animal models are employed in certain experimental designs, the rabbit is frequently utilized in ocular research due to its relevance. There exist significant physical and physiological disparities between the human and rabbit ocular systems, with the primary distinction being the atypical blinking rates, which may influence the retention of the drug on the preocular surface (Basaran et al., 2009). This approach has also been further constrained by ethical and financial considerations associated with the use of animals in therapeutic research. As an alternative, cell culture models engineered to replicate the physiological characteristics and barriers of the eye can be employed for release studies. Numerous distinct and well-defined cell culture models of ocular barriers are available, each adaptable to a variety of experimental parameters (Hornof et al., 2005).

Ocular Tolerability

At least three rabbits were used to examine the irritation tendencies of SLN preparations. In one eye, 50µl of SLNs was injected, while the other eye served as a reference, receiving an equivalent volume of physiological saline. At intervals of 0.5, 1, 3, and 12 hours, the eyes were inspected using a slit lamp, and the degree of irritability was assessed using a scoring system (Cavalli et al., 2002). The frequency of rabbit blinking five minutes after SLNs were given to the lower cul-de-sac was compared to the frequency of blinking that occurred with the reference physiological solution or a phosphate buffer of pH 7.4 to determine the level of ocular irritation (Li et al., 2008).

Corneal Hydration Levels

The corneal hydration level (HL) is a parameter commonly used to evaluate damage to cornea tissue. The normal range of HL (%) is 76-80%. A value above 83% denotes damage to the cornea epithelium or endothelium. The scleral tissues of the freshly excised corneas are carefully excised. The corneas are weighed, and the weights are recorded as Ww. Subsequently, the corneas are dried at 70°C for 12 h, after which the dry weights are recorded as Wd (Yadav et al.,2020). The level of corneal hydration (HL %) was calculated using the equation:

$$HL\ (\%) = \left(\frac{W_w - W_d}{W_w}\right) \times\ 100$$

SLNs in Ocular Disease Treatment

SLNs have garnered significant scientific interest due to their biocompatibility, biodegradability, and tolerability. Their sustained drug-release properties and enhanced permeation capabilities make them highly suitable for treating ocular diseases.

Table 3. Tabular summary of studies on SLNs for ocular delivery

Name of API	Method of SLN Preparation	Objective of study	Key Findings	Reference
Ciprofloxacin	Combination of melt-emulsion sonication and low-temperature solidification methods	To improve ocular bioavailability of topical eye drops.	SLNs coated with poly(2-ethyl-2-oxazoline) and chitosan were formulated. SLNs coated with chitosan exhibited higher mucoadhesion and retention on corneal tissues. Showed higher flux apparent permeability, and low entrapment.	Onugwu et al., 2022
Bimatoprost	Solvent evaporation method and ultrasonication technique	To improve the therapeutic effectiveness of drug for the management of glaucoma	SLNs significantly sustained drug release for up to 12 hours compared to the free drug. The formulation showed ocular tolerability and indicated no corneal tissue damage.	Satyanarayana et al., 2023
Natamycin	Hot emulsification-ultrasonication technique	To obtain sustain ocular delivery system of higher corneal penetration against deep fungal keratitis	SLNs demonstrated a prolonged drug release profile, improved corneal penetration, and enhanced antifungal efficacy, while being non-irritating and showing no cytotoxic effects on corneal tissues.	Khames et al., 2019
Atorvastatin	High pressure homogenization	To investigate local application of atorvastatin, as self-administrable eye drops for management of age-related macular degeneration	SLNs exhibited 8-fold and 12-fold higher bioavailability in aqueous and vitreous humor, respectively, compared to free drug. A 2.5-fold enhancement in corneal flux and a 13.62-fold improvement in overall stability, including significant photostability was observed.	Yadav et al., 2020
Isoniazid (INH)	Micro-emulsification method	To assess the suitability of INH-SLN for ocular drug delivery as eye drops for ocular tuberculosis treatment	INH-SLNs demonstrated extended drug release over 48 hours, a 1.6-fold enhancement in corneal permeation, and a five-fold reduction in MIC. Significant in vitro and in vivo cellular uptake was observed, leading to improved localized INH concentration in the aqueous humor, thereby ensuring effective ocular treatment.	Singh et al., 2019

Table 3. (Continued)

Name of API	Method of SLN Preparation	Objective of study	Key Findings	Reference
Metronidazole	Modified melt emulsification and ultrasonication method	To formulate chitosan-coated metronidazole-loaded SLN and evaluate their mucoadhesive properties for the treatment of ocular rosacea	Chitosan-coated metronidazole-loaded SLNs exhibited enhanced mucoadhesive properties, leading to improved ocular availability. This increased precorneal retention suggests the potential for dose reduction or longer dosing intervals, which could enhance patient compliance. Additionally, the MTZ release profile demonstrated a sustained drug release over 10 hours, supporting prolonged therapeutic effects.	Sikhondze et al., 2023
Clarithromycin	High-speed stirring and the ultra-sonication method	To explore the potential of SLNs to enhance ocular permeation and improve the therapeutic efficacy of clarithromycin in topical ocular therapy.	The optimized SLNs demonstrated superior drug permeation compared to the control solution. The SLNs exhibited ~80% drug release in 8 hours, following Weibull model kinetics. Optimized formula displayed a spherical shape, good stability, and was non-irritant for ocular administration. Pharmacokinetic analysis revealed a 150% increase in Cmax and a 2.8-fold improvement in AUC.	Nair et al., 2021

Additionally, SLNs exhibit superior physical stability and cost-effective production and serve as viable alternatives to polymeric nanoparticles, as they do not generate toxic degradation products. One of the key advantages of lipid nanocarriers is their ability to encapsulate more than one API, enabling the development of dual or multidrug-loaded lipid nanocarriers. This approach enhances therapeutic efficacy through synergistic effects, leading to improved treatment outcomes (Gugleva & Andonova, 2024). SLNs are particularly beneficial for ocular drug delivery, as they act as drug depot formulations, providing sustained drug release while enhancing corneal penetration. SLN improve ocular bioavailability by opening tight junctions between corneal epithelial cells, facilitating paracellular drug transport, and inhibiting P-glycoprotein activity, thereby reducing drug efflux. he proposed mechanisms of SLN transcorneal penetration involve endocytosis pathways, including clathrin-dependent, caveolae-dependent, and macropinocytosis-mediated uptake. The permeation and internalisation of SLNs are strongly influenced by their physicochemical properties, such as particle size, size distribution, zeta potential, and surface modifications. Generally, nanoparticles up to 200 nm have been reported to successfully penetrate the cornea. However, for periocular applications, excessive size reduction (e.g., ≈20 nm) may lead to rapid clearance, limiting their therapeutic efficacy (Gugleva & Andonova, 2024). Polydispersity index values < 0.3 signifies a homogeneous and uniform population. A lower PDI is preferred, as it improves formulation stability and reproducibility, which are essential for consistent drug delivery to the posterior segment. High zeta values (approximately ±20 mV) help overcome attractive forces between particles, preventing aggregation and ensuring colloidal stability. Cationic nanoparticles interact electrostatically with the negatively charged ocular mucosa, enhancing drug residence time and improving bioavailability (Bonilla et al., 2022). Table 3 provides a comprehensive overview of various studies conducted on SLNs for ocular drug delivery. The findings highlight the potential of SLNs in improving drug delivery and patient compliance in ocular therapies.

Therapeutic Applications of SLNs in Anterior Segment Diseases (Dry Eye, Glaucoma, Infections)

By overcoming the limitations of traditional formulations, SLNs have shown great promise in treating anterior segment disorders such as glaucoma, dry eye disease, and ocular infections. SLNs are well-suited for localized drug delivery to the anterior ocular tissues, including the cornea, conjunctiva, and aqueous humor. They enhance drug solubility, provide sustained release, and improve bioavailability. SLNs can adhere to the ocular surface, increasing the residence time of lubricating agents and anti-inflammatory drugs, and reducing the need for frequent application. Their lipid composition and nanoscale size enable strong adhesion to the mucin layer, prolonging drug retention on the ocular surface. Studies in animal models have demonstrated that cyclosporine A-loaded SLN enhance tear formation and reduces inflammation, thereby improving patient compliance by reducing dosing frequency. The biocompatibility of SLNs minimizes ocular irritation, a crucial factor in managing chronic conditions like dry eye. Given their potential to improve drug penetration through increased bioadhesion to the ocular epithelium, SLNs hold significant promise for ocular drug delivery. They are efficiently internalized by ocular tissues, potentially enhancing the bioavailability of encapsulated drugs. However, further research is needed to explore their role in supplementing the tear film and to assess the compatibility of their high-melting-point lipids with the tear film lipid layer (Garrigue et al., 2017). SLNs offer a promising approach for delivering medications that lower intraocular pressure (IOP). SLN formulations for anti-glaucoma drugs reduce systemic absorption and associated side effects by providing sustained drug release and enhancing drug penetration into the aqueous humor. SLN have been explored for anti-glaucoma medicines, such as timolol, methazolamide, bimatoprost, and latanoprost, reducing the frequency of application and minimizing systemic side effects (Gokce et al., 2009). Compared to conventional eye drops, timolol-loaded SLN has demonstrated a longer-lasting reduction in intraocular pressure in rabbits, highlighting their potential for controlled drug release and improved therapeutic outcomes. Additionally, the nanoscale properties of SLN help overcome conventional barriers increasing drug retention and availability at the targeted site. The small particle size of SLNs allows for prolonged pre-corneal residence time, controlled drug delivery, enhanced corneal absorption, and improved drug bioavailability. Several studies have demonstrated the ability of drug-loaded SLNs to effectively penetrate the corneal epithelium due to their lipophilic nature. Furthermore, SLNs can be engineered to target specific tissues within the anterior chamber, such as the trabecular meshwork, potentially enhancing drug efficacy while minimizing side effects (Wang et al., 2014; Satyanarayan et al., 2023; Dang et al., 2022). SLNs have been extensively investigated for the treatment of ocular infections due to their ability to effectively encapsulate antibiotics such as tobramycin, moxifloxacin, and levofloxacin. These formulations ensure sustained antibacterial efficacy against pathogens like *Pseudomonas aeruginosa* and *Staphylococcus aureus* while also enhancing drug stability and corneal penetration. Tobramycin-loaded SLNs have demonstrated increased ocular bioavailability and prolonged drug release, reducing dosing frequency and minimizing the risk of bacterial resistance (Severino et al., 2011). In addition to antibacterial applications, SLNs have shown potential in antifungal therapy. For instance, SLNs loaded with ketoconazole exhibited enhanced antifungal activity due to improved corneal penetration and increased bioavailability in the anterior chamber (Gokce et al., 2009). The advantages of SLNs in treating anterior segment diseases include improved precorneal retention, sustained drug release, and reduced ocular irritation, leading to better therapeutic outcomes and increased patient adherence. Moreover, SLNs offer high drug-loading capacity, prolonged drug release duration, and enhanced bioavailability, while also providing stability for chemically unstable molecules. Their biocompatibility and biodegradability make them a safe and effective delivery system, with a favourable cost-effectiveness ratio, particularly for large-scale production.

Therapeutic Applications of SLNs in Posterior Segment Diseases (Age-Related Macular Degeneration, Diabetic Retinopathy, Uveitis)

SLNs have emerged as a highly promising platform for delivering therapeutics to the posterior segment of the eye. They address critical challenges such as poor bioavailability, limited penetration of conventional pharmacological agents through the blood-retinal barrier (BRB), and the need for sustained release. Posterior segment diseases, including age-related macular degeneration (AMD), diabetic retinopathy (DR), and uveitis, require targeted and prolonged therapeutic interventions to slow disease progression and improve patient outcomes. SLNs have shown great potential in delivering atorvastatin, lutein, pentamethyl-6-chromanol for AMD treatment. Their nanoscale size and lipophilic properties enable them to penetrate the vitreous humor and reach the retina, effectively inhibiting vascular endothelial growth factor (VEGF)- a key driver of neovascularization in AMD (Mehnert & Mäder, 2012). Additionally, SLNs provide sustained drug release, reducing the need for frequent intravitreal injections, which are invasive and often associated with patient discomfort. Surface modifications, such as ligand conjugation, can enhance SLN targeting efficiency by facilitating receptor-mediated transport across the BRB, further improving drug delivery to the retinal vasculature. For diabetic retinopathy, SLNs have been explored for delivering corticosteroids such as dexamethasone, which help mitigate retinal inflammation and vascular leakage. SLN-based formulations enhance drug stability and enable controlled release, maintaining therapeutic drug concentrations in the retina over extended periods (Mehnert & Mäder, 2012). This sustained release minimizes systemic side effects and improves patient compliance. Additionally, SLNs loaded with anti-inflammatory agents have demonstrated the ability to reduce oxidative stress, a key contributor to DR progression (Gokce et al., 2009). SLNs of etoposide and epalrestat has also been formulated for diabetic retinopathy. A novel therapeutic approach involved using SLNs to deliver siRNA targeting HuR (Human antigen R) expression, a potential gene-silencing strategy for DR management (Bonilla et al., 2022). In uveitis, SLNs have been explored for the localized delivery of immunosuppressive and anti-inflammatory drugs. By delivering corticosteroids and other therapeutic agents directly to inflamed ocular tissues, SLNs minimize systemic drug exposure and associated side effects. Their ability to penetrate inflammatory barriers enhances drug bioavailability, ensuring localized and prolonged drug action in the posterior segment. Furthermore, SLNs reduce systemic toxicity by limiting drug distribution to non-target tissues while enhancing therapeutic efficacy through increased drug retention in the vitreous humor and retina (Gokce et al., 2009). The inherent versatility of SLNs, including their potential for surface modification with ligands or polymeric coatings such as polyethylene glycol (PEG), further enhances their targeting specificity. This adaptability makes SLNs a promising nanocarrier system for posterior segment drug delivery, effectively overcoming the limitations of conventional formulations while improving therapeutic outcomes in the management of AMD, DR, and uveitis.

Biocompatibility and Safety of SLNs

SLNs have attracted significant attraction due to their biocompatibility, biodegradability, and tolerability. SLNs exhibit excellent safety and biocompatibility profiles, making them particularly attractive for ocular drug delivery. Since SLNs are composed of physiological and biodegradable lipids such as tripalmitin and glyceryl behenate, they undergo natural enzymatic degradation, minimizing the risk of toxicity (Mehnert & Mäder, 2012). Their nanoscale size enables targeted drug delivery to ocular tissues while reducing systemic absorption and associated side effects. Cell viability assays have demonstrated that SLNs maintain high cell viability and exhibit minimal irritation in human corneal and conjunctival cell lines. Moreover, SLN formulations eliminate the need for preservatives such as benzalkonium chloride, which is known to cause ocular irritation in conventional eye drops (Pisella et al., 2002). Animal studies further support the safety of SLNs, showing that they maintain stable intraocular pressure (IOP) and do not induce tissue damage or inflammation (Gokce et al., 2009). However, concerns such as potential nanoparticle accumulation and surfactant-induced toxicity necessitate further optimization of formulation parameters and long-term safety evaluations. Despite these considerations, SLNs remain a promising and well-tolerated platform for ocular drug delivery, offering improved therapeutic efficacy and patient compliance. Ensuring their biocompatibility and safety is crucial for clinical translation. Currently, no SLN-based ophthalmic formulations have been successfully translated into clinical applications or commercialized. Continued research and clinical trials are essential to bridge the gap between laboratory findings and real-world applications in ophthalmology.

Challenges Associated with SLN-Based Ocular Nano-Formulations

While SLNs hold great promise for ocular drug delivery, several challenges must be addressed. One of the primary obstacles is overcoming physiological barriers such as tear film dynamics, corneal tight junctions, and ocular clearance mechanisms, which significantly limit drug retention and penetration. Additionally, formulation stability remains a critical concern, as nanoparticles may aggregate, degrade, or undergo premature drug release. A specific issue encountered in solid lipid nanoparticles (SLNs) is drug expulsion during storage, which results from the intricate crystalline behavior of lipids at the nanoscale. The transformation of lipid crystals into a more stable β-modification leads to a more refined crystalline lattice, reducing available space for drug accommodation and causing drug leakage over time. Another key challenge is burst release, a phenomenon attributed to phase separation during SLN formation. SLNs are generally considered biodegradable due to their composition of triglycerides, glycerides, and surfactants, their nanoscale size raises concerns regarding potential biotoxicity. Nanoparticles can interact with biological membranes, release toxic breakdown products, and exhibit intracellular effects. The use of surfactants in SLN formulations further complicates safety considerations, as their solubility and ionic nature influence their toxicity profile, cationic surfactants being the most hazardous. Although in vitro and in vivo studies have shown minimal toxicity, further research is needed to evaluate their long-term safety and fate after in vivo application. Additionally, manufacturing and scalability issues, including reproducibility in particle size, drug loading efficiency, and high production costs, impact the feasibility of large-scale production. Despite promising research outcomes, the transition from laboratory studies to clinical applications remains limited. Some key challenges include the ability to develop reproducible batches of lipid nanoparticles with sufficient colloidal stability during storage. In this regard, implementing a quality-by-design (QbD) approach during the (pre)formulation stage is a feasible strategy, as it enables the development of a final product with predictable quality attributes, ultimately facilitating the commercialization of SLN-based ocular formulations (Gugleva & Andonova, 2023). Other challenges posed by ocular excipients include eye discomfort, blurred vision, instability, less corneal retention, limited solubility, toxicity and incompatibilities. While preservatives play a crucial role in ensuring sterility and stability, their impact on ocular safety and packaging compatibility must be carefully evaluated. Addressing these challenges through optimized formulation strategies and careful selection of excipients is essential for the successful development and commercialization of ocular nano-formulations (Ashique et al., 2024).

Regulatory Considerations for Ocular SLNs

Regulatory considerations for ocular SLNs are critical to ensuring their safety, efficacy, and commercial viability. Several lipid-based ocular formulations have been successfully developed and evaluated in research settings. These nano-formulations offer an ideal drug delivery system by incorporating GRAS components, enabling large-scale production, and improving drug safety through lipid-based carriers. However, despite their therapeutic potential, the transition from patent innovations to commercial drug products remains limited, with most SLN-based technologies still in the research and development phase. Ongoing studies focus on assessing nanotoxicity and ocular tolerance for long-term use. Regulatory agencies, such as the U.S. Food and Drug Administration (FDA), have established stringent guidelines to ensure the quality of ophthalmic nano-formulations. The FDA's Office of Pharmaceutical Quality, under the Center for Drug Evaluation and Research

(CDER), has issued the "Quality Considerations for Topical Ophthalmic Drug Products: Guidance for Industry," which addresses microbial contamination control, particulate matter assessment, extractable and leachable from packaging materials, impurities, degradation products, and dispensing system design. Compliance with these regulations extends to the careful selection of excipients to prevent potential toxicity or adverse reactions. Navigating these regulatory challenges is essential for the successful translation of ocular SLNs from research to clinical application, ensuring their approval and commercialization while maintaining the highest standards of quality and safety (Seyfoddin et al., 2010; Ashique et al., 2024).

Future Perspectives

Advancements in ocular drug delivery are focused on exploring new surface modification strategies to enhance tissue targeting. The incorporation of targeting ligands, peptides, or aptamers can improve the pharmacokinetic properties of nanocarriers, impart mucoadhesive characteristics, prolong corneal residence time, and enhance therapeutic efficacy. More clinical trials are essential to evaluate the safety and efficacy of SLN-based ophthalmic formulations in human patients. Additionally, the development of SLNs for the co-delivery of multiple drugs offers a promising approach to treating complex ocular diseases. Another innovative strategy involves designing SLNs that release drugs in response to specific stimuli, such as pH, temperature, or enzymes, within the ocular microenvironment. Personalized medicine approaches, tailoring SLN formulations based on individual genetic profiles and disease characteristics, hold great potential for improving treatment outcomes. Intelligent drug delivery systems, including intravitreal injections, in situ gels, nanoparticles, implants, eye devices, and coils, provide novel solutions for sustained and controlled drug release. A key area of focus is the development of advanced excipients to enhance drug stability, addressing common degradation issues in the ocular environment. Future research should explore innovative polymers or protective additives to create a barrier around drug molecules, shielding them from environmental factors and extending their shelf life while maintaining therapeutic efficacy. To further improve patient safety and drug effectiveness, preservative-free formulations, single-dose packaging, and innovative self-preserving drug delivery systems are being developed to overcome the challenges associated with traditional ophthalmic formulations. These advancements will play a crucial role in optimizing ocular drug delivery, improving patient compliance, and facilitating the transition of SLN-based therapies from research to clinical applications (Ashique et al., 2024).

Conclusion

Continued research and development in the field of SLN-based ophthalmic formulations will undoubtedly lead to the creation of innovative and effective therapies, ultimately improving patient outcomes and quality of life. SLNs have demonstrated significant potential as efficient ocular drug delivery systems, as highlighted in this review. Their unique characteristics, including biodegradability, biocompatibility, and the use of GRAS lipid constituents, make them highly suitable for ophthalmic applications. Additionally, SLNs offer controlled and sustained drug release, enhance transcorneal penetration, and improve ocular bioavailability, further reinforcing their role in advancing ocular therapeutics. As research continues to refine formulation strategies and overcome existing challenges, SLN-based ophthalmic drug delivery is expected to transition from experimental studies to clinical applications, paving the way for safer and more effective treatment options in ophthalmology.

References

Achouri, D., Alhanout, K., Piccerelle, P., & Andrieu, V. (2013). Recent advances in ocular drug delivery. *Drug development and industrial pharmacy*, *39*(11), 1599-1617.

Agarwal, P., & Rupenthal, I. D. (2016). In vitro and ex vivo corneal penetration and absorption models. *Drug delivery and translational research*, *6*(6), 634–647.

Agarwal, P., Craig, J. P., & Rupenthal, I. D. (2021). Formulation Considerations for the Management of Dry Eye Disease. *Pharmaceutics*, *13*(2), 207.

Ahmed, S., Amin, M. M., El-Korany, S. M., & Sayed, S. (2022). Corneal targeted fenticonazole nitrate-loaded novasomes for the management of ocular candidiasis: Preparation, in vitro characterization, ex vivo and in vivo assessments. *Drug Delivery*, *29*(1), 2428-2441.

Ameeduzzafar, Ali, J., Fazil, M., Qumbar, M., Khan, N., & Ali, A. (2016). Colloidal drug delivery system: amplify the ocular delivery. *Drug delivery*, *23*(3), 700-716.

Anand, B. S., Dey, S., & Mitra, A. K. (2002). Current prodrug strategies via membrane transporters/receptors. *Expert opinion on biological therapy*, *2*(6), 607-620.

Ashique, S., Mishra, N., Mohanto, S., Gowda, B. H. J., Kumar, S., Raikar, A. S., Masand, P., Garg, A., Goswami, P., & Kahwa, I. (2024). Overview of processed excipients in ocular drug delivery: Opportunities so far and bottlenecks. *Heliyon, 10*(1), e23810.

Baek, J. S., & Cho, C. W. (2017). Surface modification of solid lipid nanoparticles for oral delivery of curcumin: Improvement of bioavailability through enhanced cellular uptake, and lymphatic uptake. *European Journal of Pharmaceutics and Biopharmaceutics*, *117*, 132-140.

Balyen, L. (2022). The Physiology of Cornea. *JOJ Ophthalmology*, 9(3), 555764.

Başaran, E., Demirel, M., Sırmagül, B., & Yazan, Y. (2010). Cyclosporine-A incorporated cationic solid lipid nanoparticles for ocular delivery. *Journal of microencapsulation*, *27*(1), 37-47.

Bhattacharjee, S. (2016). DLS and zeta potential–what they are and what they are not? *Journal of controlled release*, *235*, 337-351.

Bonilla, L., Espina, M., Severino, P., Cano, A., Ettcheto, M., Camins, A., García, M. L., Souto, E. B., & Sánchez-López, E. (2022). Lipid Nanoparticles for the Posterior Eye Segment. *Pharmaceutics*, *14*(1), 90.

Cavalli, R., Gasco, M. R., Chetoni, P., Burgalassi, S., & Saettone, M. F. (2002). Solid lipid nanoparticles (SLN) as ocular delivery system for tobramycin. *International journal of pharmaceutics*, *238*(1-2), 241-245.

Chantaburanan, T., Teeranachaideekul, V., Jintapattanakit, A., Chantasart, D., & Junyaprasert, V. B. (2023). Enhanced stability and skin permeation of ibuprofen-loaded solid lipid nanoparticles based binary solid lipid matrix: Effect of surfactant and lipid compositions. *International Journal of Pharmaceutics: X*, *6*, 100205.

Dang, H., Dong, C., & Zhang, L. (2022). Sustained latanoprost release from PEGylated solid lipid nanoparticle-laden soft contact lens to treat glaucoma. *Pharmaceutical development and technology*, *27*(2), 127–133.

Del Amo, E. M., Rimpelä, A.-K., Heikkinen, E., Kari, O. K., Ramsay, E., Lajunen, T., Schmitt, M., Pelkonen, L., Bhattacharya, M., Richardson, D., Subrizi, A., Turunen, T., Reinisalo, M., Itkonen, J., Toropainen, E., Casteleijn, M., Kidron, H., Antopolsky, M., Vellonen, K.-S., Ruponen, M., & Urtti, A. (2017). Pharmacokinetic aspects of retinal drug delivery. *Progress in retinal and eye research*, *57*, 134-185.

DelMonte, D. W., & Kim, T. (2011). Anatomy and physiology of the cornea. *Journal of Cataract & Refractive Surgery*, *37*(3), 588-598.

Dhanya, C. S., Paul, W., Rekha, M. R., & Joseph, R. (2023). Solid lipid nanoparticles of lauric Acid: a prospective drug carrier for oral drug delivery. *Journal of Molecular Liquids*, *380*, 121738.

Ficarrotta, K. R., & Passaglia, C. L. (2020). Intracranial pressure modulates aqueous humour dynamics of the eye. *The Journal of physiology*, *598*(2), 403-413.

Ganesan, P., & Narayanasamy, D. (2017). Lipid nanoparticles: Different preparation techniques, characterization, hurdles, and strategies for the production of solid lipid nanoparticles and nanostructured lipid carriers for oral drug delivery. *Sustainable Chemistry and Pharmacy*, *6*, 37–56.

Garrigue, J. S., Amrane, M., Faure, M. O., Holopainen, J. M., & Tong, L. (2017). Relevance of Lipid-Based Products in the Management of Dry Eye Disease. *Journal of ocular pharmacology and therapeutics*, *33*(9), 647–661.

Gökçe, E. H., Sandri, G., Eğrilmez, S., Bonferoni, M. C., Güneri, T., & Caramella, C. (2009). Cyclosporine a-loaded solid lipid nanoparticles: ocular tolerance and in vivo drug release in rabbit eyes. *Current eye research*, *34*(11), 996-1003.

Gugleva, V., & Andonova, V. (2023). Recent Progress of Solid Lipid Nanoparticles and Nanostructured Lipid Carriers as Ocular Drug Delivery Platforms. *Pharmaceuticals*, *16*(3), 474.

Gupta, D. K., Sharma, S. K., Gaur, P. K., & Singh, A. P. (2022). Lovastatin loaded solid lipid nanoparticles for transdermal delivery: In vitro characterization. *Research Journal of Pharmacy and Technology*, *15*(3), 1085-1089.

Heiati, H., Tawashi, R., & Phillips, N. C. (1998). Drug retention and stability of solid lipid nanoparticles containing azidothymidine palmitate after autoclaving, storage and lyophilization. *Journal of microencapsulation*, *15*(2), 173-184.

Hernández-Esquivel, R. A., Navarro-Tovar, G., Zárate-Hernández, E., & Aguirre-Bañuelos, P. (2022). Solid lipid nanoparticles (SLN). In *Nanocomposite materials for biomedical and energy storage applications*. Intechopen.

Hornof, M., Toropainen, E., & Urtti, A. (2005). Cell culture models of the ocular barriers. *European journal of pharmaceutics and biopharmaceutics*, *60*(2), 207-225.

Ita, K. (2022). Anatomy and physiology of the eye. In *Elsevier eBooks* (pp. 73–92).

Jain, S. K., Chourasia, M. K., Masuriha, R., Soni, V., Jain, A., Jain, N. K., & Gupta, Y. (2005). Solid lipid nanoparticles bearing flurbiprofen for transdermal delivery. *Drug delivery*, *12*(4), 207-215.

Jakubiak, P., Cantrill, C., Urtti, A., & Alvarez-Sánchez, R. (2019). Establishment of an in vitro–in vivo correlation for melanin binding and the extension of the ocular half-life of small-molecule drugs. *Molecular pharmaceutics*, *16*(12), 4890-4901.

Kakkar Thukral, D., Dumoga, S., & K. Mishra, A. (2014). Solid lipid nanoparticles: promising therapeutic nanocarriers for drug delivery. *Current drug delivery*, *11*(6), 771-791.

Khames, A., Khaleel, M. A., El-Badawy, M. F., & El-Nezhawy, A. O. (2019). Natamycin solid lipid nanoparticles–sustained ocular delivery system of higher corneal penetration against deep fungal keratitis: preparation and optimization. *International journal of nanomedicine*, 2515-2531.

Larkins, M. C., Zubair, M., & Thombare, A. (2024). Osmometer. StatPearls Publishing.

Lee, V. H. L., & Robinson, J. R. (1979). Mechanistic and quantitative evaluation of precorneal pilocarpine disposition in albino rabbits. *Journal of pharmaceutical sciences*, *68*(6), 673-684.

Li, X., Nie, S. F., Kong, J., Li, N., & Ju, C. Y. (2008). A controlled-release ocular delivery system for ibuprofen based on nanostructured lipid carriers. *International journal of pharmaceutics*, *363*(1-2), 177-182.

Lou, X., Hu, Y., Zhang, H., Liu, J., & Zhao, Y. (2021). Polydopamine nanoparticles attenuate retina ganglion cell degeneration and restore visual function after optic nerve injury. *Journal of Nanobiotechnology*, *19*, 1-15.

Lüpke, N. P., The hen's egg test (HET)—an alternative toxicity test. *British Journal of Dermatology*, 115(s31), 133-135.

Malekpour-Galogahi, F., Hatamian-Zarmi, A., Ganji, F., Ebrahimi-Hosseinzadeh, B., Nojoki, F., Sahraeian, R., & Mokhtari-Hosseini, Z. B. (2018). Preparation and optimization of rivastigmine-loaded tocopherol succinate-based solid lipid nanoparticles. *Journal of liposome research*, *28*(3), 226-235.

Malhotra, A., Minja, F. J., Crum, A., & Burrowes, D. (2011, February). Ocular anatomy and cross-sectional imaging of the eye. In *Seminars in Ultrasound, CT and MRI* (Vol. 32, No. 1, pp. 2-13). WB Saunders.

Manjunath, K., Reddy, J. S., & Venkateswarlu, V. (2005). Solid lipid nanoparticles as drug delivery systems. *Methods Find Exp Clin Pharmacol*, *27*(2), 127-144.

Mehnert, W., & Mäder, K. (2012). Solid lipid nanoparticles: production, characterization and applications. *Advanced drug delivery reviews*, *64*, 83-101.

Mishra, D., Gade, S., Glover, K., Sheshala, R., & Singh, T. R. R. (2023). Vitreous humor: composition, characteristics and implication on intravitreal drug delivery. *Current eye research*, *48*(2), 208-218.

Mukherjee, S., Ray, S., & Thakur, R. S. (2009). Solid lipid nanoparticles: a modern formulation approach in drug delivery system. *Indian journal of pharmaceutical sciences*, *71*(4), 349.

Nair, A. B., Shah, J., Al-Dhubiab, B. E., Jacob, S., Patel, S. S., Venugopala, K. N., Morsy, M. A., Gupta, S., Attimarad, M., Sreeharsha, N., & Shinu, P. (2021). Clarithromycin Solid Lipid Nanoparticles for Topical Ocular Therapy: Optimization, Evaluation and In Vivo Studies. Pharmaceutics, 13(4), 523.

Naylor, A., Hopkins, A., Hudson, N., & Campbell, M. (2019). Tight junctions of the outer blood retina barrier. *International journal of molecular sciences*, *21*(1), 211.

Onugwu, A. L., Attama, A. A., Nnamani, P. O., Onugwu, S. O., Onuigbo, E. B., & Khutoryanskiy, V. V. (2022). Development and optimization of solid lipid nanoparticles coated with chitosan and poly (2-ethyl-2-oxazoline) for ocular drug delivery of ciprofloxacin. *Journal of Drug Delivery Science and Technology*, *74*, 103527.

Pandey, S., Shaikh, F., Gupta, A., Tripathi, P., & Yadav, J. S. (2021). A recent update: solid lipid nanoparticles for effective drug delivery. *Advanced Pharmaceutical Bulletin*, *12*(1), 17.

Pawar, P., Yadav, A., & Gharge, V. (2018). Ocular drug delivery and novel formulations: A review. *Asian Journal of Pharmacy and Technology*, 8(4), 248–254.

Pisella, P. J., Pouliquen, P., & Baudouin, C. (2002). Prevalence of ocular symptoms and signs with preserved and preservative free glaucoma medication. *British Journal of Ophthalmology*, *86*(4), 418-423.

Qushawy, M., Prabahar, K., Abd-Alhaseeb, M., Swidan, S., & Nasr, A. (2019). Preparation and evaluation of carbamazepine solid lipid nanoparticle for alleviating seizure activity in pentylenetetrazole-kindled mice. *Molecules*, *24*(21), 3971.

Ruan, X., Liu, Z., Luo, L., & Liu, Y. (2020). Structure of the lens and its associations with the visual quality. *BMJ Open Ophthalmology*, *5*(1), e000459.

Saad, S., Ahmad, I., Kawish, S. M., Khan, U. A., Ahmad, F. J., Ali, A., & Jain, G. K. (2020). Improved cardioprotective effects of hesperidin solid lipid nanoparticles prepared by supercritical antisolvent technology. *Colloids and Surfaces B: Biointerfaces*, *187*, 110628.

Santana, C. P., Matter, B. A., Patil, M. A., Silva-Cunha, A., & Kompella, U. B. (2023). Corneal Permeability and Uptake of Twenty-Five Drugs: Species Comparison and Quantitative Structure-Permeability Relationships. *Pharmaceutics*, *15*(6), 1646.

Satyanarayana, S. D., Abu Lila, A. S., Moin, A., Moglad, E. H., Khafagy, E.-S., Alotaibi, H. F., Obaidullah, A. J., & Charyulu, R. N. (2023). Ocular Delivery of Bimatoprost-Loaded Solid Lipid Nanoparticles for Effective Management of Glaucoma. *Pharmaceuticals*, *16*(7), 1001.

Sawant, K. K., & Dodiya, S. S. (2008). Recent advances and patents on solid lipid nanoparticles. *Recent patents on drug delivery & formulation*, *2*(2), 120-135.

Schermer, A., Galvin, S., & Sun, T. T. (1986). Differentiation-related expression of a major 64K corneal keratin in vivo and in culture suggests limbal location of corneal epithelial stem cells. *The Journal of cell biology*, *103*(1), 49-62.

Severs, N. J. (2007). Freeze-fracture electron microscopy. *Nature protocols*, *2*(3), 547-576.

Seyfoddin, A., Shaw, J., & Al-Kassas, R. (2010). Solid lipid nanoparticles for ocular drug delivery. *Drug Delivery*, *17*(7), 467–489.

Shirodkar, R. K., Kumar, L., Mutalik, S., & Lewis, S. (2019). Solid lipid nanoparticles and nanostructured lipid carriers: Emerging lipid based drug delivery systems. *Pharmaceutical Chemistry Journal*, *53*(5), 440–453.

Sikhondze, S. S., Makoni, P. A., Walker, R. B., & Khamanga, S. M. M. (2023). Chitosan-Coated SLN: A Potential System for Ocular Delivery of Metronidazole. *Pharmaceutics*, *15*(7), 1855.

Singh, M., Bharadwaj, S., Lee, K. E., & Kang, S. G. (2020). Therapeutic nanoemulsions in ophthalmic drug administration: Concept in formulations and characterization techniques for ocular drug delivery. *Journal of Controlled Release*, *328*, 895-916.

Singh, M., Guzman-Aranguez, A., Hussain, A., Srinivas, C. S., & Kaur, I. P. (2019). Solid lipid nanoparticles for ocular delivery of isoniazid: evaluation, proof of concept and in vivo safety & kinetics. *Nanomedicine (London, England)*, *14*(4), 465–491.

Spadaro, A., Pasquinucci, L., Lorenti, M., Santagati, L. M., Sarpietro, M. G., Turnaturi, R., Parenti, C., & Montenegro, L. (2021). Solid lipid nanoparticles as carriers for the synthetic opioid LP2: characterization and in vitro release. *Applied Sciences*, *11*(21), 10250.

Szabadi, E. (2018). Functional organization of the sympathetic pathways controlling the pupil: light-inhibited and light-stimulated pathways. *Frontiers in Neurology*, *9*, 1069.

Tóth-Molnár, E., & Ding, C. (2020). New insight into lacrimal gland function: Role of the duct epithelium in tear secretion. *The Ocular Surface*, *18*(4), 595-603.

Vaneev, A., Tikhomirova, V., Chesnokova, N., Popova, E., Beznos, O., Kost, O., & Klyachko, N. (2021). Nanotechnology for topical drug delivery to the anterior segment of the eye. *International journal of molecular sciences*, *22*(22), 12368.

Vicente-Pascual, M., Gómez-Aguado, I., Rodríguez-Castejón, J., Rodríguez-Gascón, A., Muntoni, E., Battaglia, L., Del Pozo-Rodríguez, A., & Solinís Aspiazu, M. Á. (2020). Topical Administration of SLN-Based Gene Therapy for the Treatment of Corneal Inflammation by De Novo IL-10 Production. *Pharmaceutics*, *12*(6), 584.

Viera, L. M. D. A. L., Silva, R. S., da Silva, C. C., Presgrave, O. A. F., & Boas, M. H. S. V. (2022). Comparison of the different protocols of the Hen's Egg Test-

Chorioallantoic Membrane (HET-CAM) by evaluating the eye irritation potential of surfactants. *Toxicology in Vitro*, *78*, 105255.

Wallig, M. A., Bolon, B., Haschek, W. M., & Rousseaux, C. G. (Eds.). (2017). *Fundamentals of toxicologic pathology*. Academic press.

Wang, F., Chen, L., Zhang, D., Jiang, S., Shi, K., Huang, Y., Li, R., & Xu, Q. (2014). Methazolamide-loaded solid lipid nanoparticles modified with low-molecular weight chitosan for the treatment of glaucoma: vitro and vivo study. *Journal of drug targeting*, *22*(9), 849–858.

Xu, X., & Zuo, Y. Y. (2022). Nanomedicine for ocular drug delivery. In *Nanomedicine* (pp. 1-32). Singapore: Springer Nature Singapore.

Yadav, M., Schiavone, N., Guzman-Aranguez, A., Giansanti, F., Papucci, L., Perez de Lara, M. J., Singh, M., & Kaur, I. P. (2020). Atorvastatin-loaded solid lipid nanoparticles as eye drops: proposed treatment option for age-related macular degeneration (AMD). *Drug delivery and translational research*, *10*(4), 919–944.

Yang, S. C., & Zhu, J. B. (2002). Preparation and characterization of camptothecin solid lipid nanoparticles. *Drug development and industrial pharmacy*, *28*(3), 265-274.

Chapter 5

Polymeric Nanoparticles for Ocular Therapies

Shahzar Khan
Nadia Ilyas
Humaira Nawaz
Adnan Khan
and Samiullah Khan*
Department of Microbiology, Faculty of Biological Sciences, Quaid-i-Azam University, Islamabad, Pakistan

Abstract

The emergence of polymeric nanoparticles has revolutionized ocular drug delivery by providing innovative solutions to longstanding challenges such as poor bioavailability, rapid drug clearance, and limited permeability across ocular barriers. These nanoscale carriers offer enhanced drug targeting, controlled release profiles, and improved drug stability, making them highly suitable for treating both anterior and posterior segment ocular diseases. This chapter explores various types of polymeric nanoparticles, including biodegradable and non-biodegradable systems, along with their therapeutic applications in conditions such as glaucoma, macular degeneration, and diabetic retinopathy. Particular emphasis is placed on surface modifications and ligand conjugation techniques that enhance mucoadhesion and site-specific targeting. Furthermore, the development of stimuli-responsive polymers has enabled on-demand drug release tailored to the ocular microenvironment. The chapter also addresses the clinical potential, regulatory considerations, and future prospects of polymeric

* Corresponding Author's Email: samikhan@qau.edu.pk

In: Advances in Nanotechnology for Ocular Drug Delivery
Editors: Anjali Sharma, Ajmer Singh Grewal, Geeta Deswal et al.
ISBN: 979-8-89530-814-1
© 2025 Nova Science Publishers, Inc.

nanoparticles in ophthalmology. By integrating advances in nanotechnology, materials science, and ocular pharmacology, polymeric nanoparticles continue to evolve as a promising platform for effective, targeted, and patient-friendly treatment of ocular disorders.

Keywords: biodegradable polymers, controlled release, diabetic retinopathy, glaucoma, mucoadhesive nanoparticles, nanotechnology in ophthalmology, ocular drug delivery, polymeric nanoparticles, stimuli-responsive polymers

Abbreviations

API: Active Pharmaceutical Ingredient
BRB: Blood-Retinal Barrier
CRISPR: Clustered Regularly Interspaced Short Palindromic Repeats
FDA: Food and Drug Administration
NPs: Nanoparticles
PCL: Polycaprolactone
PEG: Polyethylene Glycol
PLA: Polylactic Acid
PLGA: Poly(lactic-co-glycolic acid)
PNPs: Polymeric Nanoparticles
siRNA: Small Interfering RNA
VEGF: Vascular Endothelial Growth Factor

Introduction

Ocular diseases currently affect millions across the world. Three primary eye conditions which require medical attention are glaucoma, age-related macular degeneration, and diabetic retinopathy. Ocular diseases affect both patients' quality of life and generate massive financial costs for healthcare systems (Sahoo et al., 2008). The eye possesses numerous barriers that prevent drug substances from penetrating its tissues. Entry of drugs into the eye is hindered by three protective barriers, including the cornea, conjunctiva, and blood-retinal barrier. The bioavailability of traditional eye drops and ointments remains poor (Agarwal et al., 2018). Nanotechnology can be an innovative technique to enhance drug delivery to the eyes. Drugs become more effective through nanoscale carrier delivery systems. Biological barriers become penetrable when nanoparticles provide controlled drug release (Sultana et al., 2011). Polymeric nanoparticles (PNPs) function as flexible nanocarrier systems and present a powerful delivery platform for drugs intended for the eye. These nanoparticles consist of biocompatible and biodegradable polymers. PNPs preserve drugs within the system and enable precise delivery to distinct parts of the eye (Nagarwal et al., 2009). Advancements are being made in the utilization of PNPs in therapeutic applications for ocular diseases. Various studies aim to develop novel formulations that enhance drug loading and provide gradual drug release mechanisms (Rajan et al., 2024). This chapter aims to highlight the limitations of traditional ocular drug delivery systems. Moreover, the chapter sheds light on the importance of PNPs in revolutionizing ocular drug delivery.

Anatomy of the Eye and Barriers to Drug Delivery

The human eye has a very complex structure, possessing many biological barriers that hinder the delivery of drugs. The cornea, aqueous humor, iris, and lens make up the anterior segment of the eye, while the vitreous humor, retina, and choroid form the posterior segment. Each segment of the eye presents a unique obstacle to drug penetration. There are many static and dynamic barriers in the human eye, and these barriers make ocular drug delivery very complicated (Gorantla et al., 2020). Topically applied drugs are rapidly eliminated from the ocular surface due to blinking and tear turnover. In addition, the tight junctions of the corneal epithelium make it a major barrier restricting the passage of hydrophilic drugs. The conjunctiva and blood-retinal barrier (BRB) also block the diffusion of systemic drugs into intraocular tissues. The BRB, formed by the retinal capillary endothelium and retinal pigment epithelium, tightly restricts molecule entry into the retina, making drug delivery to the posterior segment highly challenging (Suri et al., 2020). Traditional drug delivery methods, like eye drops and ointments, possess low bioavailability and cannot easily pass the eye's barriers. For example, less than 5% of an eye drop reaches the intraocular tissues. These drugs are rapidly cleared, and limited penetration occurs through the cornea. Likewise, systemic administration is ineffective in treating diseases of the posterior segment. Intravitreal injections are effective for delivering drug contents to the retina; however, this process can cause infection, retinal detachment, and discomfort. These challenges highlight the need for, and importance of, alternative drug delivery routes (Akhter et al., 2022). PNPs have been developed as a potential solution. They improve drug retention, penetration, and controlled release. They can cross the corneal epithelium. Moreover, mucoadhesive nanoparticles that adhere to the ocular mucosa improve drug retention in the precorneal area. Further surface modifications, including PEG coating, enhance drug stability and reduce clearance, facilitating drug release to the retina. Nanoparticle-based drug carriers improve drug absorption and prolong drug release. PNPs can provide better therapeutic outcomes in treating both anterior and posterior segment disorders (Naik et al., 2020). PNPs also increase permeability while reducing systemic toxicity. Nanocarriers like liposomes, dendrimers, and solid lipid nanoparticles have superior capabilities in bypassing ocular barriers. Positively charged nanoparticles strongly interact with the negatively charged corneal surface, enabling better drug penetration (Tsung et al., 2023). In general, PNPs represent a significant advancement in ocular drug delivery (Baig et al., 2024).

Polymeric Nanoparticles: An Overview

PNPs constitute a large class of nanomaterials that have attracted significant interest as platforms for drug delivery and diagnostics. Generally, these submicron-sized particles (10 to 1,000 nm) are made from natural or synthetic polymers and readily allow the binding and adsorption of an active pharmaceutical ingredient (API). Synthetic polymers offer a route for controlled release, targeted delivery, and improved drug stability (Feron et al., 2012). Two basic types of PNPs are termed nanospheres and nanocapsules. Nanospheres, which are solid matrix systems, distribute drug molecules homogeneously within their polymeric core and support sustained drug release (Kumar et al., 2015; Pinto Reis et al., 2006). Nanocapsules are based on a core–shell structure. The drug is loaded into the core, while the outer shell controls drug release, allowing regulation of release kinetics and minimizing premature drug degradation (Couvreur et al., 2002). Preparation methods for PNPs are diverse. Emulsion polymerization is a commonly used method that involves the polymerization of monomers in an emulsified system composed of water, surfactants, and initiators. This technique produces monodisperse nanoparticles, which are more efficient for encapsulating both hydrophilic and hydrophobic drugs (Guterres et al., 2007). Solvent evaporation or extraction is the most widely used method for synthesizing PNPs. In this process, the polymers and drugs are dissolved in a volatile organic solvent and then emulsified into an aqueous phase. Nanoparticle formation occurs upon solvent removal. This method helps optimize drug loading efficiency (Quintanar et al., 1998). In the case of natural polymers such as chitosan, ionic gelation is the preferred method. This process involves ionic interactions between the polymer and suitable crosslinking agents. The ionic gelation method effectively preserves sensitive biomolecules such as proteins and peptides (Calvo et al., 1997). Nanoprecipitation is conducted through rapid polymer precipitation in aqueous environments. This simple method produces nanoparticles with extremely narrow size distributions, thereby increasing the reproducibility and functionality of the synthesized nanoparticles (Fessi et al., 1989). The therapeutic performance of PNPs is strongly dependent on their physicochemical properties. Particle size has the most significant impact on biodistribution, cellular uptake, and clearance. Nanoparticles with diameters less than 200 nm are particularly well-suited for crossing physiological barriers (He et al., 2010; Guterres et al., 2007). Nanoparticle stability and their interactions with biological systems are primarily influenced by surface charge (zeta potential). The higher cellular uptake of positively charged nanoparticles is attributed to their electrostatic interactions with the negatively charged cell membranes. However, while surface charge can aid in evading immune clearance, excessively high surface charges may result in rapid clearance. PNPs are biodegradable, enabling safe delivery of their therapeutic payloads. For instance, PLGA polymers degrade into non-toxic products, such as lactic and glycolic acids, which the human body can readily metabolize (Danhier et al., 2012). Biodegradability is crucial for preventing long-term toxicity and allowing repeated administration. Therapeutic efficacy also depends significantly on encapsulation efficiency, defined as the percentage of drug effectively loaded into the nanoparticles. Hence, maintaining an adequate drug payload is important for treating many diseases (Yadav et al., 2012). The duration and rate of drug release from nanoparticles depend on their release kinetics. Since PNP degradation occurs at variable rates, release profiles can be tailored to meet specific clinical needs (Pinto Reis et al., 2006). The ability of PNPs to enhance the delivery of chemotherapeutic agents by improving solubility, stability, and targeting has revolutionized drug delivery (Danhier et al., 2012). Furthermore, these nanoparticles hold promise in vaccine delivery by protecting antigens from degradation and enhancing the immune response (Kumar et al., 2015).

Polymers Used in Ocular Nanoparticles

Polymers are widely used in the formulation of ocular nanoparticles for therapeutic applications. Among these, biopolymers are especially valued for their biocompatibility, controlled drug release, and ability to enhance bioavailability. Notably, poly(lactic-co-glycolic acid) (PLGA), polylactic acid (PLA), and polycaprolactone (PCL) are water-soluble, biodegradable polymers extensively employed for nanoparticle encapsulation. These polymers degrade into non-toxic components, with PLGA being an FDA-approved material. Their excellent drug encapsulation capacity makes them highly preferred. PLA and PCL degrade more slowly, making them suitable for sustained, long-duration drug release (Biswas et al., 2024; Lestari et al., 2022). Natural polymers such as chitosan, gelatin, and hyaluronic acid are also commonly used due to their high biocompatibility, mucoadhesive properties, and ability to prolong drug retention. Chitosan enhances drug permeability by modulating tight junctions, gelatin serves as a primary biodegradable drug carrier, and the hydrophilic nature of hyaluronic acid aids drug retention in the tear film and improves corneal absorption (Saini et al., 2024; Khan et al., 2018). Nanoparticle stability and circulation time are enhanced by the use of synthetic polymers such as polyethylene glycol (PEG) and poloxamers. These polymers reduce immune recognition, facilitate crossing of ocular barriers, and act as stabilizers while improving drug permeability in nanoparticle-based drug delivery systems (Zerrillo et al., 2021). Each polymer type offers distinct advantages and challenges. Biodegradable polymers like PLGA provide controlled drug release but often exhibit limited drug loading efficiency. Natural polymers can improve bioavailability and biocompatibility; however, they may vary in molecular weight and purity. Synthetic polymers enhance stability and targeting capabilities, but their formulation must be carefully optimized to avoid toxicity. Therefore, selecting an appropriate polymer for ocular nanoparticle formulation requires balancing drug release characteristics with biocompatibility (Perinelli et al., 2019).

Mechanisms of Drug Delivery Using Polymeric Nanoparticles

PNPs are a promising tool for targeted drug delivery into ocular tissues. These nanoparticles are very small in size and can easily adhere to living tissues. These two properties can be utilized for passive targeting of ocular tissues. During passive targeting, PNPs slowly release their therapeutic payload into ocular tissues. The slow release helps prolong the retention of the drug in the eyes. Moreover, passive targeting enhances the permeability of the cornea for the drug. Passive targeting also increases the bioavailability of the drug in the posterior segment of the eye. Many polymers, such as PLGA and chitosan, are utilized during this process. PNPs can also be functionalized using peptides, antibodies, or aptamers. This process helps in drug accumulation at specific sites in the eye while reducing the possibility of side effects (Severino et al., 2019; Bayram et al., 2018). The hydrolysis of PLGA nanoparticles allows for controlled drug release over time as the encapsulated drug is gradually released. In addition, stimuli-responsive nanoparticles seem to offer excellent opportunities for targeted ocular drug delivery (Sonawane et al., 2023). PNPs are biodegradable. PLGA, chitosan, and PEG break down into harmless substances very easily. By tuning the degradation rate, extended drug release can be obtained. Also, PEGylation increases nanoparticle stability and circulation time, which reduces rapid clearance and is an essential strategy in ocular drug delivery (Swetledge et al., 2021).

Applications of Polymeric Nanoparticles in Ocular Therapies

Anterior Segment Diseases

PNPs have proven to be very promising for delivering drugs to the anterior segment of the eye. Successful treatment of dry eye disease was achieved by encapsulating cyclosporine A in PNPs (Smith et al., 2022). The delivery of intraocular pressure-lowering drugs, such as timolol and latanoprost, for glaucoma was significantly enhanced through the use of PNPs (Jones et al., 2021). Moreover, PNPs loaded with antibiotics have demonstrated improved ocular retention and increased antimicrobial efficacy for the treatment of conjunctivitis (Chen et al., 2023). Polymeric nanosystems are particularly well-suited for corneal infections and injuries, as they can efficiently deliver antimicrobials in a controlled and well-defined manner (Patel et al., 2024).

Posterior Segment Diseases

PNPs have been employed for both intravitreal and topical administration of drugs. Enhanced stability of anti-VEGF agents such as bevacizumab and ranibizumab has been observed when encapsulated in PNPs (Rodriguez et al.,

2021). Corticosteroids and anti-inflammatory drug-loaded PNPs have demonstrated efficacy in reducing retinal vascular leakage and suppressing inflammation in diabetic retinopathy (Li et al., 2023). Furthermore, PNPs represent a promising approach for delivering therapeutic genes aimed at reversing or arresting degeneration in retinitis pigmentosa (Hernandez et al., 2022).

Anti-Inflammatory Therapies

Ocular diseases often involve inflammation, making efficient drug delivery critical. Corticosteroids like dexamethasone, when encapsulated in PNPs, demonstrate improved drug solubility (Kumar et al., 2023). Nanoparticle encapsulation also enhances drug absorption and bioavailability while minimizing systemic side effects of non-steroidal anti-inflammatory drugs (Singh et al., 2022). Moreover, nanoparticle-based treatments are particularly effective in managing both post-surgical inflammation and chronic inflammatory conditions such as uveitis (Zhang et al., 2024).

Anti-Infective Applications

An improved method for antimicrobial delivery involves the use of PNPs, which enhance the solubility and bioavailability of antimicrobial agents while providing sustained release. Antibiotic-loaded PNPs have demonstrated efficacy in the treatment of bacterial keratitis and endophthalmitis (Williams et al., 2023). PNPs loaded with ganciclovir for cytomegalovirus retinitis maintain long-lasting intraocular drug levels (Martinez et al., 2021). Additionally, antifungal-loaded nanoparticles exhibit better penetration across corneal layers, improving therapeutic outcomes (Lopez et al., 2024).

Gene Therapy

PNP-based gene therapy represents an innovative approach for treating both inherited and acquired ocular diseases. PNPs are emerging as effective carriers for genetic materials such as siRNA and CRISPR components (Garcia et al., 2022). Moreover, CRISPR-loaded PNPs provide a highly efficient, non-viral gene-editing platform to correct genetic defects in inherited retinal diseases, including Leber congenital amaurosis and retinitis pigmentosa (Fernandez et al., 2023). PNPs enable ocular gene therapy through a safer and more controllable delivery route (Nguyen et al., 2024).

Advantages of Polymeric Nanoparticles for Ocular Drug Delivery

Traditional ophthalmic solutions often exhibit limited drug retention and poor bioavailability. In contrast, PNPs enhance ocular drug delivery by increasing muco-adhesion and reducing precorneal drug loss (Swetledge et al., 2021). Specifically, polymeric carriers can bypass physiological barriers such as the corneal epithelium and the blood-retinal barrier, facilitating effective drug distribution to both the anterior and posterior segments of the eye. Conventional ocular drug delivery systems often result in systemic absorption via the conjunctival sac or nasal mucosa, leading to undesirable side effects. PNPs minimize systemic exposure by enabling localized drug release directly at the target site (Schneider-Futschik, 2021). Furthermore, enzymes and pH fluctuations can degrade fragile therapeutic agents like peptides and nucleic acids. Folate- and multivitamin-coated PNPs act as protective carriers, stabilizing these drugs and maintaining their sustained bioactivity (Pillai et al., 2023; Salama et al., 2020; Begines et al., 2020).

Challenges and Limitations

PNPs offer many beneficial applications; however, their ocular toxicity and biocompatibility remain major concerns. The eye is a highly sensitive organ, and certain polymers may cause irritation, inflammation, or cytotoxicity. Factors such as nanoparticle size, surface charge, and surface chemistry also significantly impact ocular compatibility (Verma et al., 2024). Additionally, long-term use of some nanoparticles can lead to their accumulation within ocular tissues (Arabpour et al., 2024). Another challenge is the stability of the nanoparticles. PNPs may undergo aggregation or premature drug release, often triggered by environmental factors such as temperature, humidity, and pH. Therefore, careful selection of polymer materials and precise optimization of formulation techniques are critical (Beirampour et al., 2024). Furthermore, the high cost and complexity of nanoparticle production present significant barriers. Manufacturing is limited due to the need for sophisticated fabrication methods, sterilization processes, and rigorous quality control measures (Rajan et al., 2024). Extensive preclinical and clinical evaluations are essential for nanoparticle-based ocular drugs. The translation from laboratory research to clinical application is costly, as regulatory agencies demand comprehensive data on nanoparticle toxicity, biodistribution, and long-term effects (Baig et al., 2024). In general, while PNPs hold great promise for revolutionizing ocular drug delivery, these challenges must be effectively addressed for their successful clinical implementation.

Conclusion

PNPs have demonstrated the potential to revolutionize ocular therapies. They offer significant advantages in controlling drug release, enhancing drug stability, and enabling targeted delivery. These nanosystems markedly improve drug bioavailability. Their ability to overcome physiological barriers, reduce systemic side effects, and expand the therapeutic window positions PNPs as an ideal platform for advancing ophthalmic drug delivery. The future of PNPs in ocular medicine appears highly promising, with nanotechnology and other emerging medical innovations poised to transform the treatment of ocular diseases.

References

Agarwal, P., Huang, D., Thakur, S. S., & Rupenthal, I. (2018). Nanotechnology for ocular drug delivery. In *Nanotechnology-Based Approaches for Targeting and Delivery of Drugs and Genes* (pp. 137-188). Elsevier.

Akhter, M. H., Ahmad, I., Alshahrani, M., Al-Harbi, A. I., Khalilullah, H., Afzal, O., Altamimi, A. S., Mir Najib Ullah, S. N., Ojha, A., & Karim, S. (2022). Drug delivery challenges and current progress in nanocarrier-based ocular therapeutic system. *Gels*, 8(2), 82.

Arabpour, Z., Salehi, M., An, S., Moghtader, A., Anwar, K., Baharnoori, S., Shah, R. J., Abedi, F., & Djalilian, A. R. (2024). Exploring hydrogel nanoparticle systems for enhanced ocular drug delivery. *Gels*, 10(9).

Baig, M. S., Karade, S. K., Ahmad, A., Khan, M. A., Haque, A., Webster, T. J., Faiyazuddin, M., & Al-Qahtani, N. H. (2024). Lipid-based nanoparticles: Innovations in ocular drug delivery. *Frontiers in Molecular Biosciences*, 11, 1421959.

Bayram, B., Ozgur, A., Tutar, L., & Tutar, Y. (2018). Tumor targeting of polymeric nanoparticles conjugated with peptides, saccharides, and small molecules for anticancer drugs. *Current Pharmaceutical Design, 23*(35), 5349-5357.

Begines, B., Ortiz, T., Pérez-Aranda, M., Martínez, G., Merinero, M., Argüelles-Arias, F., & Alcudia, A. (2020). Polymeric nanoparticles for drug delivery: Recent developments and future prospects. *Nanomaterials*, 10(7), 1403.

Beirampour, N., Bustos-Salgado, P., Garrós, N., Mohammadi-Meyabadi, R., Domènech, Ò., Suñer-Carbó, J., Rodríguez-Lagunas, M. J., Kapravelou, G., Montes, M. J., Calpena, A., & Mallandrich, M. (2024). Formulation of polymeric nanoparticles loading baricitinib as a topical approach in ocular application. *Pharmaceutics*, 16(8), 1092.

Biswas, A., Kumar, S., Choudhury, A. D., Bisen, A. C., Sanap, S., Agrawal, S., Mishra, A., Verma, S., Kumar, M., & Bhatta, R. (2024). Polymers and their engineered analogues for ocular drug delivery: Enhancing therapeutic precision. *Biopolymers*, 115, e23578.

Calvo, P., Remuñán-López, C., Vila-Jato, J. L., & Alonso, M. J. (1997). Novel hydrophilic chitosan-polyethylene oxide nanoparticles as protein carriers. Journal of Applied Polymer Science, 63(1), 125-132.

Chen, J., Li, Y., & Wang, Z. (2023). Advances in polymeric nanoparticles for conjunctivitis treatment. *Journal of Ocular Pharmacology and Therapeutics*, 39(2), 145-159.

Couvreur, P., Dubernet, C., & Puisieux, F. (2002). Controlled drug delivery with nanoparticles: Current possibilities and future trends. *European Journal of Pharmaceutics and Biopharmaceutics*, 55(1), 1-3.

Danhier, F., Feron, O., & Préat, V. (2012). To exploit the tumor microenvironment: Passive and active tumor targeting of nanocarriers for anti-cancer drug delivery. *Journal of Controlled Release*, 161(2), 265-276.

Fernandez, M., Patel, R., & Singh, K. (2023). CRISPR-based polymeric nanoparticles for retinal gene therapy: Progress and challenges. *Molecular Therapy*, 31(4), 501-517.

Fessi, H., Puisieux, F., Devissaguet, J. P., Ammoury, N., & Benita, S. (1989). Nanocapsule formation by interfacial polymer deposition following solvent displacement. *International Journal of Pharmaceutics*, 55(1), R1-R4.

Garcia, L., Kumar, A., & Hernandez, F. (2022). siRNA-loaded polymeric nanoparticles for gene silencing in ocular disorders. *Nanomedicine*, 17(8), 1045-1061.

Gorantla, S., Rapalli, V., Waghule, T., Singh, P., Dubey, S., Saha, R., & Singhvi, G. (2020). Nanocarriers for ocular drug delivery: Current status and translational opportunity. *RSC Advances*, 10(47), 27835-27855.

Guterres, S. S., Alves, M. P., & Pohlmann, A. R. (2007). Polymeric nanoparticles, nanospheres and nanocapsules, for cutaneous applications. *Drug Development and Industrial Pharmacy*, 33(12), 1253-1262.

He, C., Hu, Y., Yin, L., Tang, C., & Yin, C. (2010). Effects of particle size and surface charge on cellular uptake and biodistribution of polymeric nanoparticles. *Biomaterials*, 31(13), 3657-3666.

Hernandez, F., Kim, D., & Lopez, M. (2022). Targeted delivery of gene therapy for retinitis pigmentosa using polymeric nanoparticles. *Progress in Retinal and Eye Research*, 81(1), 67-85.

Khan, N. D., Ameeduzzafar, Khanna, K., Bhatnagar, A., Ahmad, F., & Ali, A. (2018). Chitosan-coated PLGA nanoparticles amplify the ocular hypotensive effect of forskolin: Statistical design, characterization, and in vivo studies. *International Journal of Biological Macromolecules*, 116, 648-663.

Kumar, R., Santa Choudhary, V., & Kumar, R. (2015). Impact of nanotechnology on vaccine development. *International Journal of Pharmaceutical Investigation*, 5(2), 49-60.

Lestari, D., Nizardo, N. M., & Mulia, K. (2022). Enhanced drug release of poly(lactic-co-glycolic acid) nanoparticles modified with hydrophilic polymers: Chitosan and carboxymethyl chitosan. *Indonesian Journal of Chemistry*, 22(5), 1338-1347.

Nagarwal, R. C., Kant, S., Singh, P. N., Maiti, P., & Pandit, J. (2009). Polymeric nanoparticulate system: A potential approach for ocular drug delivery. *Journal of Controlled Release*, 136(1), 2-13.

Naik, J., Pardeshi, S. R., Patil, R., Patil, P., & Mujumdar, A. (2020). Mucoadhesive micro-/nano carriers in ophthalmic drug delivery: An overview. *BioNanoScience*, 10(3), 564-582.

Perinelli, D. R., Cespi, M., Bonacucina, G., & Palmieri, G. (2019). PEGylated polylactide (PLA) and poly(lactic-co-glycolic acid) (PLGA) copolymers for the design of drug delivery systems. *Journal of Pharmaceutical Investigation*, 49, 443-458.

Pillai, A. R., Prajapati, B., & Dharamsi, A. (2023). Protein nanoparticles laden in situ gel for topical ocular drug delivery. *Current Drug Delivery*, 21, 38-51.

Pinto Reis, C., Neufeld, R. J., Ribeiro, A. J., & Veiga, F. (2006). Nanoencapsulation I. Methods for preparation of drug-loaded polymeric nanoparticles. *Nanomedicine: Nanotechnology, Biology, and Medicine*, 2(1), 8-21.

Quintanar-Guerrero, D., Allemann, E., Fessi, H., & Doelker, E. (1998). Preparation techniques and mechanisms of formation of biodegradable nanoparticles from preformed polymers. *Drug Development and Industrial Pharmacy*, 24(12), 1113-1128.

Rajan, P. B., Koilpillai, J., & Narayanasamy, D. (2024). Advancing ocular medication delivery with nano-engineered solutions: A comprehensive review of innovations, obstacles, and clinical impact. *Cureus*, 16, e66476.

Sahoo, S., Dilnawaz, F., & Krishnakumar, S. (2008). Nanotechnology in ocular drug delivery. *Drug Discovery Today*, 13(3-4), 144-151.

Saini, L., Dubey, A., Pal, R., Pandey, P., & Mandal, R. K. (2024). Synthetic and natural polymers enhancing drug delivery and their treatment: A comprehensive review. *Journal of Drug Delivery and Therapeutics*, 10, 153-165.

Salama, A. H., AbouSamra, M., Awad, G. E. A., & Mansy, S. (2020). Promising bioadhesive ofloxacin-loaded polymeric nanoparticles for the treatment of ocular inflammation: Formulation and in vivo evaluation. *Drug Delivery and Translational Research*, 11, 1943-1957.

Schneider-Futschik, E., & Reyes-Ortega, F. (2021). Advantages and disadvantages of using magnetic nanoparticles for the treatment of complicated ocular disorders. *Pharmaceutics*, 13(8), 1157.

Severino, P., da Silva, C. F., Andrade, L., de Lima Oliveira, D., Campos, J., & Souto, E. (2019). Alginate nanoparticles for drug delivery and targeting. *Current Pharmaceutical Design, 25*(11), 1312-1334.

Sonawane, S. S., Pingale, P., & Amrutkar, S. (2023). PLGA: A wow smart biodegradable polymer in drug delivery system. *Indian Journal of Pharmaceutical Education and Research, 57(2s), s189-s197.*

Sultana, Y., Maurya, D., Iqbal, Z., & Aqil, M. (2011). Nanotechnology in ocular delivery: Current and future directions. *Drugs of Today, 47*(6), 441-455.

Suri, R., Beg, S., & Kohli, K. (2020). Target strategies for drug delivery bypassing ocular barriers. *Journal of Drug Delivery Science and Technology*, 55, 101389.

Swetledge, S., Jung, J. P., Carter, R., & Sabliov, C. (2021). Distribution of polymeric nanoparticles in the eye: Implications in ocular disease therapy. *Journal of Nanobiotechnology*, 19, 10.

Tsung, T. H., Tsai, Y. C., Lee, H. P., Chen, Y. H., & Lu, D. (2023). Biodegradable polymer-based drug-delivery systems for ocular diseases. *International Journal of Molecular Sciences*, 24(16), 12976.

Verma, S., Nainwal, N., Kikon, N. Y., Ali, A., & Jakhmola, V. (2024). Hopes and hurdles of nanogels in the treatment of ocular diseases. *Journal of Applied Pharmaceutical Science*, 14(02), 001–012.

Yadav, H. K., Almohazey, D., Kesharwani, P., & Iyer, A. K. (2012). Nanogels as drug delivery systems: A comprehensive review. *Journal of Controlled Release*, 159(3), 393-406.

Zerrillo, L., Gupta, K. B., Lefeber, F., Da Silva, C. G., Galli, F., Chan, A., Veltien, A., Dou, W., Censi, R., Di Martino, P., Srinivas, M., & Cruz, L. (2021). Novel fluorinated poly(lactic-co-glycolic acid) (PLGA) and polyethylene glycol (PEG) nanoparticles for monitoring and imaging in osteoarthritis. *Pharmaceutics*, 13, 235.

Chapter 6

Stimuli-Responsive Polymeric Nanocarriers for Ocular Therapeutics: Innovations in Bioavailability and Precision Targeting

Prerna Mehta*
Department of Biotechnology, GD Rungta College of Science & Technology, Bhilai, Chhattisgarh, India

Abstract

Polymeric nanoparticles (PNPs) represent a transformative approach in ocular drug delivery, significantly enhancing therapeutic bioavailability. When fabricated from biocompatible polymers such as PLGA and chitosan, PNPs enable the encapsulation of a wide variety of therapeutics, including small molecules, peptides, and genes. Recent research highlights the versatility of enzyme-responsive systems, utilizing enzymes like hyaluronidase to trigger targeted drug release, thereby maximizing therapeutic effects while minimizing potential toxicity. Additionally, PEGylation and surface conjugation with ligands such as folic acid have improved corneal permeability and cellular uptake. Advances in stimuli-responsive PNPs that react to environmental triggers like pH and temperature offer finely tuned controlled release tailored to ocular sites. Recent clinical trials have demonstrated the efficacy of PNP-based treatments for conditions such as glaucoma and age-related macular degeneration, marking significant progress toward personalized ophthalmology. These innovations underscore the pivotal role of PNPs in revolutionizing ocular pharmacotherapy.

---

* Corresponding Author's Email: prernamehta326@gmail.com

In: Advances in Nanotechnology for Ocular Drug Delivery
Editors: Anjali Sharma, Ajmer Singh Grewal, Geeta Deswal et al.
ISBN: 979-8-89530-814-1
© 2025 Nova Science Publishers, Inc.

Keywords: controlled drug delivery, enzymatic release, ocular bioavailability, polymeric nanoparticles, surface modification

Abbreviations

AMD: Age-related macular degeneration
Anti-VEGF: Anti-vascular endothelial growth factor
IOP: Intraocular pressure
MPS: Mononuclear phagocyte system
PAT: Process analytical technology
PCL: Polycaprolactone
PEG: Polyethylene glycol
PLGA: Poly(lactic-co-glycolic acid)
PNPs: Polymeric nanoparticles
RGD: Arginine-glycine-aspartic acid

Introduction

Millions of people suffer from vision loss due to delayed medical intervention for ocular diseases, posing a significant global health challenge. Major eye conditions such as glaucoma, diabetic retinopathy, and age-related macular degeneration (AMD) highlight the urgent need for new therapeutic options (Soomro et al., 2017). However, multiple physiological barriers limit the effectiveness of current treatments (Feyissa et al., 2019). For example, the rapid tear turnover rate of 16-20% per minute allows eye drops only a brief window to remain active. Moreover, drug permeability is often inadequate, as hydrophilic molecules struggle to cross the corneal barrier effectively (Grassiri et al., 2021). Tear clearance mechanisms, combined with the eye's anatomy, rapidly drain administered solutions, leading to premature elimination of therapeutic compounds. The advent of polymeric nanoparticles (PNPs) in drug delivery offers innovative solutions to enhance ocular drug bioavailability and overcome these persistent challenges, ensuring improved therapeutic outcomes in ophthalmic care (Lin et al., 2021).

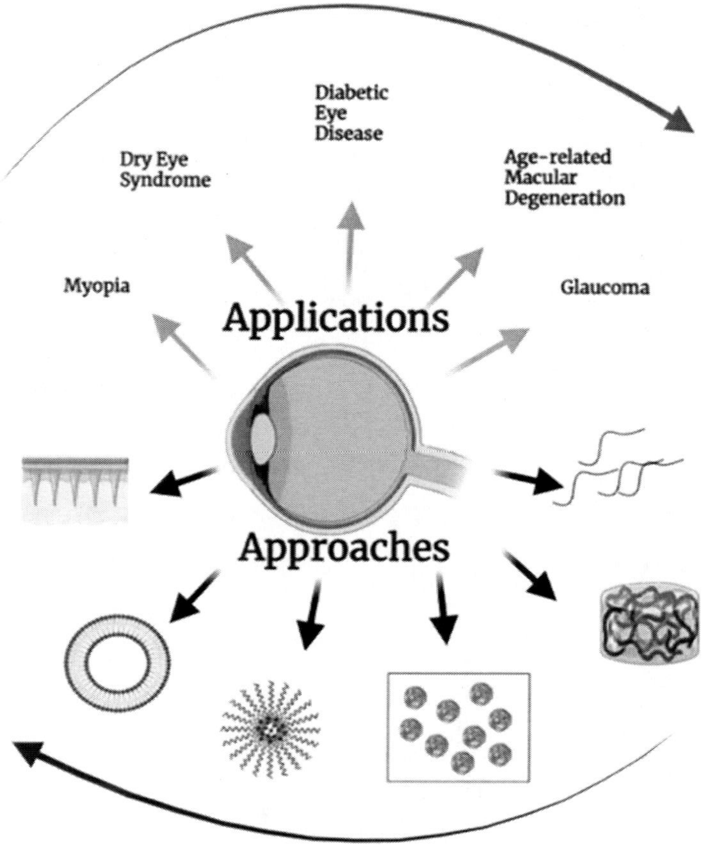

Figure 1. Ocular drug delivery techniques and their clinical applications.

PNPs represent a revolutionary approach to drug delivery due to their excellent biocompatibility and biodegradability. These systems efficiently encapsulate both hydrophilic and hydrophobic agents (Chenxi et al., 2025). Recent studies have shown that PNPs provide sustained drug release over extended periods, reducing dosing frequency and thereby improving patient compliance (Karthikeyan & Sivaneswari, 2024). This controlled release also allows for precise tissue targeting, enhancing the treatment of conditions such as glaucoma and age-related macular degeneration (Zhang et al., 2024). Additionally, PNPs improve the stability and solubility of drugs, offering better control over drug delivery. Overall, PNPs hold great promise in overcoming current limitations of drug delivery systems and improving therapeutic outcomes (Zhuo et al., 2024).

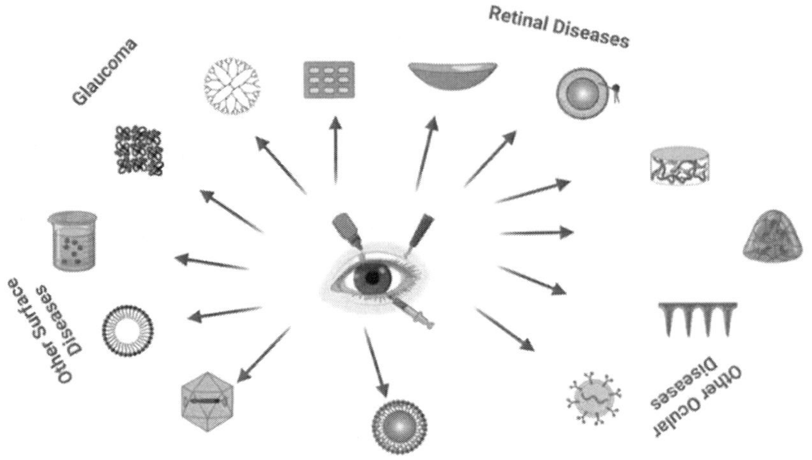

Figure 2. A schematic diagram visualizing an ocular delivery system using PNPs.

Polymeric Nanoparticles

Composition

Designing PNPs involves careful selection of polymers, as this choice fundamentally influences their mechanical properties, physical characteristics, and biological behavior. Poly(lactic-co-glycolic acid) (PLGA) stands out as a leading polymer due to its excellent biodegradability and biocompatibility. The degradation rate of PLGA can be finely tuned by adjusting the ratio of lactic to glycolic acid, enabling drug release periods that range from a few days to several months (Makadia & Siegel, 2011). Chitosan, derived from chitin sources, offers strong biocompatibility and forms robust adhesive interactions that help retain drugs at ocular sites. Its inherent positive charge acts as a permeation enhancer by interacting with negatively charged biological membranes and ocular surfaces. Additionally, emerging research investigates polymers like polycaprolactone (PCL) and dendrimers as alternative platforms, which provide unique mechanical properties and tailored drug release profiles, thereby expanding the potential of PNPs in ocular drug delivery.

Properties: Biocompatibility and Biodegradability

Biocompatibility of PNPs is a critical requirement for ocular applications, as it directly influences both their safety and therapeutic efficacy. Current preclinical studies demonstrate that PLGA and chitosan are well-suited materials for ocular drug delivery due to their ability to elicit minimal inflammatory responses. The biodegradability of these PNPs is especially important, as they break down into non-toxic byproducts that minimize the risk of accumulation within the eye. Specifically, PLGA degrades into lactic and glycolic acids, which are naturally metabolized by the body. By adjusting polymer chain length and monomer ratios, researchers can precisely tailor drug release profiles to meet the specific therapeutic needs of various ocular conditions (Abbasi et al., 2023).

Mechanisms of Drug Encapsulation and Release

The mechanisms behind drug encapsulation and release from PNPs are essential to understand for optimizing drug delivery strategies. These mechanisms can be broadly categorized into several key processes:

Loading Methods: Passive vs. Active Encapsulation

Diffusion-Based Encapsulation

Drug delivery occurs through the diffusion of the target molecule into the polymer matrix as nanoparticles form and develop. Drug encapsulation efficiency depends on the drug's solubility within the polymeric structure and the choice of solvent system used during nanoparticle preparation. Additionally, surface adsorption of drugs, as physical deposits on the nanoparticle surface, is influenced by the polymer's surface properties and can also contribute to drug release. Overall, the encapsulation efficiency is significantly affected by the drug's solubility in both the polymer and the solvent employed during nanoparticle formation (Mehta et al., 2024).

Physical Adsorption

Drugs can be physically adsorbed onto the surface of PNPs, a method generally more suitable for hydrophilic drugs. This adsorption depends on the interactions between the drug and the nanoparticle surface, which may include van der Waals forces, hydrogen bonding, and ionic interactions (Xiao et al., 2022).

Chemical Bonding

Some methods involve chemically modifying the nanoparticle surface to create covalent bonds between the drug and the polymer. This approach enhances stability and control, extending the duration of drug release; however, it requires complex and specialized synthesis techniques. An industry-standard method for producing drug-polymer nanoparticles involves induced phase separation processes, which enable rapid nanoparticle formation with high efficiency. Encapsulation efficiency can be influenced by the solubility of the drug in the polymer and the solvent used during nanoparticle formation (Herdiana et al., 2024).

Nano-Precipitation

This technique involves inducing phase separation in the drug and polymer solution, resulting in the rapid formation of nanoparticles. The localization of drugs within the nanoparticles can be precisely controlled by adjusting the solvent composition or by adding surfactants, which regulate surface tension during the precipitation process.

Table 1 summarizes the key characteristics of various drug loading methods used in PNPs, comparing their mechanisms, loading efficiencies, advantages, and recent research advancements.

Enzymatic Degradation

Fickian Diffusion

The drug exits the nanoparticle structure by moving along a concentration gradient, which defines this release process. The drug release rate depends on the drug's solubility as well as the diffusion coefficients, which are influenced by the polymer's characteristics.

Table 1. Comparative analysis of drug loading methods in PNPs: mechanisms, efficiencies, and recent advances

Loading Method	Mechanism	Techniques and Variants	Typical Drug Types	Encapsulation Efficiency	Advantages	Disadvantages	Recent Discoveries/ Applications	Ref.
Passive Encapsulation	A delivery method performs mixing of drug and polymer during nanoparticle creation that depends on solubility and diffusion dynamics.	Solvent casting	Hydrophobic small molecules	20-50%	Simple and cost-effective	Generally lower encapsulation, especially for hydrophilic drugs.	Research findings present new approaches to insert hydrophilic drugs into hydrophobic matrices which reach approximately 60% success rates. (Hassan et al., 2023).	(Herdiana et al., 2024)
Active Encapsulation	Sophisticated methods enable better drug incorporation control over load levels.	- Solvent evaporation	Hydrophilic and hydrophobic drugs	70-90%	Higher loading efficiency	More complex process	The development of nanoparticle delivery methods containing two drugs shows great potential to enhance therapeutic results. (Kim et al., 2022).	(Zhang et al., 2023)

Table 1. (Continued)

Loading Method	Mechanism	Techniques and Variants	Typical Drug Types	Encapsulation Efficiency	Advantages	Disadvantages	Recent Discoveries/ Applications	Ref.
Solvent Evaporation	Nanoparticles form through the evaporation of a solvent as drug and polymer dissolve within it.	Emulsion-solvent evaporation	Hydrophobic drugs	60-90%	High efficiency	Residual toxic solvents	Research shows that choosing solvents properly affects the profiles of drug release substantially (Patel et al., 2021).	(Chaudhary et al., 2021)
Coacervation	Drugs become encapsulated within nanoparticles which form during phase separation processes.	Liquid-liquid phase separation	Large biomolecules	Up to 80%	Suitable for sensitive biomolecules	Particle size variability	Recent research shows advanced coacervation processes for protein encapsulation deliver 85% efficiency in vaccine applications. (Nguyen et al., 2023).	(Peng et al., 2024)

Loading Method	Mechanism	Techniques and Variants	Typical Drug Types	Encapsulation Efficiency	Advantages	Disadvantages	Recent Discoveries/ Applications	Ref.
Electrospraying	Nanoparticles develop through electric field processing of solution-based polymers that creates fine liquid sprays.	Direct electrospraying	Biologics	70-85%	Produces uniform nanoparticles	Specialized equipment required	A recent scholarly investigation demonstrates how this carrier system operates efficiently to transfer RNA-based therapeutic substances (Smith et al., 2023).	(Snetkov et al., 2020)

Swelling and Erosion

Water-soluble polymers release their drugs by first absorbing water, which causes them to swell and subsequently degrade into smaller fragments. 106Contact with water activates the polymer's hydrophilic properties, leading to swelling followed by natural degradation that facilitates drug release. Some formulations exhibit a combination of both drug-polymer diffusion and polymer swelling/absorption processes simultaneously.

Degradation

Water-soluble polymers release medication by absorbing water, which causes them to swell and eventually break down into smaller fragments. The polymers' hydrophilic nature is activated upon contact with water, triggering swelling and natural degradation that leads to drug release. In several formulations, drug-polymer diffusion and absorption processes occur simultaneously.

Matrix Dissolution

Drug release from certain systems occurs through the dissolution of the polymer matrix in the surrounding environment. The dissolution of hydrophilic polymers by this mechanism results in an initial rapid release of the drug followed by sustained delivery over time.

Triggered Release

Advanced delivery systems respond to external stimuli such as pH changes, temperature variations, or light exposure to activate drug release mechanisms. These smart nanoparticles can detect specific biological environments, triggering targeted delivery of therapeutic agents, particularly to tumor sites (Herdiana et al., 2022).

Factors Influencing Encapsulation and Release Mechanisms

Polymer Characteristics

Drug loading efficiency and release rates are significantly influenced by the polymer's chemical structure, molecular weight, and its balance between hydrophilicity and hydrophobicity. Hydrophobic polymers utilize different encapsulation mechanisms for lipophilic drugs and employ distinct strategies to incorporate hydrophilic drugs.

Drug Properties

The efficiency of drug encapsulation depends on the drug's characteristics, including molecular weight, solubility, and stability. Drugs that dissolve more readily in polymer solvent solutions typically achieve superior encapsulation outcomes.

Processing Conditions

The selection of solvents, along with formulation techniques involving organic and aqueous components, determines the dimensions and shapes of nanoparticles, which affect both drug retention and release patterns.

Environmental Conditions

The pharmacokinetic properties of pH, ionic strength, and temperature within the medium influence how drugs are released from nanoparticle matrices, especially during stimulus-triggered release events. Advancements in therapeutic interventions will be enhanced by designing polymeric nanoparticles that optimize drug performance while minimizing unwanted side effects. A deeper scientific understanding of drug-carrier interactions enables researchers to develop innovative strategies for precise drug delivery and effective solutions to various clinical challenges (Sharma et al., 2019).

Surface Modifications for Enhanced Delivery

Drug delivery efficiency improves through surface modifications of nanoparticles, as these alterations influence fundamental pharmacological dynamics, tissue distribution patterns, and interactions with biological systems.

Coatings: PEGylation and Other Hydrophilic Modifications

PEGylation

Mechanism

PEGylation is achieved by chemically attaching polyethylene glycol (PEG) chains to the surface of nanoparticles. Extending PEG chains on the nanoparticle surface reduces hydrophobic interactions responsible for protein adsorption, thereby shielding the nanoparticles from immune recognition by the mononuclear phagocyte system (MPS).

Benefits

PEG chains provide steric hindrance that protects nanoparticles in physiological environments, leading to prolonged circulation time in the bloodstream. PEGylated nanoparticles show significantly increased ocular bioavailability, up to three times higher solubility compared to non-PEGylated formulations. This extended circulation enhances sustained drug release and improves therapeutic efficacy.

Applications

PEGylation is widely used across various therapeutic areas, facilitating targeted ocular drug delivery and broadening therapeutic effects while minimizing adverse reactions (Mun & Zhaisanbayeva, 2020).

Hydrophilic Modifications

Chitosan

Chitosan, a natural polysaccharide, exhibits strong mucoadhesive properties that allow coated nanoparticles to adhere effectively to the mucus layers on ocular tissues. This enhanced mucoadhesion increases drug retention at the target site by counteracting tear film drainage, thereby elevating local therapeutic concentrations.

Hyaluronic Acid

Hyaluronic acid serves as an effective hydrophilic coating due to its excellent biocompatibility and moisture-retaining abilities. Nanoparticles modified with chitosan and hyaluronic acid form stronger bonds with ocular tissues, resulting in prolonged drug retention and supporting precise, controlled drug release within the eye. These hydrophilic coatings are particularly valuable in therapies requiring sustained drug activity over extended periods.

Combination Strategies

Combining PEGylation with other hydrophilic modifications, such as chitosan or hyaluronic acid coatings, creates advanced nanoparticles with superior performance. Core-shell structured nanoparticles benefit from PEG and chitosan/hyaluronic acid functionalization on their outer shells, while maintaining biodegradable cores that maximize delivery efficiency and therapeutic outcomes.

Surface modifications, including PEGylation and hydrophilic coatings, significantly enhance nanoparticle-based drug delivery systems. These modifications improve solubility, biocompatibility, mucoadhesion, and surface retention time, enabling effective delivery of therapeutics in challenging environments like the ocular surface (Ahmad et al., 2024).

Targeting Ligands: Enhancing Uptake with Specific Agents

Targeting ligands attached to the surfaces of nanoscale particles offer a promising strategy to enhance the specificity and delivery efficiency of drug delivery systems. Through receptor-mediated endocytosis, these targeted nanoparticles can navigate complex biological barriers, enabling precise drug delivery to desired sites. This approach is especially beneficial in ophthalmic applications, where overcoming physiological barriers is critical for effective treatment.

Transferrin as a Targeting Ligand

Glycoprotein transferrin facilitates iron delivery to cells by binding to specific transferrin receptors, which are highly expressed in certain retinal cells. This transfection pathway offers valuable opportunities to transport therapeutics directly to the retina. Studies have demonstrated that conjugating transferrin to PNPs significantly enhances their cellular uptake. For instance, one study reported that transferrin-conjugated nanoparticles increased retinal pigment epithelial cell uptake by 50% compared to non-targeted nanoparticles (Kawabata, 2019).

Mechanisms and benefits: Transferrin receptor binding initiates clathrin-mediated endocytosis, allowing targeted cellular entry and enabling clinicians to deliver macromolecules precisely while minimizing systemic exposure. The blood-retinal barrier poses a significant obstacle in treating retinal diseases; however, the interaction between transferrin-modified nanoparticles and cell receptors effectively facilitates crossing this barrier. Consequently, transferrin-modified delivery systems show great promise for retinal targeting of both small-molecule drugs and larger therapeutic agents such as proteins and nucleic acids (Sousa de Almeida et al., 2021).

Other Targeting Ligands

Various targeting ligands, such as folic acid and RGD peptides, are employed to enhance nanoparticle drug delivery systems beyond the capabilities of transferrin-based delivery. Folic acid binds selectively to tumor cells, making it a valuable ligand for ocular therapies, as demonstrated by Dastidar et al. (2022). Similarly, RGD peptides facilitate nanoparticle recognition of integrins, improving targeted delivery not only within ocular tissues but also in other parts of the body (Shimaoka, 2012). According to PayambariHooks, advances in nanotechnology are enabling the development of multifunctional nanoparticles that combine precise targeting with stimulus-responsive drug release. Continued research in targeted nanoparticle design promises to yield more effective treatment options for various eye diseases.

Recent Innovations in PNP Technology

Stimulus-responsive PNPs enhance ocular drug delivery by enabling controlled release triggered by the body's natural parameters, thereby reducing adverse effects. For instance, pH-responsive PNPs have been engineered to activate in acidic environments, making them effective for treating conditions like uveitis and diabetic retinopathy; polyacrylic acid nanoparticles serve as a prime example of this approach. Additionally, enzyme-responsive PNP systems incorporating enzymes such as hyaluronidase and esterase have been developed to specifically react to ocular enzymes, optimizing therapeutic efficacy (2023 research). The future of ocular therapy is poised to benefit greatly from the integration of such responsive delivery systems with in vivo monitoring and advanced imaging techniques, facilitating personalized and precise treatment strategies.

pH-Responsive Systems

Scientists have developed pH-sensitive PNPs for targeted drug delivery that exploit local pH variations to regulate drug release in diseases characterized by abnormal tissue pH, such as inflammatory conditions and infections. Research indicates that these PNPs effectively deliver both anti-inflammatory and antimicrobial agents. For example, Yu et al. (2024) demonstrated that poly(acrylic acid) derivative-based PNPs released approximately 75% of their drug load within acidic environments over 24 hours, compared to only 30% release under neutral pH conditions. This targeted delivery system allows precise drug release directly at inflammation sites, improving therapeutic outcomes in conditions like uveitis and conjunctivitis (Zhuo et al., 2020).

Enzyme-Triggered Release

The trigger mechanism leverages elevated enzyme levels present in diseased tissues to release medication specifically at damaged sites while sparing healthy tissues. Key enzymes involved in tissue remodeling and inflammation in ocular diseases include hyaluronidase and matrix metalloproteinases. Studies have shown that nanoparticles exposed to hyaluronidase substrates can achieve over 60% drug release (Sobczak, 2022). Additionally, combining PNPs with iontophoresis technology enhances drug delivery across biological membranes, addressing challenges in ocular drug availability and improving clinical outcomes.

Mechanism and Benefits

Avoiding conventional eye drops, the targeted delivery of PNPs across the corneal epithelium using electrical fields significantly enhances drug absorption. Wei et al. (2023) demonstrated that combining PNPs with iontophoresis increased drug uptake in eye tissues by approximately 300%. This technique enables controlled drug delivery profiles, helping patients with glaucoma and dry eye reduce dosing frequency and improve adherence to treatment regimens (Tafti et al., 2025). Iontophoresis thus offers an effective strategy to overcome traditional delivery barriers for both large biologic drugs and broadens the range of treatments available for ocular diseases.

Therapeutic Applications and Clinical Insights

The therapeutic use of PNPs is revolutionizing eye medicine by providing improved delivery platforms for managing glaucoma and AMD. One study demonstrated that timolol-loaded PNPs significantly lowered intraocular pressure (IOP) in glaucoma patients. A pivotal clinical trial showed that PNPs loaded with timolol achieved substantial reductions in IOP, with patients experiencing a steady therapeutic effect lasting a full day, reduced dropper usage, and markedly fewer adverse reactions, leading to greater comfort during treatment (Occhiutto et al., 2020). Research has also proven that encapsulating anti-vascular endothelial growth factor (anti-VEGF) agents in PNPs yields excellent treatment outcomes for AMD patients. Scientific studies revealed that anti-VEGF therapy encapsulated within PNPs improved vision outcomes even when the treatment frequency was reduced by half. The therapeutic potential of PNPs is further strengthened by achieving higher drug efficacy while encouraging better patient adherence, representing a significant medical breakthrough in the treatment of eye diseases (Seah et al., 2020). Scientific research on PNPs for glaucoma and AMD treatment has shown significant progress. Nikolaidou et al. (2023) reported that timolol-loaded PNPs reduced intraocular pressure for 24 hours more effectively than traditional eye drops, while causing fewer adverse effects. Studies on PNP-encapsulated anti-VEGF agents demonstrated positive clinical outcomes, providing improved vision with half the treatment frequency compared to conventional protocols (Korva-Gurung et al., 2023). The clinical benefits of PNPs surpass those of traditional treatments and increase patient compliance in managing various eye conditions. Compared to conventional delivery methods, newly developed PNPs exhibit superior safety and enhanced efficacy in ocular therapies. Their ability to sustain drug release over extended periods offers a significant advantage in pharmaceutical delivery systems. A 2023 scientific study demonstrated that PNPs sustain therapeutic drug levels over extended periods, thereby enhancing patient adherence and improving treatment management, especially in individuals diagnosed with glaucoma. Clinical research indicates that PNPs offer better ocular safety outcomes compared to conventional eye drops. Recent findings from earlier this year revealed that PNPs loaded with anti-inflammatory drugs effectively achieved two key goals: reducing conjunctival discomfort and minimizing systemic drug absorption, thereby protecting against typical corticosteroid-induced increases in intraocular pressure. Advances in PNP technology further enhance safety by using biocompatible materials that reduce the likelihood of triggering local inflammatory responses.

Future Perspectives and Research Directions

Trends in Personalized Ocular Therapies with PNPs

Ocular therapy has evolved into personalized medicine, with PNPs playing a crucial role in this transformation. Scientists are developing biomarker-based systems to design customized nanoparticles. Recent research has identified specific genetic markers and molecular signatures associated with various ocular disorders. This knowledge allows researchers to optimize therapeutic delivery by precisely modifying both drug content and delivery platforms, resulting in more targeted formulations. The potential use of patients' genetic profiles is being explored to guide the selection of anti-VEGF agents for the most effective treatment of age-related macular degeneration. Current PNP-based ocular drug delivery prioritizes adapting drug release timing as a key advancement. New designs focus on creating systems that automatically adjust drug release in response to specific biological signals or changes in medical condition. Smart PNPs studied recently demonstrate adaptive capabilities by triggering drug release based on pH variations and inflammatory markers, enabling truly personalized therapeutic delivery. This promising approach in ocular medicine aims to improve patient compliance and reduce side effects by tailoring treatments to the individual biological differences of each patient.

Ongoing Challenges and Potential Improvements in Nanoparticle Design

Clinical researchers recognize PNP technology as a valuable approach for ocular therapeutics, offering enhanced drug delivery and improved patient outcomes. However, several significant challenges hinder its widespread acceptance and implementation in clinical practice. Addressing these obstacles is essential to advancing research, ensuring patient safety, and maximizing therapeutic effectiveness.

Manufacturing Consistency

The medical application of PNPs faces challenges because their therapeutic success relies heavily on maintaining consistent particle size, drug loading, and electrical properties, yet manufacturing processes remain unstable. Manufacturing reliability can be enhanced through the use of Process Analytical Technology (PAT), which enables real-time monitoring of production parameters and helps ensure regulatory compliance. Environmental factors, particularly temperature and humidity, pose stability issues that can degrade PNPs and reduce drug efficacy. Researchers are actively developing improved stabilization methods, including protective coatings and optimized storage conditions, to extend the therapeutic durability of PNP formulations.

Regulatory Hurdles

The transition of PNPs from the laboratory to clinical application faces significant roadblocks, notably the challenges in navigating complex regulatory requirements. The intricate nature of current regulatory frameworks creates difficulties for researchers and manufacturers working in this field. Regulatory bodies such as the U.S. Food and Drug Administration recognize the need for specific standards tailored to the unique properties of nanotechnology. Streamlining and clarifying approval processes for PNPs would accelerate the marketing of innovative therapies and encourage greater investment in research and development. The innovation ecosystem benefits when regulatory agencies maintain ongoing dialogue with academic institutions and industry stakeholders. To address these challenges effectively, multiple organizations should establish dedicated task forces focused on evaluating the safety and efficacy of nanoparticles for targeted applications.

Collaborative Research Initiatives

An effective solution to the challenges facing clinical translation of PNPs lies in fostering research investment programs and coordinated interagency efforts within the field of ocular therapeutics. Multidisciplinary teams, comprising material scientists, pharmacologists, regulatory professionals, and clinical practitioners, should collaboratively drive the advancement of nanoparticle technologies by identifying designs that balance safety with therapeutic efficacy. Integrating research and development efforts through collaborative testing and shared resources can enhance the creation of best practices and harmonized clinical trial protocols. Such partnerships promote collective problem-solving and ensure that patient safety remains a top priority throughout the development process. Advancing clinical applications of PNPs will also depend on the innovation of novel materials and distinctive nanoparticle architectures. Future therapeutic systems that incorporate biological targeting mechanisms and smart nanoparticle platforms offer the potential to significantly enhance treatment precision, minimize side effects, and ultimately lead to more effective ocular therapies.

Conclusion

Ocular drug delivery systems have undergone a significant transformation with the introduction of PNPs for the treatment of various eye diseases. PNPs offer numerous advantages over traditional methods, including enhanced bioavailability, improved corneal permeability, and targeted delivery mechanisms. Their versatility allows for the encapsulation of diverse therapeutic agents, making them suitable for addressing a broad spectrum of ocular conditions such as glaucoma, diabetic retinopathy, and AMD. Looking ahead, future research must prioritize the development of advanced PNP formulations tailored to individual patient needs. Precision treatment strategies will increasingly rely on genomic and proteomic analyses to guide the dynamic optimization of drug formulations and delivery profiles. This personalized approach promises to improve both the safety and therapeutic efficacy of ocular treatments, marking a pivotal shift toward more effective and patient-specific ophthalmic care. The forthcoming 2024 FDA regulatory guidelines for nanotechnology offer a promising opportunity to streamline approval pathways and accelerate the clinical adoption of nanoparticle-based therapies. These clearer, more structured frameworks are expected to help translate laboratory breakthroughs into tangible patient applications with greater efficiency. However, realizing this potential hinges on robust interdisciplinary collaboration, uniting researchers, clinicians, and regulatory experts to drive innovation while ensuring safety, efficacy, and compliance. Such cooperation will be vital in guiding the development and successful integration of advanced nanoparticle systems into routine clinical practice. New PNPs must be supported by robust real-world evidence obtained through well-designed clinical trials to validate their efficacy across diverse patient populations. The ocular drug delivery landscape is on the verge of a significant transformation, driven by the innovation in PNP technologies that are reshaping therapeutic approaches. Advancements in PNPs promise improved treatment outcomes, precise and controlled drug delivery, and enhanced patient adherence, paving the way for a future marked by personalized and highly effective ocular healthcare. Continued research and the successful clinical translation of PNPs offer a powerful opportunity to revolutionize treatment strategies for a wide range of eye diseases, representing a major leap forward in therapeutic innovation.

References

Abbasi, R., Shineh, G., Mobaraki, M., Doughty, S., & Tayebi, L. (2023). Structural parameters of nanoparticles affecting their toxicity for biomedical applications: A review. *Journal of Nanoparticle Research*, 25(3), 43.

Ahmad, K., Zhang, Y., Chen, P., Yang, X., & Hou, H. (2024). Chitosan interaction with stomach mucin layer to enhance gastric retention and mucoadhesive properties. *Carbohydrate Polymers*, 333, 121926.

Chaudhary, S. A., Patel, D. M., Patel, J. K., & Patel, D. H. (2021). Solvent emulsification evaporation and solvent emulsification diffusion techniques for nanoparticles. In *Emerging Technologies for Nanoparticle Manufacturing* (pp. 287–300). Springer International Publishing.

Chenxi, Z., Hemmat, A., Thi, N., & Afrand, M. (2025). Nanoparticle-enhanced drug delivery systems: An up-to-date review. *Journal of Molecular Liquids*, 424, 126999.

Dastidar, D. G., Ghosh, D., & Das, A. (2022). Recent developments in nanocarriers for cancer chemotherapy. *OpenNano*, 8, 100080.

Feyissa, G. T., Woldie, M., Munn, Z., & Lockwood, C. (2019). Exploration of facilitators and barriers to the implementation of a guideline to reduce HIV-related stigma and discrimination in the Ethiopian healthcare settings: A descriptive qualitative study. *PLOS ONE*, 14(5), e0216887.

Grassiri, B., Zambito, Y., & Bernkop-Schnürch, A. (2021). Strategies to prolong the residence time of drug delivery systems on ocular surface. *Advances in Colloid and Interface Science*, 288, 102342.

Herdiana, Y., Febrina, E., Nurhasanah, S., Gozali, D., Elamin, K. M., & Wathoni, N. (2024). Drug loading in chitosan-based nanoparticles. *Pharmaceutics*, 16(8), 1043.

Herdiana, Y., Wathoni, N., Shamsuddin, S., & Muchtaridi, M. (2022). Drug release study of the chitosan-based nanoparticles. Heliyon, 8(1), e08674.

Karthikeyan, E., & Sivaneswari, S. (2024). Advancements in transdermal drug delivery systems: Enhancing medicine with pain-free and controlled drug release. *Intelligent Pharmacy*, Available online 11 October 2024.

Kawabata, H. (2019). Transferrin and transferrin receptors update. *Free Radical Biology and Medicine*, 133, 46–54.

Korva-Gurung, I., Kubin, A.-M., Ohtonen, P., & Hautala, N. (2023). Visual outcomes of anti-VEGF treatment on neovascular age-related macular degeneration: A real-world population-based cohort study. *Pharmaceuticals*, 16(7), 927.

Lin, X., Wu, X., Chen, X., Wang, B., & Xu, W. (2021). Intellective and stimuli-responsive drug delivery systems in eyes. *International Journal of Pharmaceutics*, 602, 120591.

Makadia, H. K., & Siegel, S. J. (2011). Poly Lactic-co-Glycolic Acid (PLGA) as biodegradable controlled drug delivery carrier. *Polymers*, 3(3), 1377–1397.

Mehta, P., Chelike, D. K., & Rathore, R. K. (2024). Adsorption-based approaches for exploring nanoparticle effectiveness in wastewater treatment. *ChemistrySelect*, 9(25), e202400959.

Mun, E. A., & Zhaisanbayeva, B. A. (2020). The role of nanoparticle PEGylation in drug delivery. *Advanced Materials & Technologies*, 2(18), 010–018.

Nikolaidou, A., Spyratou, E., Sandali, A., Gianni, T., Platoni, K., Lamprogiannis, L., & Efstathopoulos, E. P. (2025). Utilization of nanoparticles for treating age-related macular degeneration. *Pharmaceuticals*, 18(2), 162.

Occhiutto, M. L., Maranhão, R. C., Costa, V. P., & Konstas, A. G. (2020). Nanotechnology for medical and surgical glaucoma therapy—A review. *Advances in Therapy*, 37(1), 155–199.

Peng, Q., Wang, T., Yang, D., Peng, X., Zhang, H., & Zeng, H. (2024). Recent advances in coacervation and underlying noncovalent molecular interaction mechanisms. *Progress in Polymer Science*, 153, 101827.

Seah, I., Zhao, X., Lin, Q., Liu, Z., Su, S. Z. Z., Yuen, Y. S., Hunziker, W., Lingam, G., Loh, X. J., & Su, X. (2020). Use of biomaterials for sustained delivery of anti-VEGF to treat retinal diseases. *Eye*, 34(8), 1341–1356.

Sharma, G., Parchur, A. K., Jagtap, J. M., Hansen, C. P., & Joshi, A. (2019). Hybrid nanostructures in targeted drug delivery. In *Hybrid Nanostructures for Cancer Theranostics* (pp. 139–158). Elsevier.

Shimaoka, M. (Ed.). (2012). *Integrin and Cell Adhesion Molecules* (Vol. 757). Humana Press.

Snetkov, P., Zakharova, K., Morozkina, S., Baranov, M., Olekhnovich, R., & Uspenskaya, M. (2020). Electrosprayed nanoparticles based on hyaluronic acid: Preparation and characterization. *Technologies*, 8(4), 71.

Sobczak, M. (2022). Enzyme-responsive hydrogels as potential drug delivery systems—State of knowledge and future prospects. *International Journal of Molecular Sciences*, 23(8), 4421.

Soomro, T. A., Khan, M. A. U., Gao, J., Khan, T. M., & Paul, M. (2017). Contrast normalization steps for increased sensitivity of a retinal image segmentation method. *Signal, Image and Video Processing*, 11(8), 1509–1517.

Sousa de Almeida, M., Susnik, E., Drasler, B., Taladriz-Blanco, P., Petri-Fink, A., & Rothen-Rutishauser, B. (2021). Understanding nanoparticle endocytosis to improve targeting strategies in nanomedicine. *Chemical Society Reviews*, 50(9), 5397–5434.

Tafti, M. F., Fayyaz, Z., Aghamollaei, H., Jadidi, K., & Faghihi, S. (2025). Drug delivery strategies to improve the treatment of corneal disorders. *Heliyon*, 11(2), e41881.

Wei, D., Pu, N., Li, S.-Y., Wang, Y.-G., & Tao, Y. (2023). Application of iontophoresis in ophthalmic practice: An innovative strategy to deliver drugs into the eye. *Drug Delivery*, 30(1), 2165736.

Xiao, X., Teng, F., Shi, C., Chen, J., Wu, S., Wang, B., Meng, X., Essiet Imeh, A., & Li, W. (2022). Polymeric nanoparticles—Promising carriers for cancer therapy. *Frontiers in Bioengineering and Biotechnology*, 10, 1024143.

Yu, H., Gao, R., Liu, Y., Fu, L., Zhou, J., & Li, L. (2024). Stimulus-responsive hydrogels as drug delivery systems for inflammation-targeted therapy. *Advanced Science*, 11(1), 2306152.

Zhang, H., Yang, J., Sun, R., Han, S., Yang, Z., & Teng, L. (2023). Microfluidics for nano-drug delivery systems: From fundamentals to industrialization. *Acta Pharmaceutica Sinica B*, 13(8), 3277–3299.

Zhang, Y., Watson, S., Ramaswamy, Y., & Singh, G. (2024). Intravitreal therapeutic nanoparticles for age-related macular degeneration: Design principles, progress, and opportunities. *Advances in Colloid and Interface Science*, 329, 103200.

Zhuo, S., Zhang, F., Yu, J., Zhang, X., Yang, G., & Liu, X. (2020). pH-sensitive biomaterials for drug delivery. *Molecules*, 25(23), 5649.

Zhuo, Y., Zhao, Y.-G., & Zhang, Y. (2024). Enhancing drug solubility, bioavailability, and targeted therapeutic applications through magnetic nanoparticles. *Molecules*, 29(20), 4854.

Chapter 7

Liposomes in Ocular Drug Delivery

Prabhjot Kaur[1,*]
Abhinav[2]
Ajay Bilandi[3]
and Kumar Guarve[1]
[1]Guru Gobind Singh College of Pharmacy, Yamuna Nagar, Haryana, India
[2]Amar Shaheed Baba Ajit Singh and Jujhar Singh Memorial College of Pharmacy, Bela, Ropar, Punjab, India
[3]Bhai Gurdas College of Pharmacy, Sangrur, Punjab, India

Abstract

In the pharmaceutical domain, ocular drug delivery remains one of the most challenging areas due to the eye's intricate anatomy and the presence of multiple physiological barriers. Despite these complexities, local administration continues to be the preferred route, largely due to improved patient compliance. However, local delivery also presents significant boundaries, including low bioavailability, systemic toxicity, and the need for frequent dosing. Intravitreal injections are commonly used to deliver drugs to the posterior segment of the eye, yet they are often associated with poor patient compliance and potential complications from repeated parenteral administration. To overcome these challenges, ocular nanoparticles have been developed to enhance bioavailability, prolong drug release, and improve drug stability. Among these, liposomes have emerged as the most widely applied nanocarriers in clinical practice. Renowned for their ability to bypass ocular barriers and deliver therapeutic agents to targeted sites effectively, liposomes

* Corresponding Author's Email: prabhjotkaur@gmail.com

In: Advances in Nanotechnology for Ocular Drug Delivery
Editors: Anjali Sharma, Ajmer Singh Grewal, Geeta Deswal et al.
ISBN: 979-8-89530-814-1
© 2025 Nova Science Publishers, Inc.

offer the unique advantage of encapsulating both hydrophilic and lipophilic drugs. This chapter compiles the controlled release properties, therapeutic advantages, and recent advancements in liposomal technology for the treatment of various ocular diseases.

Keywords: ocular drug delivery, eye diseases, nanoparticles, liposomes, bioavailability, nanotechnology, drug stability

Abbreviations

AMD: Age related macular degeneration
GUVs: Giant unilamellar vesicles
LUVs: Large unilamellar vesicles
MLVs: Multilamellar vesicles
ODDS: Ocular drug delivery system
P-gp: P glycoprotein
PVA: Poly vinyl acryladine
RPE: Retinal pigment epithelium
SUVs: Small unilamellar vesicles
ULVs: Unilamellar vesicles
VEGF: Vascular endothelial growth factor

Introduction

Nearly 80% of all sensory input is received through the eyes, making them the most important sensory organ in the human body. The eye is a highly sophisticated, specialized, and isolated organ. Anatomically, it is divided into anterior and posterior segments, which contributes to its complexity and also renders it highly susceptible to disease and injury. Despite the presence of several defense mechanisms, the eye's direct communication with the external environment makes it vulnerable to infections, trauma, and other injuries (Madal et al., 2024). According to the World Health Organization (WHO), at least 2.2 billion people worldwide suffer from some form of vision impairment (Li et al., 2023). Diabetic retinopathy, cataracts, glaucoma, keratitis, age-related macular degeneration (AMD), and other ocular disorders can significantly impair visual acuity and diminish a patient's quality of life. Pharmacological intervention remains the primary approach for managing most eye diseases. A major focus of current research is determining how to effectively deliver therapeutic agents to targeted ocular tissues at optimal concentrations while minimizing damage to healthy cells (Wei et al., 2023). Ocular drug delivery systems are designed to overcome anatomical and physiological barriers, enhance drug stability and therapeutic efficacy, prolong ocular retention time, and reduce the frequency of dosing. These systems also facilitate the delivery of combination therapies, improve patient adherence, and minimize drug-related side effects (Silva et al., 2021). Traditional delivery methods commonly employed in clinical practice, including topical eye drops, conjunctival and scleral administration, intracameral and intravitreal injections, retrobulbar injection, and systemic administration, have shown therapeutic benefits in managing ocular diseases. However, as previously noted, the presence of complex ocular barriers significantly limits drug penetration to target sites and hinders sustained drug retention. Consequently, the bioavailability of these conventional treatments often remains below 5%. The emergence of nanotechnology has led to dynamic advancements in ocular drug delivery, enabling the development of innovative therapeutic strategies with improved efficacy and precision (Gholizadeh et al., 2021). Nanocarriers offer numerous advantages over conventional drug delivery methods in ocular therapy. These include the ability to overcome ocular barriers, enhance transcorneal permeability, extend drug residence time, reduce drug degradation, lower dosing frequency, improve patient compliance, and enable sustained or controlled drug release. Additionally, nanocarriers facilitate targeted drug delivery and gene therapy. Among the novel drug delivery systems, liposomes have been extensively investigated for the treatment of both anterior and posterior segment ocular diseases (Onugwo et al., 2023). This chapter discusses the role and therapeutic potential of such advanced drug carriers in improving ocular drug delivery outcomes.

Absorption Barriers for Ocular Drug Delivery

Due to numerous physiological and anatomical challenges, ophthalmologists and pharmaceutical scientists have long encountered significant obstacles in effective ocular drug delivery (Li et al., 2023). Both static (e.g., corneal epithelium and blood-ocular barriers) and dynamic (e.g., tear turnover and nasolacrimal drainage) ocular barriers impede the absorption of therapeutic agents and obstruct the administration of exogenous drugs. These barriers contribute to poor drug bioavailability, as a substantial portion of the topically administered dose is either rapidly drained through the nasolacrimal system or absorbed systemically via the conjunctival sac, reducing the amount that reaches the intended ocular tissues (Leclercq et al., 2022). The cornea acts as a formidable barrier to drug penetration. Due to the presence of tight intercellular junctions (zonula occludens) in the lipophilic corneal epithelium, only small, lipophilic molecules can penetrate transcellularly, while paracellular diffusion of larger macromolecules is effectively blocked. In contrast, the hydrophilic corneal stroma permits the passage of hydrophilic drugs with molecular sizes up to approximately 500 kDa, while restricting the entry of most hydrophobic compounds. Additionally, the corneal endothelium allows the diffusion of chemicals and particles as small as 20 nm (Suri et al., 2020). Beyond molecular weight, the drug's lipophilicity, often expressed as the log P (octanol-water partition coefficient), significantly influences transcorneal penetration. Drugs with log P values less than 0 tend to be hydrophilic and favor diffusion through the stroma, whereas those with log P values greater than 0 are lipophilic and more readily cross the epithelial layer (Durairaj, 2017). Drugs can also be absorbed directly into the vitreous humor through the sclera and conjunctiva following topical administration. This occurs via passive diffusion through scleral water channels, which range in size from 30 nm to 300 nm (Varela-Fernández et al., 2020). Additionally, protein binding, particularly to melanin pigments in the iris and ciliary body, and enzyme-mediated metabolism within the vitreous, involving enzymes such as cytochrome P450 reductase and esterases, significantly influence the pharmacokinetics of intraocular drugs (Argikar et al., 2027). The choroid, retina, vitreous, and sclera are key components of the eye's posterior segment. While molecular size has a greater impact on drug permeability through the sclera than lipophilicity, the cornea is generally less permeable than the sclera. The sclera primarily restricts the transport of lipophilic drugs. The choroid, in particular, serves as a significant barrier to macromolecule penetration. Therapeutic compounds are often unable to pass through the pigmented choroid into the vitreous humor due to efflux transporters such as P-glycoprotein (P-gp) and multidrug resistance proteins. The vitreous humor consists mostly of water (98-99% w/w), with only a small fraction of solids like collagen and glycosaminoglycans. According to Leclercq et al. (2017), the vitreous humor acts as a diffusional barrier to therapeutic molecules administered via injection, particularly affecting high molecular weight drugs and suspended particles. However, factors such as aging, vitreoretinal surgery, and multiple injection procedures can influence drug distribution and intravitreal pharmacokinetics (Shastri et al., 2023). A major obstacle to ocular absorption of substances administered via the trans-scleral or systemic route is the blood-retinal barrier. This barrier is primarily composed of tight junctions between the retinal pigment epithelium (RPE) and the retinal endothelial blood vessels. The RPE serves as a rate-limiting permeation barrier, especially for hydrophilic compounds, where permeation lag time increases with molecular weight. Additionally, melanin within the RPE can bind to small, lipophilic, and cationic compounds, further restricting their penetration in this region (O'Leary and Campbell, 2023). Various strategies have been proposed to overcome these challenges in ocular drug delivery, among which nanodrug delivery systems are considered promising (Irimia et al., 2018).

The Role of Nanotechnology in Enhancing Ocular Drug Permeation

Novel drug delivery technologies have been developed to enhance drug bioavailability and overcome the challenges associated with ocular drug delivery. Over the past decades, research has focused on creating and implementing innovative delivery methods to improve mucoadhesion, corneal penetration, and ocular retention time. The development of nanocarriers offers numerous advantages, including overcoming ocular barriers, enhancing trans-corneal permeability, prolonging drug residence time, reducing dosing frequency, enabling targeted drug delivery, and facilitating controlled or sustained drug release (Qamar et al., 2019). Various nanocarriers, such as liposomes, niosomes, micelles, nanoemulsions, cubosomes, and nano-wafers, have been designed specifically for ocular administration. Nanotechnology-based delivery systems are expected to transform the current landscape of ocular drug delivery.

Liposomal Carriers in Ocular Drug Delivery

Typical liposomes (Figure 1) are simple vesicles consisting of an aqueous core enclosed by a lipid bilayer. Their unique structure allows them to encapsulate both hydrophilic and hydrophobic drugs effectively (Koutsoviti et al., 2021; Kaur and Kriplani, 2024). Besides this key advantage, liposomes are non-toxic, biocompatible, and biodegradable (Jagrati and Sengar, 2014).

Structurally, liposomes are classified into unilamellar vesicles (ULVs) and multilamellar vesicles (MLVs). ULVs are further subdivided by size into small unilamellar vesicles (SUVs), large unilamellar vesicles (LUVs), and giant unilamellar vesicles (GUVs). The aqueous core of ULVs is enclosed by a single lipid bilayer composed of phosphatidylglycerol or lecithin. In contrast, MLVs consist of multiple concentric lipid bilayers, each separated by an aqueous compartment. Hydrophobic drugs can be encapsulated within the lipid bilayers, while hydrophilic drugs are contained in the aqueous core. Due to their larger aqueous volume, multilamellar liposomes have a higher capacity to encapsulate hydrophilic drugs compared to unilamellar liposomes (Basu et al., 2024). Several studies have demonstrated the efficacy of liposomal ocular delivery for both anterior and posterior segments of the eye (Das et al., 2022; Ameeduzzafar et al., 2016). Liposomal technology effectively enhances bioavailability and biocompatibility while enabling sustained, controlled, or personalized drug release. Liposomes offer targeted delivery, reduce the required drug dosage, and improve pharmacokinetics and drug solubility. Additionally, they can mitigate the ocular toxicity of certain medications (Das et al., 2022; López-Cano et al., 2021). However, liposomal systems also have some limitations, such as low entrapment efficiency for hydrophilic drugs and challenges related to storage stability (Das et al., 2022; Tan et al., 2017). Liposomes are generally biocompatible with the tear film owing to their structural characteristics (Das et al., 2022). Their unique physicochemical properties make liposomes efficient carriers for ocular drug delivery by enhancing transcorneal absorption, reducing drug clearance, and providing exceptional stability (Lajunen et al., 2016).

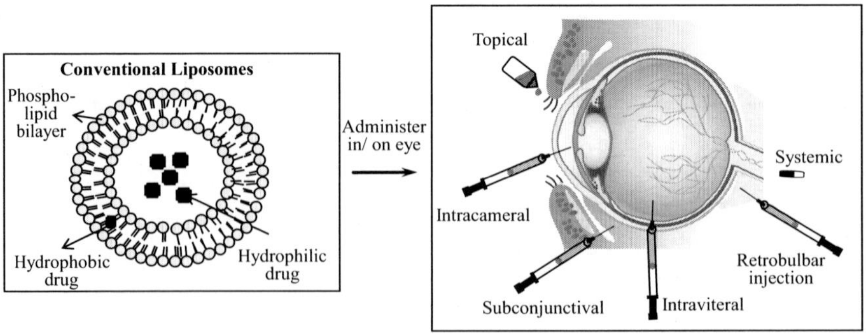

Figure 1. Liposomal based drug delivery system for ocular application from various routes of eye.

Mechanisms of Liposomal Permeation via Ocular Surfaces

The permeability mechanisms of liposomes have been extensively studied but are still not completely understood. Four main mechanisms have been identified:

Adsorption

One of the key steps in intracellular drug delivery is the adsorption of liposomes onto the cell membrane. When cell surface proteins are present, adsorbed liposomes become leaky, releasing their contents near the cell membrane. This increases the local drug concentration, facilitating drug uptake by passive diffusion or active transport (Agarwal et al., 2016).

Endocytosis

After adsorption, liposomes are engulfed by the cell membrane and internalized into endosomes. They are subsequently transported to lysosomes, where lysosomal enzymes degrade the lipids, releasing the encapsulated drug into the cytoplasm.

Fusion

Liposomes can deliver their contents directly into the cytoplasm by fusing their lipid bilayer with the lipoidal cell membrane. This fusion involves intermixing and lateral diffusion of lipids between the liposomal and cellular membranes (Tasharrofi et al., 2022).

Lipid Exchange

Lipid transfer proteins on the cell membrane recognize liposomes due to the similarity of liposomal lipids to cellular phospholipids. This triggers lipid exchange, destabilizing the liposomal membrane and releasing drug molecules intracellularly (López-Cano et al., 2021).

Table 1. Some recent examples of pre-clinical studies related to liposomal drug delivery systems for ocular diseases

Drug	Therapeutic use	Method	Particle Size	Entrapement Efficiency	Key points	Reference
Conbersept	Neovascular age related mascular degeneration	Thin film dispersion method	152.4 ± 1.3 nm	45.8% ± 2.6	11.55-fold enhanced intra-ocular concentration (than free drug) was reported.	(Sun et al., 2024)
Curcumin	Dry eyes; Meibomian gland dysfunction	Thin film hydration method	~150 nm	93%	Compared to clinically approved cyclosporin A liposomes, formulation enhanced tear breakup time, and improved signs of inflammation.	(Baranauskas et al., 2024)
Latanaprost	Glaucoma	Thin film hydration method	150-200 nm	-	3 times higher relative bioavailability than marketed formulation, offered sustained release of drug for 24 hours.	(Brugnera et al., 2024)
Timol maleate	Glaucoma	Thin film hydration method	340.8 nm	-	Improved stability and controlled release of drug, improving patient compliance.	(Bose et al., 2025)
Citicoline	Glaucoma	Thin film hydration method	118 ± 4 nm	40 ± 11%	Improving their bioavailability and therapeutic potential.	(Bonechi et al., 2023)
Dexamethasone, bevacizumab	Neovascular diseases	Thin film hydration method	139± 2 nm	97± 5%	Sustained release with improved bioavailability was reported.	(Farooq et al., 2024)
Epoitin-β	Neuroprotective action	Thin film hydration method	144± 1.3 nm	-	1.9 times higher efficiency compared to free drug.	(Pakian et al., 2025)
Astaxanthin	Retinal disease therapy	Self formimg method	274 ± 4 nm	98%	Enhanced permeation, tissue accumulation and thereby increased overall efficacy.	(Luviano et al., 2025)

Pre-Clinical Studies Involving a Liposomal Ocular Drug Delivery System

Liposomes have been extensively studied for ocular drug delivery. Therefore, this chapter focuses on recent studies evaluating liposomal applications. The most commonly used technique for liposome preparation is the thin-film hydration method. Table 1 summarizes the applications of liposomes in various ocular diseases (Sun et al., 2024; Baranauskas et al., 2024; Brugnera et al., 2024; Bose et al., 2025; Farooq et al., 2024; Pakian et al., 2025; Luviano et al., 2025).

Neovascular Age Related Macular Degeneration

AMD is a leading cause of blindness worldwide and represents a significant public health concern. It is characterized by the formation of pathological choroidal neovascular membranes, which results in retinal exudation and hemorrhage. Vascular endothelial growth factor (VEGF), a key mediator of vascular permeability and angiogenesis, is upregulated during this abnormal vascular proliferation (Shirian et al., 2024). The primary treatment for AMD involves repeated intravitreal injections of anti-VEGF agents. However, these injections are highly invasive, can lead to ocular complications, and often suffer from poor patient compliance. To address the limitations of conventional formulations, dual ophthalmic liposomes encapsulating penetration enhancers and hyaluronic acid have been developed (Sun et al., 2024). Additionally, liposomal formulations of dexamethasone and bevacizumab have been explored to improve drug delivery efficacy (Farooq et al., 2024). Epoetin-β has also been encapsulated in liposomes to enhance its therapeutic effectiveness (Pakian et al., 2025).

Dry Eyes

Tear film instability and decreased tear production initiate a vicious cycle, where hyperosmolarity, ocular inflammation, and apoptosis contribute to keratitis and other surface damages, collectively characterizing the widespread condition known as dry eye disease (Leonardi et al., 2025). For instance, biomimetic liposomes encapsulating curcumin have been reintroduced to treat meibomian gland dysfunction and dry eye symptoms. This treatment significantly reduced ocular surface inflammation. Histological studies further demonstrated increased goblet cell density and corneal epithelial thickness, indicating reversal of ocular surface damage caused by dry eye (Baranauskas et al., 2024).

Glaucoma

Glaucoma is the leading cause of blindness worldwide, posing a significant societal burden due to lost productivity and rising healthcare costs. Elevated intraocular pressure is the major modifiable risk factor for glaucoma and can be managed using topical eye drops, oral medications, laser therapy, or surgical interventions (Yuan et al., 2025). Among these, topical prostaglandin analogues such as latanoprost are commonly prescribed as the first-line treatment for lowering intraocular pressure. Topical ophthalmic medications often suffer from limited bioavailability due to their rapid clearance from the ocular surface. To address this, synthetic liposomes loaded with latanoprost along with hyaluronic acid, betaine, and leucine have been developed, demonstrating a sustained effect lasting up to 24 hours (Brugnera et al., 2024). Additionally, liposomal delivery has been explored to mitigate the adverse effects associated with timolol maleate eye drops (Bose et al., 2025). Moreover, citicoline incorporated in liposomal formulations has shown promise as an adjunct therapy for glaucoma management (Bonechi et al., 2023).

Retinal Disease Therapy

Astaxanthin, a potent antioxidant carotenoid with notable anti-inflammatory and neuroprotective properties, shows promise for treating retinal disorders such as age-related macular degeneration. However, its clinical application is limited by poor ocular permeability, low stability, and poor water solubility, highlighting the need for effective delivery strategies. To address these challenges, astaxanthin has been successfully encapsulated in liposomal systems, enhancing its therapeutic potential (Luviano et al., 2025).

Methods of Preparation Employed for Liposomes Formulation

The thin-film hydration method is predominantly employed for the formulation of liposomes targeting ocular delivery (Sun et al., 2024; Baranauskas et al., 2024; Brugnera et al., 2024; Bose et al., 2025). Other notable formulation techniques include the self-forming method (Luviano et al., 2025), ether injection method (Jain et al., 2025), and coacervation phase separation technique (Badran et al., 2024).

The Thin-Film Hydration Method

Thin-film hydration is the most commonly preferred method for liposomal preparation due to several advantages, including higher yield, improved entrapment efficiency, production of nano-sized particles, ease of use, and reduced processing time. However, this method has certain limitations, notably issues with scalability (Kraft et al., 2014). The process involves dissolving lipophilic components such as surfactants and cholesterol in an organic solvent within a round-bottom flask. A rotary vacuum evaporator is then used to remove the solvent, resulting in the formation of a thin, dry lipid film on the interior surface of the flask. Subsequently, an aqueous drug-containing solution is added to hydrate the lipid film, leading to the formation of multilamellar vesicles (Moammeri et al., 2023).

The Self-Forming Method

The lipid phase is mixed with the aqueous phase, followed by filtration to ensure homogeneity, after which the drug is incorporated into the formulated liposomes (Luviano et al., 2025).

Limitations of a Liposomal Drug Delivery System in Ocular Drug Delivery

Despite their numerous advantages, liposome-based formulations have faced limited therapeutic applications historically due to several inherent constraints. One of the primary challenges is their stability. Liposomes composed of unsaturated lipids are prone to chemical instability caused by hydrolysis and oxidation. Additionally, the encapsulated drug can contribute to physical instability of the vesicles. Liposomes may also aggregate, forming larger particles that are more readily recognized and cleared by phagocytic cells, which in turn reduces ocular absorption. Instability can also arise from the partitioning of lipophilic drugs from the lipid bilayer into the surrounding solvent. Conventional liposomes lack the ability to distinguish target cells from non-target cells, posing a significant hurdle in achieving targeted drug delivery. Moreover, the requirement for prolonged, continuous drug administration necessitates the modification of liposomal properties to enhance their therapeutic performance (López-Cano et al., 2021).

Conclusion and Future Perspectives

Because of its distinct anatomical and physiological makeup and the low absorption of traditional medications, the eye is considered a difficult subject for researchers. Topical medications have the lowest permeability and bioavailability because direct ocular contact with the external environment provides a high level of protection against exogenous toxins. Drugs' physicochemical properties are crucial, just like in other drug administration routes. Furthermore, when administering drugs to the eyes, more stringent requirements including sterility and isotonicity need to be considered. It should also have a pH close to that of the eye to prevent irritation or blurring of vision. Despite the fact that many people throughout the world suffer from ocular disorders, the anatomy of the eye makes it impossible to administer drugs effectively. While suprachoroidal injections also have a number of adverse effects when used systemically, topically administered dosage forms including ointments and drops retain for a shorter period of time in the eye; as a result, bioavailability is minimal. The earliest class of vesicular systems to be produced were liposomes. They are known more than other vesicular systems because of this. Commercial vesicular ocular drug delivery systems are liposomal; however, there is no denying that additional vesicular systems might be commercialized. The pharmaceutical industry is more likely to produce systems that are very useful in the cosmetics sector. There is no doubt that the possibility of commercial product development will spur additional research in this area. In the upcoming years, vesicular systems will undoubtedly be utilized more frequently for ocular medication targeting. With alterations, novel vesicular systems with distinct structures from those already recognized can also be created. Compared to traditional drug delivery, these systems also offer the patient a longer period of effective drug release, which improves patient compliance. Vesicular carriers are expected to replace traditional systems as more commercial preparations become accessible.

References

Ahmed, M. S., Khalil, I. A., Hassanein, K., El-Gendy, G. A., & Ismail, S. (2023). Celecoxib niosomal in-situ gel as a valuable drug delivery system for ocular inflammation. *Bulletin of Pharmaceutical Sciences Assiut University*, *46*(2), 757-783.

Agarwal, R., Iezhitsa, I., Agarwal, P., Abdul Nasir, N. A., Razali, N., Alyautdin, R., & Ismail, N. M. (2016). Liposomes in topical ophthalmic drug delivery: an update. *Drug delivery*, *23*(4), 1075-1091.

Argikar, U. A., Dumouchel, J. L., Dunne, C. E., & Bushee, A. J. (2017). Ocular non-P450 oxidative, reductive, hydrolytic, and conjugative drug metabolizing enzymes. *Drug metabolism reviews*, *49*(3), 372-394.

Basu, B., Prajapati, B., Dutta, A., & Paliwal, H. (2024). Medical application of liposomes. *Journal of Exploratory Research in Pharmacology*, *9*(1), 13-22.

Baranauskas, V., Jaruseviciene, R., Grigalavicius, M., Galgauskas, S., Karabanovas, V., & Steponkiene, S. (2024). Biomimetic curcumin-loaded liposomes for the treatment of dry eyes and meibomian gland dysfunction: an in vivo study. *Journal of Clinical Medicine*, *13*(21), 6436.

Bonechi, C., Mahdizadeh, F. F., Talarico, L., Pepi, S., Tamasi, G., Leone, G., Consumi, M., Donati, A. and Magnani, A. (2023). Liposomal encapsulation of citicoline for ocular drug delivery. *International Journal of Molecular Sciences*, *24*(23), 16864.

Bose, A., Majumdar, S., & Halder, A. (2025). Development and statistical optimization of timolol maleate encapsulated liposome using 32 full factorial design. *Proceedings of the Indian National Science Academy*, 1-15.

Brugnera, M., Vicario-de-la-Torre, M., González-Cela Casamayor, M. A., López-Cano, J. J., Bravo-Osuna, I., Huete-Toral, F., González Rubio, M. L., Carracedo, G., Molina-Martínez, I. T., Andrés-Guerrero, V. and Herrero-Vanrell, R. (2024). Enhancing the hypotensive effect of latanoprost by combining synthetic phosphatidylcholine liposomes with hyaluronic acid and osmoprotective agents. *Drug Delivery and Translational Research*, 14, 2804-2822.

Das, B., Nayak, A. K., & Mallick, S. (2022). Lipid-based nanocarriers for ocular drug delivery: An updated review. *Journal of Drug Delivery Science and Technology*, *76*, 103780.

Durairaj, C. (2017). Ocular pharmacokinetics. *Pharmacologic Therapy of Ocular Disease*, 31-55.

Farooq, U., O'Reilly, N. J., Ahmed, Z., Gasco, P., Singh, T. R. R., Behl, G., Fitzhenry, L. & McLoughlin, P. (2024). Design of liposomal nanocarriers with a potential for combined dexamethasone and bevacizumab delivery to the eye. *International Journal of Pharmaceutics*, *654*, 123958.

Gholizadeh, S., Wang, Z., Chen, X., Dana, R., & Annabi, N. (2021). Advanced nanodelivery platforms for topical ophthalmic drug delivery. *Drug discovery today*, *26*(6), 1437-1449.

Irimia, T., Ghica, M. V., Popa, L., Anuţa, V., Arsene, A. L., & Dinu-Pîrvu, C. E. (2018). Strategies for improving ocular drug bioavailability and corneal wound healing with chitosan-based delivery systems. *Polymers*, *10*(11), 1221.

Jagrati, K. M., & Sengar, A. (2024). Liposomal vesicular delivery system: An innovative nano carrier. *World Journal of Pharmaceutical Research*, *13*(13), 1155-1169.

Jain, N., Verma, A., Rastogi, V., Jain, N., & Sharma, A. K. (2025). Niosomal gel containing dipivefrin hydrochloride for glaucoma treatment: Development, characterization and In vitro-In vivo assessment. *Food & Pharma International.*

Kaur, P., & Kriplani, P. (2024). Quality by design for Niosome-Based nanocarriers to improve transdermal drug delivery from lab to industry. *International Journal of Pharmaceutics*, 666, 124747.

Koutsoviti, M., Siamidi, A., Pavlou, P., & Vlachou, M. (2021). Recent advances in the excipients used for modified ocular drug delivery. *Materials*, *14*(15), 4290.

Kraft, J. C., Freeling, J. P., Wang, Z., & Ho, R. J. (2014). Emerging research and clinical development trends of liposome and lipid nanoparticle drug delivery systems. *Journal of Pharmaceutical Sciences*, *103*(1), 29-52.

Lajunen, T., Nurmi, R., Kontturi, L., Viitala, L., Yliperttula, M., Murtomäki, L., & Urtti, A. (2016). Light activated liposomes: Functionality and prospects in ocular drug delivery. *Journal of Controlled Release*, *244*, 157-166.

Leclercq, B., Mejlachowicz, D., & Behar-Cohen, F. (2022). Ocular barriers and their influence on gene therapy products delivery. *Pharmaceutics*, *14*(5), 998.

Leonardi, A., Di Zazzo, A., Cutrupi, F., & Iaccarino, L. (2025). Dry eye and systemic diseases. *Saudi Journal of Ophthalmology*, 39, 5-13.

Li, S., Chen, L., & Fu, Y. (2023). Nanotechnology-based ocular drug delivery systems: recent advances and future prospects. *Journal of Nanobiotechnology*, *21*(1), 232.

López-Cano, J. J., González-Cela-Casamayor, M. A., Andrés-Guerrero, V., Herrero-Vanrell, R., & Molina-Martínez, I. T. (2021). Liposomes as vehicles for topical ophthalmic drug delivery and ocular surface protection. *Expert Opinion on Drug Delivery*, *18*(7), 819-847.

Luviano, M., Zúñiga-González, O. G., López-Naranjo, E. J., Navarro-Partida, J., Orozco-Guareño, E., Ramírez-Arreola, D. E., & González-Ortiz, L. J. (2025). Efficient liposomal encapsulation and intraocular diffusion of astaxanthin in a topical ophthalmic delivery system. *Materials Letters*, 383, 138034.

Mandal, S., Vishvakarma, P., & Bhumika, K. (2024). Developments in Emerging Topical Drug Delivery Systems for Ocular Disorders. *Current Drug Research Reviews*, *16*(3), 251-267.

Moammeri, A., Chegeni, M. M., Sahrayi, H., Ghafelehbashi, R., Memarzadeh, F., Mansouri, A., Akbarzadeh, I., Abtahi, M. S., Hejabi, F. and Ren, Q. (2023). Current advances in niosomes applications for drug delivery and cancer treatment. *Materials Today Bio*, *23*, 100837.

O'Leary, F., & Campbell, M. (2023). The blood–retina barrier in health and disease. *The FEBS journal*, *290*(4), 878-891.

Onugwu, A. L., Nwagwu, C. S., Onugwu, O. S., Echezona, A. C., Agbo, C. P., Ihim, S. A., Emeh, P., Nnamani, P. O., Attama, A. A., & Khutoryanskiy, V. V. (2023). Nanotechnology based drug delivery systems for the treatment of anterior segment eye diseases. *Journal of Controlled Release*, *354*, 465-488.

Qamar, Z., Qizilbash, F. F., Iqubal, M. K., Ali, A., Narang, J. K., Ali, J., & Baboota, S. (2019). Nano-based drug delivery system: recent strategies for the treatment of ocular disease and future perspective. *Recent Patents on Drug Delivery & Formulation*, *13*(4), 246-254.

Pakian, S., Nabid, M. R., Satarian, L., Abandansari, H. S., & Mirkani, A. (2025). Topical Administration of Mucoadhesive Liposomes–Epoetin-β for Targeting the Ocular Posterior Segment. *Molecular Pharmaceutics*, 4, 2287-2305.

Shastri, D. H., Silva, A. C., & Almeida, H. (2023). Ocular delivery of therapeutic proteins: a review. *Pharmaceutics*, *15*(1), 205.

Shirian, J. D., Shukla, P., & Singh, R. P. (2025). Exploring new horizons in neovascular age-related macular degeneration: Novel mechanisms of action and future therapeutic avenues. *Eye*, *39*(1), 40-44.

Silva, B., São Braz, B., Delgado, E., & Gonçalves, L. (2021). Colloidal nanosystems with mucoadhesive properties designed for ocular topical delivery. *International Journal of Pharmaceutics*, *606*, 120873.

Sun, C., Zhang, S., Xu, N., Liu, K., Wei, F., Zhang, X., Zhang, J., Gao, S., Yu, Y., & Ding, X. (2024). Topical Ophthalmic Liposomes Dual-Modified with Penetratin and Hyaluronic Acid for the Noninvasive Treatment of Neovascular Age-Related Macular Degeneration. *International Journal of Nanomedicine*, 19, 1887-1908.

Suri, R., Beg, S., & Kohli, K. (2020). Target strategies for drug delivery bypassing ocular barriers. *Journal of Drug Delivery Science and Technology*, *55*, 101389.

Tan, G., Yu, S., Pan, H., Li, J., Liu, D., Yuan, K., Yang, X. and Pan, W. (2017). Bioadhesive chitosan-loaded liposomes: A more efficient and higher permeable ocular delivery platform for timolol maleate. *International Journal of Biological Macromolecules*, *94*, 355-363.

Tasharrofi, N., Nourozi, M., & Marzban, A. (2022). How liposomes pave the way for ocular drug delivery after topical administration. *Journal of Drug Delivery Science and Technology*, *67*, 103045.

Varela-Fernández, R., Díaz-Tomé, V., Luaces-Rodríguez, A., Conde-Penedo, A., García-Otero, X., Luzardo-Álvarez, A., Fernández-Ferreiro, A. and Otero-Espinar, F. J. (2020). Drug delivery to the posterior segment of the eye: biopharmaceutic and pharmacokinetic considerations. *Pharmaceutics*, *12*(3), 269.

Wei, J., Mu, J., Tang, Y., Qin, D., Duan, J., & Wu, A. (2023). Next-generation nanomaterials: advancing ocular anti-inflammatory drug therapy. *Journal of Nanobiotechnology*, *21*(1), 282.

Yuan, P. H. S., Dorling, M., Shah, M., Panarelli, J. F., & Durr, G. M. (2025). Combined microinvasive glaucoma surgery with phacoemulsification in open-angle glaucoma: a systematic review and meta-analysis. *American Journal of Ophthalmology*, *270*, 154-163.

Chapter 8

Niosomes in Ocular Drug Delivery

Shailesh Sharma[1,*]
Neelam Sharma[1]
Punam Gaba[1,2]
Rahul Kumar Sharma[1]
and Shruti[1]
[1]Amar Shaheed Baba Ajit Singh Jujhar Singh Memorial College of Pharmacy, Bela, Punjab, India
[2]Maharaja Agrasen School of Pharmacy, Maharaja Agrasen University, Kalujhanda, Himachal Pradesh, India

Abstract

Ocular drug delivery presents considerable challenges due to the eye's unique anatomical and physiological barriers, including the corneal epithelium, tear turnover, and blood-ocular barriers. Niosomes, vesicular carriers composed of non-ionic surfactants and cholesterol, have emerged as a promising strategy to overcome these obstacles. They offer numerous advantages such as enhanced drug stability, controlled release, and improved permeation across ocular barriers. Due to their excellent biocompatibility and ability to encapsulate both hydrophilic and lipophilic drugs, niosomes represent highly adaptable delivery systems for a broad range of therapeutic agents. Furthermore, surface modifications, such as the addition of ligands for active targeting or mucoadhesive polymers, can further improve their pharmacological properties and site-specific delivery. In ophthalmic therapy, these self-assembled vesicles have been explored for the treatment of both anterior

[*] Corresponding Author's Email: shailesh.bela@gmail.com

In: Advances in Nanotechnology for Ocular Drug Delivery
Editors: Anjali Sharma, Ajmer Singh Grewal, Geeta Deswal et al.
ISBN: 979-8-89530-814-1
© 2025 Nova Science Publishers, Inc.

segment disorders (e.g., glaucoma and conjunctivitis) and posterior segment diseases (e.g., diabetic retinopathy and macular degeneration). Research has underscored their potential to enhance the therapeutic efficacy of antibiotics, anti-inflammatory agents, and anti-glaucoma medications. In summary, niosomes offer a novel and promising platform for improving drug effectiveness, safety, and patient compliance, thereby paving the way for advanced strategies in ocular drug delivery.

Keywords: ocular, surfactant, vesicular, retinopathy, glaucoma, nano-carriers, therapeutic, conjunctivitis, hydrophilic, mucoadhesive, antibiotic

Abbreviations

AMD: Age-related Macular Degeneration
AS: Anterior Segment
BB: Blood-Ocular Barriers
BC: Biocompatibility
CE: Corneal Epithelium
CR: Controlled Release
DDS: Drug Delivery System
DS: Drug Stability
IP: Improved Permeation
LHP: Lipophilic and Hydrophilic Phases
LT: Ligand Targeting
MT: Mucoadhesive Targeting
NIO: Niosomes
ODDS: Ocular Drug Delivery System
PM: Patient Compliance
PS: Posterior Segment
SA: Surface Alteration
SAA: Surface Active Agents
VDDS: Vascular Drug Delivery Systems
VE: Vesicular Encapsulation

Introduction

In recent years, pharmaceutical research has increasingly focused on niosomes, a novel vesicular drug delivery system, due to their potential to enhance drug stability, targeted delivery, and bioavailability. This innovative vesicular system primarily consists of non-ionic surfactants and cholesterol as an excipient. Compared to liposomes, niosomes demonstrate greater chemical stability, longer shelf life, and improved resistance to oxidation and degradation. The inclusion of surfactants in their synthetic bilayer structure renders them non-immunogenic, biodegradable, and medically safe (Allen, 1998). Several methods have been developed for niosome preparation, including sonication, reverse-phase evaporation, thin-film hydration, and transmembrane pH gradient drug loading (Baillie et al., 1985). These methods vary depending on the desired vesicle size and lamellarity. Among them, sonication is the most recommended technique for producing uniformly small unilamellar vesicles (Hu & Rhodes, 1999). Current trends in nanomaterial development emphasize the integration of biotechnology and nanotechnology to establish environmentally sustainable methodologies grounded in green chemistry (Kaur et al., 2004). In addition to niosomes, various nanocarrier systems, such as dendrimers, solid lipid nanoparticles (SLNs), nanostructured lipid carriers (NLCs), nanomicelles, polymeric nanoparticles, mesoporous silica nanoparticles, and gold and silver nanoparticles, have emerged as essential tools in nanomedicine. Nanomedicine is a multidisciplinary field that combines genetic engineering, pharmaceutical science, nanotechnology, and artificial intelligence to advance healthcare. These interdisciplinary approaches aim to enable targeted delivery of pharmacological agents for treating diverse diseases and to facilitate the development of novel drug delivery platforms for drugs with poor bioavailability (Hu & Rhodes, 1999; Kaur et al., 2004). Drug delivery via transcytosis through M cells in Peyer's patches of the intestinal lymphatic tissue enhances metabolic availability by overcoming the anatomical barriers of the gastrointestinal tract. Surfactant vesicles are taken up by the reticuloendothelial system (Azmin et al., 1985), enabling localized drug accumulation, which is particularly beneficial for treating diseases such as leishmaniasis, where parasites target the liver and spleen. Besides the reticuloendothelial system, immunoglobulins and other non-reticuloendothelial components can also recognize the lipid surfaces of these delivery systems (Szoka & Papahadjopoulos, 1980). Encapsulation of various antineoplastic drugs within these carrier vesicles has been shown to reduce drug-induced adverse effects while maintaining, and in some cases enhancing, antitumor efficacy. However, liposomes and synthetic carriers differ significantly in composition: liposomes are composed of double-chain phospholipids (which may be charged or neutral), whereas non-vesicular carriers such as niosomes consist of uncharged single-chain surfactants and cholesterol (Jadon et al., 2009). Phospholipid vesicles generally contain a higher concentration of steroid alcohol compared to other nanocarriers, which contributes to their lower drug entrapment efficiency relative to niosomes. Additionally, the production of encapsulated drug carriers such as liposomes is costly. Components like glycerophospholipids are chemically unstable due to their susceptibility to oxidative degradation. Moreover, their formulation demands specific handling and storage conditions, and the purity of naturally sourced phospholipids can vary significantly (Baillie et al., 1986). Due to their structural similarities with other carrier systems, self-assembled vesicles such as niosomes have been developed as alternatives to overcome challenges related to the mass production, stability, and sterilization of liposomes. These vesicles can encapsulate both hydrophilic and hydrophobic drugs within their aqueous compartments and bilayer structures, respectively. Moreover, they have the potential to improve the pharmacokinetics and biodistribution of therapeutics, thereby enabling targeted drug delivery and enhancing the absorption and uptake efficiency of the encapsulated drugs (Gregoriadis, 1981). By prolonging the circulation time of drugs in the bloodstream, protecting active molecules from the biological environment, and restricting their activity to specific target cells, such systems can significantly improve therapeutic outcomes. A specific type of niosome, known as a bola-surfactant niosome, is composed of α,ω-hexadecyl-bis(1-aza-18-crown-6) (bola-surfactant), Span 80, and cholesterol in a 2:3:1 molar ratio. The design of these nanosystems depends on surfactants that are non-immunogenic, biodegradable, and biocompatible. The eye poses a unique challenge for drug delivery due to its complex anatomical and physiological barriers, including the mucoaqueous tear film and the corneal epithelium. The corneal epithelium is reinforced by tight junctions and desmosomes, which protect the eye from external insults (Baillie et al., 1985). Internally, the retinal and ciliary epithelium, along with the endothelial cells of the iris blood vessels, constitute the blood-aqueous barrier. Additionally, the blood-retinal barrier, together with the sclera and conjunctiva, isolates the eye from systemic circulation. These protective structures significantly restrict the effectiveness of both topical and systemic ocular drug delivery. For example, eye drops typically achieve less than 3% bioavailability, as the majority of the medication is cleared from the ocular surface within 2–3 minutes (Jadon et al., 2009).

Delivering therapeutic agents to the posterior segment of the eye via topical administration remains particularly challenging, thereby limiting effective treatment of diseases in this region.

The History of Niosomes

Initial Advancements (1970s-1980s)

The discovery of liposomal nanoparticles in the 1960s marked a significant milestone in pharmaceutical research, laying the foundation for vesicular drug delivery. However, the instability of artificial vesicles, primarily caused by phospholipid oxidation and hydrolysis, posed considerable challenges (Hu & Rhodes, 1999). This led researchers to explore more stable alternatives, with non-ionic surfactants emerging as promising candidates for vesicle formation.

1970s: During this decade, the potential of surfactant-based carriers was first proposed. Researchers observed that non-ionic surfactants could spontaneously self-assemble into bilayer structures in aqueous solutions, forming vesicles that closely resembled liposomes (Cummings et al., 1984).

1980s: This period witnessed the formal introduction of non-ionic surfactant vesicles. Notably, the cosmetics company L'Oréal patented a method to enhance skin delivery using these vesicles and pioneered the first colloidal formulations, primarily for cosmetic purposes. Subsequent studies demonstrated that these vesicles could effectively encapsulate polar (hydrophilic) drugs, making them suitable for pharmaceutical applications (Cummings et al., 1984).

1990s: The use of synthetic vesicles expanded significantly into medical applications. These systems were explored as carriers for vaccines, as well as for topical and transdermal drug delivery. Research during this era confirmed their biocompatibility and effectiveness in encapsulating both hydrophilic and lipophilic drugs (Cummings et al., 1984).

Technological Advancements (2000s-2010s)

The use of amphiphilic vesicles in drug formulation has significantly expanded with advancements in nanotechnology and pharmaceutical sciences. Several innovative vesicular systems have been developed to enhance drug delivery efficiency. For example, proniosomes are dehydrated blends that form niosomes upon hydration, offering improved stability and ease of storage. Ethosomes and transferosomes are specially modified vesicular carriers designed to enhance transdermal drug delivery by improving skin penetration. Additionally, polymeric niosomes represent hybrid systems that enable sustained and controlled drug release. The therapeutic applications of these vesicular systems have broadened to include antifungal and antibacterial treatments, gene therapy, and targeted drug delivery for cancer (Baillie et al., 1986).

Current Progress (2020s)

In recent years, research has increasingly focused on bilayer vesicles that respond to external stimuli, serving as smart nanocarriers for controlled drug release. These advanced delivery systems have shown great potential in the development of peptide and mRNA vaccines. Additionally, combination therapy, co-delivering multiple drugs using niosomal carriers, has attracted attention for its ability to enhance therapeutic efficacy while minimizing side effects (Allen, 1998). The optimization of these complex formulations is being further propelled by artificial intelligence (AI) and machine learning, which enable more precise design and prediction of drug delivery behavior.

Benefits of Target Delivery System

In recent years, vesicular systems have emerged as vital components in drug delivery research, especially within the fields of targeted therapeutics, personalized medicine, and nanomedicine.

Improved Medicinal Stability

One of the major advantages of niosomes over liposomes is their superior chemical and physical stability. Unlike phospholipid-based encapsulation systems, nanocarriers composed of non-ionic surfactants exhibit greater resistance to oxidation and hydrolysis, thereby enhancing shelf life and reducing drug degradation (Jayaraman et al., 1996). Furthermore, their rigid bilayer structure minimizes premature drug leakage and improves entrapment efficiency, making them highly suitable for commercial pharmaceutical formulations (Azmin et al., 1985).

Controlled and Prolonged Drug Release

Colloidal vesicular systems can prolong the release of therapeutic agents, thereby reducing dosing frequency and enhancing patient compliance. Their bilayer structure facilitates gradual drug release, maintaining therapeutic levels over extended periods while minimizing the risk of toxicity and adverse effects (Jayaraman et al., 1996). This makes these systems particularly valuable for managing chronic conditions such as cancer, diabetes, and hypertension. For instance, niosomal formulations of NSAIDs like ibuprofen and diclofenac have demonstrated prolonged pain relief accompanied by fewer gastrointestinal side effects.

Targeted Drug Delivery

Niosomes can be engineered for site-specific drug delivery, minimizing systemic toxicity and adverse immune responses. Their cost-effectiveness and ease of scalable production make them attractive alternatives to liposomes, particularly due to the use of stable and inexpensive non-ionic surfactants (Jadon et al., 2009).

Versatile Drug Encapsulation

These carriers can encapsulate both hydrophilic and lipophilic drugs, with aqueous-soluble drugs contained in the core and lipid-soluble drugs incorporated within the bilayer. Niosomes enhance the stability of sensitive molecules such as peptides and proteins by protecting them from enzymatic degradation. Additionally, they are being explored as delivery vehicles for gene therapy, effectively carrying DNA and RNA-based therapeutics to improve cellular uptake and gene expression efficiency (Jadon et al., 2009).

Benefits of Non-Ionic Surfactants

Compared to phospholipid-based liposomes, niosomal vesicles demonstrate greater resistance to hydrolysis and oxidation. Additionally, their cost-effective production supports wider application in therapeutic development (Baillie et al., 1986).

Reduced Side Effects and Drug Toxicity

Encapsulation of drugs in niosomes reduces systemic toxicity by preventing premature interaction with healthy tissues. Controlled release enables the use of smaller dosages while maintaining therapeutic efficacy. For instance, anticancer drugs such as doxorubicin have shown decreased cardiotoxicity when delivered through niosomal carriers (Gregoriadis, 1981).

Multiple Routes of Administration

Niosomes are adaptable to various administration routes:

- Oral: Enhance intestinal absorption and protect drugs from enzymatic degradation in the gastrointestinal tract.
- Parenteral: Enable both immediate and sustained systemic drug delivery.
- Topical/Transdermal: Commonly used in dermatology and cosmetics to achieve localized effects and enhance skin penetration.
- Ophthalmic: Improve ocular bioavailability and extend drug retention on the eye surface.
- Pulmonary/Nasal: Facilitate effective mucosal absorption, making them ideal for vaccines and respiratory treatments (Allen, 1998).

Customizable Formulation Properties

The properties of niosomes can be customized by modifying the types and ratios of cholesterol and surfactants used. This adaptability enables precise optimization of the carrier system to meet specific therapeutic objectives.

Furthermore, niosomes are compatible with other drug delivery platforms, increasing their potential for use in combination therapies (Kaur et al., 2004).

Types of Niosomes

Niosomes are classified based on the number of bilayers and their size (Table 1), with these structural differences influencing their preparation techniques and therapeutic applications.

Small Unilamellar Vesicles (SUVs)

These are spherical vesicles with a single lipid bilayer surrounding an aqueous core, typically measuring 10-100 nm in diameter. SUVs are well-suited for rapid drug delivery and are frequently used in biological research due to their resemblance to natural cell membranes (Szoka & Papahadjopoulos, 1980).

Large Unilamellar Vesicles (LUVs)

Slightly larger at 100-300 nm, LUVs also possess a single bilayer but provide greater drug encapsulation capacity. This makes them appropriate for complex delivery systems and membrane-related studies.

Multilamellar Vesicles (MLVs)

Table 1. Composition, method of preparation, and applications of niosomes

Type	Composition	Preparation Method	Applications
SUVs	Amphiphilic lipids	i. Sonication ii. Extrusion	i. Biomedical research ii. Drug delivery
LUVs	Aqueous core and amphiphilic lipids	i. Reverse evaporation ii. Ethanol injection	i. Studying membrane properties ii. Synthetic biology
MLVs	Phospholipids (e.g., phosphatidylcholine, phosphatidylglycerol) and cholesterol (for stability)	i. Thin film hydration ii. Reverse phase evaporation iii. Freeze-thaw cycling	i. Sustained release drug delivery ii. Lipophilic drug transport iii. Vaccine delivery

These consist of multiple concentric lipid bilayers, often exceeding 300 nm in size. MLVs allow for the sustained and controlled release of both hydrophilic and lipophilic drugs, making them particularly advantageous for long-term drug delivery and vaccine formulations.

Mechanisms of Drug Release from Niosomes

The release of drugs from niosomes is governed by multiple mechanisms, which depend on the physicochemical properties of the drug and the structural features of the vesicles. These mechanisms include:

Passive Diffusion through the Lipid Bilayer

Lipophilic drugs can diffuse gradually through the lipid bilayer into the surrounding environment. This passive diffusion is primarily affected by factors such as the type of surfactant used, cholesterol concentration, and membrane rigidity. For instance, methotrexate, used in cancer therapy, is released from niosomal carriers via this mechanism.

Vesicle Degradation and Leakage

Over time, niosomal vesicles may undergo oxidative degradation or hydrolysis, compromising membrane integrity and resulting in drug leakage. Enzymatic activity, pH changes, and interactions with biological components such as plasma proteins accelerate this breakdown, facilitating drug release in vivo (Baillie et al., 1986).

Fusion with Biological Membranes

Non-ionic surfactants in niosomes can fuse with cellular membranes, enabling direct transfer of the encapsulated drug into the cytoplasm. This mechanism significantly enhances intracellular drug delivery, making it especially useful for targeted therapies (Baillie et al., 1985).

Intracellular Drug Release via Endocytosis

Niosomes can be internalized by cells through endocytic pathways such as phagocytosis. Once internalized, they localize within lysosomes where the acidic pH and enzymatic activity facilitate vesicle degradation, leading to intracellular drug release (Jadon et al., 2009).

Release Induced by Swelling and Osmotic Pressure

In environments with variable pH or ionic strength, such as the gastrointestinal tract or bloodstream, water influx into the vesicle may cause swelling. This increase in osmotic pressure can eventually rupture the vesicle, releasing the drug.

Membrane Deterioration or Erosion

Over time, artificial lipid bilayers can degrade due to hydrolytic or enzymatic activity. Surfactant metabolism or the presence of enzymes such as lipases in biological fluids can hasten this process, resulting in the dissolution of the vesicle and release of the drug (Hu & Rhodes, 1999).

Factors Influencing Ocular Drug Absorption

The efficiency of ocular drug absorption is affected by a combination of physiological, anatomical, physicochemical, and formulation-related factors. A thorough understanding of these variables is crucial for optimizing drug delivery and enhancing therapeutic outcomes in ocular treatments.

Formulation-Related Factors

Tear Film Dynamics

The tear film, with a turnover rate of approximately 0.5-2.2 μL/min, rapidly dilutes and eliminates administered drugs. Reflex tearing, often induced by ocular irritation, further decreases drug residence time on the ocular surface, thereby limiting absorption.

Tear Volume and Drainage

The normal tear volume within the conjunctival sac is around 7-10 μL, whereas conventional eye drop formulations typically range from 30-50 μL. The surplus fluid is rapidly drained via the nasolacrimal duct, leading to systemic absorption rather than ocular retention, and reducing the effective dose available at the target site.

Corneal Barrier

Topically applied drugs are primarily absorbed through the cornea, which comprises three distinct layers with differing permeability characteristics:

- Lipophilic epithelium: Facilitates penetration of lipid-soluble drugs while acting as a barrier to hydrophilic compounds.
- Hydrophilic stroma: Permits diffusion of water-soluble drugs but impedes the passage of lipophilic agents.
- Lipophilic endothelium: Although it plays a lesser role, it allows the entry of small, lipophilic molecules into the anterior chamber.

pH and Buffering Capacity

The tear film maintains a physiological pH of approximately 7.4 and possesses inherent buffering capacity to restore homeostasis following drug instillation. Drugs with extreme pH values can cause irritation, triggering reflex tearing and further reducing absorption. Additionally, protein binding to tear proteins such as albumin, globulins, and lysozyme can sequester active drug molecules, lowering their bioavailability.

Blood-Ocular Barriers

- Blood-aqueous barrier: Formed by tight junctions in the ciliary epithelium, this barrier limits drug entry into the aqueous humor and anterior chamber.
- Blood-retinal barrier: Comprising tight junctions within the retinal pigment epithelium and retinal capillary endothelium, it restricts drug access to the posterior segment of the eye.

Nasolacrimal Drainage

Drugs that exceed the conjunctival sac's holding capacity are drained through the nasolacrimal duct into the systemic circulation. This process not only reduces ocular bioavailability but may also result in systemic side effects, while simultaneously minimizing the contact time between the drug and ocular tissues.

Physicochemical Properties of the Drug

Molecular Size and Shape

Smaller molecules diffuse more readily across the corneal epithelium. In contrast, large macromolecules such as proteins and peptides exhibit poor permeability due to their size and structural complexity, often requiring specialized delivery systems.

pKa and Ionization

The corneal epithelium favors the absorption of unionized drug molecules. Since ionization is influenced by the pH of the surrounding environment, drugs must be formulated near their pKa to maximize the proportion of unionized species, thereby enhancing permeability and solubility.

Solubility

Drugs with poor aqueous solubility face significant absorption challenges. Incorporating solubility enhancers, such as surfactants, co-solvents, or cyclodextrins, can increase drug availability at the absorption site, improving therapeutic efficacy.

Lipophilicity and Hydrophilicity

Lipophilic drugs are better suited for corneal penetration due to the lipophilic nature of the epithelial layer. Conversely, hydrophilic drugs, which are restricted by the lipophilic epithelium, may be absorbed through alternative routes such as the conjunctiva or sclera.

Clinical Applications of Niosomes in Treating Ocular Diseases

Niosomes, vesicular drug delivery systems composed of non-ionic surfactants, represent a promising platform for enhancing ocular drug delivery. Their capacity to encapsulate both hydrophilic and lipophilic drugs, combined with their affordability, stability, and formulation flexibility, makes them superior alternatives to conventional eye drops and, in some cases, even liposomes.

Glaucoma Therapy

Glaucoma is a chronic optic neuropathy characterized by elevated intraocular pressure (IOP), which can lead to progressive vision loss. Conventional eye drops suffer from low bioavailability due to rapid tear drainage and systemic absorption. Niosomal nanocarriers improve drug penetration and prolong ocular retention. Timolol maleate niosomes enhance corneal permeability and sustain IOP reduction, thereby reducing dosing frequency. Brimonidine tartrate niosomes increase ocular absorption and minimize systemic side effects such as bradycardia and hypotension.

Dry Eye Syndrome (DES)

DES is caused by tear film instability, leading to ocular discomfort, dryness, and inflammation. Niosomes enable controlled and sustained drug release, reducing the frequency of administration. Cyclosporine-loaded niosomes enhance ocular surface retention and decrease inflammation. Hyaluronic acid-based vesicles stabilize the tear film and provide prolonged hydration.

Ocular Infections (Bacterial, Viral, Fungal)

Local delivery is crucial for maintaining effective drug concentrations at the infection site while minimizing systemic exposure. Niosomes improve drug penetration and extend the duration of antimicrobial activity. Moxifloxacin niosomes provide prolonged antibacterial effects for treating bacterial conjunctivitis. Acyclovir niosomes enhance corneal penetration for the treatment of viral keratitis.

Cataract Prevention and Treatment

Cataracts, primarily caused by oxidative damage to the lens, result in visual impairment. Niosomes improve antioxidant delivery and prolong ocular retention. Carnosine-loaded nano-lamellar vesicles serve as antioxidants that slow lens opacification. Curcumin-loaded vesicles reduce oxidative stress and inflammation with lower toxicity than conventional formulations.

Retinal Disorders (e.g., AMD, Diabetic Retinopathy)

The blood-retinal barrier (BRB) poses a major obstacle to drug delivery in posterior eye diseases. Niosomes offer targeted retinal delivery with extended therapeutic action. Dexamethasone-loaded niosomes reduce inflammation in diabetic retinopathy. Bevacizumab niosomes enhance intraocular retention and reduce the frequency of intravitreal injections in AMD.

Postoperative Drug Delivery (e.g., LASIK, Cataract Surgery)

Post-surgical inflammation and infection require sustained therapy. Niosomes maintain therapeutic drug levels, reducing the need for repeated dosing and preserving ocular sterility. Tromethamine-ketorolac niosomes provide prolonged anti-inflammatory effects after surgery. Cefuroxime-loaded vesicles effectively prevent postoperative infections, particularly following cataract procedures.

Inflammation and Uveitis

Uveitis and other ocular inflammations often necessitate corticosteroid or NSAID therapy. Niosomes extend drug action while minimizing adverse effects. Dexamethasone niosomes offer prolonged anti-inflammatory benefits. Flurbiprofen niosomes improve drug retention and reduce inflammation with fewer side effects. linical trials are underway for niosomal formulations of cyclosporine, timolol, and anti-VEGF agents to further enhance ocular therapeutic outcomes.

Conclusion

Based on the findings discussed, niosomes represent a highly promising and adaptable platform for overcoming the eye's complex physiological and anatomical barriers to drug delivery. Their unique composition, primarily comprising cholesterol and non-ionic surfactants, allows for the effective encapsulation of both hydrophilic and lipophilic drugs, thereby improving drug stability, enhancing bioavailability, and enabling controlled and sustained release. These vesicular carriers have shown significant therapeutic benefits in the management of various ocular disorders, including glaucoma, dry eye syndrome, infections, and retinal diseases. While certain formulation and regulatory challenges persist, ongoing advances in nanotechnology, surface engineering, and the incorporation of biocompatible materials continue to expand the potential of niosomes. As research progresses, niosomal systems are poised to play a transformative role in ocular therapeutics, ultimately improving treatment outcomes and patient adherence.

References

Allen, T. M. (1998). Liposomal Drug Formulations. *Drugs*, *56*(5), 747–756.

Azmin, M. N., Florence, A. T., Handjani-Vila, R. M., Stuart, J. F. B., Vanlerberghe, G., & Whittaker, J. S. (1985). The effect of non-ionic surfactant vesicle (niosome) entrapment on the absorption and distribution of methotrexate in mice. *Journal of Pharmacy and Pharmacology*, *37*(4), 237–242.

Baillie, A. J., Coombs, G. H., Dolan, T. F., & Laurie, J. (1986). Non-ionic surfactant vesicles, niosomes, as a delivery system for the anti-leishmanial drug, sodium stibogluconate. *Journal of Pharmacy and Pharmacology*, *38*(7), 502–505.

Baillie, A. J., Florence, A. T., Hume, L. R., Muirhead, G. T., & Rogerson, A. (1985). The preparation and properties of niosomes—non-ionic surfactant vesicles. *Journal of Pharmacy and Pharmacology*, *37*(12), 863–868.

Cummings, J., Stuart, J. F. B., & Calman, K. C. (1984). Determination of adriamycin, adriamycinol and their 7-deoxy aglycones in human serum by high-performance liquid chromatography. *Journal of Chromatography B: Biomedical Sciences and Applications*, *311*, 125–133.

Gregoriadis, G. (1981). Targeting of drugs: implications in medicine. *The Lancet*, *318*(8240), 241–247.

Hu, C., & Rhodes, D. G. (1999). Proniosomes: A Novel Drug Carrier Preparation. *International Journal of Pharmaceutics*, *185*(1), 23–35.

Jadon, P. S., Gajbhiye, V., Jadon, R. S., Gajbhiye, K. R., & Ganesh, N. (2009). Enhanced Oral Bioavailability of Griseofulvin via Niosomes. *AAPS PharmSciTech*, *10*(4), 1186.

Jayaraman, S. C., Ramachandran, C., & Weiner, N. (1996). Topical Delivery of Erythromycin from Various Formulations: An in Vivo Hairless Mouse Study. *Journal of Pharmaceutical Sciences*, *85*(10), 1082–1084.

Kaur, I. P., Garg, A., Singla, A. K., & Aggarwal, D. (2004). Vesicular systems in ocular drug delivery: an overview. *International Journal of Pharmaceutics*, *269*(1), 1–14.

Szoka, F., & Papahadjopoulos, D. (1980). Comparative Properties and Methods of Preparation of Lipid Vesicles (Liposomes). *Annual Review of Biophysics and Bioengineering*, *9*(1), 467–508.

Chapter 9

Micelles for Ocular Therapeutics

Neelam Singh[1]
Ritu Chauhan[1]
Deepti Aggrawal[1]
Reenu Chauhan[1]
and Chaitanya Vinayak Narayan[2,*]
[1]HR Institute of Pharmacy, HRIT University, Uttar Pradesh, India
[2]I.T.S College of Pharmacy, Ghaziabad, Uttar Pradesh, India

Abstract

Conventional ocular drug delivery methods face several significant challenges due to three primary physiological barriers: (a) the complex, protective structure of the cornea that limits drug permeability; (b) the metabolic activity of ocular cells that can degrade therapeutic agents; and (c) the tear film's rapid clearance mechanism, which reduces drug retention time. These factors often result in poor drug bioavailability and limited therapeutic efficacy. In recent years, micellar drug delivery systems, particularly polymeric micelles, have emerged as a promising approach to overcome these limitations. Micelles offer multiple advantages for ocular therapy, including enhanced solubilization of poorly water-soluble drugs, controlled and sustained drug release, minimal ocular irritation, and excellent biocompatibility. Their nanoscale size and optical transparency further make them well-suited for ocular administration without impairing vision. This review explores recent advances in micelle-based drug delivery for targeting both the anterior and posterior segments of the eye, and discusses the benefits, ongoing

* Corresponding Author's Email: chaitanyavinayak@gmail.com

In: Advances in Nanotechnology for Ocular Drug Delivery
Editors: Anjali Sharma, Ajmer Singh Grewal, Geeta Deswal et al.
ISBN: 979-8-89530-814-1
© 2025 Nova Science Publishers, Inc.

challenges, and future prospects of these formulations in ocular pharmacotherapy.

Keywords: controlled release, micelles, nanomicelles, ocular bioavailability, ocular drug delivery, polymeric micelles, tear clearance barriers

Abbreviations

API: Active Pharmaceutical Ingredient
BAB: Blood-Aqueous Barrier
BC: Bruch's-Choroid
BRB: Blood-Retinal Barrier
CMC: Critical Micelle Concentration
CNV: Choroidal Neovascularization
DES: Dry Eye Syndrome
DME: Diabetic Macular Edema
DMF: Dimethylformamide
DR: Diabetic Retinopathy
ERG: Electroretinography
KCS: Keratoconjunctivitis Sicca
NPDR: Non-Proliferative Diabetic Retinopathy
OSD: Ocular Surface Disease
PACG: Primary Angle-Closure Glaucoma
PCL: Polycaprolactone
PEG: Polyethylene Glycol
PLA: Polylactic Acid
PLGA: Poly(lactic-co-glycolic acid)
POAG: Primary Open-Angle Glaucoma
PVA: Polyvinyl Alcohol
PVP: Polyvinylpyrrolidone
RP: Retinitis Pigmentosa
RPE: Retinal Pigment Epithelium
siRNA: Small Interfering RNA

Introduction

Ocular drug delivery presents unique challenges due to the eye's intricate anatomical and physiological barriers, including the corneal epithelium, tear turnover, and blood-ocular barriers. Conventional formulations, such as eye drops and ointments, often exhibit low bioavailability, rapid drug clearance, and limited penetration into deeper ocular tissues. To address these limitations, micelle-based drug delivery systems have emerged as a promising strategy in ocular therapeutics. Micelles are nano-sized, amphiphilic structures formed by the self-assembly of surfactant molecules in aqueous environments (Binkhathlan et al., 2023). Comprising a hydrophobic core and a hydrophilic shell, micelles are ideal for encapsulating poorly water-soluble drugs, thereby enhancing solubility and enabling sustained release. In ocular applications, micelles offer numerous benefits, including improved corneal permeability, prolonged retention time, reduced toxicity, and better patient compliance. Recent developments in polymeric micelles, stimuli-responsive systems, and targeted delivery techniques have further broadened their applicability for treating ocular disorders such as glaucoma, dry eye syndrome, uveitis, and retinal diseases. By protecting drugs from enzymatic degradation, enhancing stability, and providing controlled release, micelles represent a cutting-edge approach in modern ocular pharmacotherapy (Binkhathlan et al., 2023; Assiri et al., 2024).

Anatomy of the Eye

The human eye is a highly specialized organ that functions similarly to a camera, capturing and processing light to enable vision. It comprises multiple layers and structures, each essential to visual function. The outermost layer includes the cornea, a transparent dome that focuses incoming light, and the sclera, the white, protective outer covering. The middle layer, or uvea, consists of the iris, which regulates light entry by adjusting pupil size; the ciliary body, which controls lens shape for focusing; and the choroid, a vascular layer that supplies oxygen and nutrients. The innermost layer is the retina, which contains photoreceptor cells: rods, responsible for vision in low light, and cones, responsible for color perception. Visual signals from the retina are transmitted to the brain via the optic nerve. The eye also contains two fluid-filled chambers: the aqueous humour, which maintains intraocular pressure and nourishes the anterior segment, and the vitreous humour, which helps maintain the eye's shape and provides structural support. In addition, extraocular muscles control eye movements, while lacrimal glands produce tears that lubricate and protect the ocular surface (Willoughby et al., 2010). Together, this intricate architecture enables precise and dynamic visual perception.

Barriers of Ocular Drug Delivery Systems

Ophthalmic medications are most commonly administered through topical formulations such as solutions, suspensions, and ointments. However, alternative delivery routes, including subconjunctival, intravitreal, periocular, intravenous, oral administration, and implantable devices, are also employed to address specific therapeutic needs (Li et al., 2023). A major challenge in ocular drug delivery lies in overcoming the eye's complex anatomical and physiological barriers to ensure efficient drug penetration into target tissues. After crossing the tear film and lacrimal fluid, drugs must traverse multiple membrane barriers located in the cornea, conjunctiva, iris–ciliary body, and retina (Diwan et al., 2022). These barriers are primarily composed of tightly joined epithelial and endothelial cells, which significantly restrict paracellular drug transport. Based on their anatomical location and functional role, these barriers are typically categorized into different classes, each posing unique challenges to effective ocular drug delivery.

Corneal Barrier

The cornea is a transparent, avascular, and multilayered structure composed of the corneal epithelium, basement membrane, Bowman's layer, stroma, Descemet's membrane, and endothelium. It functions as both a mechanical and biochemical barrier, significantly limiting the penetration of topically administered drugs (Wu et al., 2023). The epithelial cells, especially those at the apical surface, are interconnected by tight junctions that restrict paracellular transport. Consequently, lipophilic drugs generally demonstrate higher corneal permeability compared to hydrophilic compounds. Despite the barrier properties of the corneal epithelium, transcorneal permeation remains the primary route for drug absorption from the lacrimal fluid into the aqueous humor (Thakur et al., 2023).

Conjunctiva

The conjunctiva is a thin, transparent membrane that covers the anterior sclera and lines the inner surface of the eyelids. It consists of three main layers: (i) the outer epithelium, which serves as a permeability barrier; (ii) the substantia propria, which contains blood vessels, lymphatics, and nerves; and (iii) the submucosa, which loosely connects to the underlying Tenon's capsule. Due to its rich vascularization, the presence of goblet cells, and its potential for transdifferentiation, the conjunctiva plays a vital role in non-corneal drug delivery, particularly for macromolecular therapeutics. The conjunctival epithelium contains tight junctions with a bioelectrical resistance exceeding 1500 $\Omega \cdot cm^2$, providing robust protection while regulating the passage of hydrophilic substances (Wu et al., 2023).

Tear Film

The tear film, regulated by orbital glands and ocular surface secretions, plays a critical role in maintaining the health of the cornea and conjunctiva. It is composed of proteins, lipids, electrolytes, and mucins, forming a dynamic barrier and protective interface. Topical administration into the conjunctival cul-de-sac remains the most common route for ocular drug delivery. However, precorneal drainage into the nasolacrimal ducts leads to the loss of approximately 90% of the administered drug (Wu et al., 2023). Additional factors such as drug dilution, protein binding, and rapid tear turnover further limit drug bioavailability. Although commercial droppers typically dispense 20-50 μL of solution, the precorneal space can retain only 7-10 μL, resulting in substantial drug loss within 15-30 seconds of instillation (Li et al., 2023).

Reflex Blinking

The human eye can temporarily retain up to 30 μL of fluid; however, any excess volume is rapidly eliminated through reflex blinking, occurring approximately 5 to 7 times per minute, and nasolacrimal drainage, significantly reducing the amount of drug available for therapeutic action (Li et al., 2023).

Metabolism in Ocular Tissues

Enzymatic metabolism within ocular tissues also plays a significant role in drug clearance. Drugs containing aromatic hydrocarbons are particularly susceptible to extensive enzymatic degradation. Hayakawa et al. demonstrated that the high metabolic activity of the conjunctiva leads to poor absorption of insulin and peptide-based drugs in rabbits. Similarly, metabolic clearance can exceed the rate of aqueous humor turnover, further diminishing drug bioavailability (Ahmed et al., 2023).

Sclera and Bruch's-Choroid Complex

The sclera, primarily composed of collagen fibrils and glycoproteins, covers a surface area of approximately 16.3 cm^2. Drug permeability through the sclera is significantly influenced by molecular weight (MW), with macromolecules demonstrating lower permeability compared to smaller molecules. However, molecular radius is often considered a more accurate predictor of diffusion for macromolecular compounds. Beyond the sclera, the Bruch's-choroid (BC) complex, which consists of fenestrated capillaries and elastic connective tissue layers, presents an even more formidable barrier, particularly to positively charged lipophilic drugs. These compounds tend to bind within the BC complex, forming slow-release depots that delay the onset of therapeutic action (Li et al., 2023).

Retina and Blood-Retinal Barrier (BRB)

The retina poses a significant barrier to the delivery of large molecules, primarily due to BRB, which restricts systemic diffusion into retinal tissues. The BRB is composed of two distinct layers:

- The inner BRB, formed by retinal capillary endothelial cells connected by tight intercellular junctions, serves to protect the retina by limiting the passage of hydrophilic compounds and macromolecules.
- The outer BRB, constituted by the retinal pigment epithelium (RPE), is located between the photoreceptors and the choriocapillaries. Lipophilic drugs typically traverse the RPE via transcellular

pathways, whereas hydrophilic drugs rely on paracellular transport through tight junctions (Ahmed et al., 2023).

Blood-Aqueous Barrier (BAB)

Located in the anterior segment of the eye, BAB is formed by the non-pigmented ciliary epithelium and the endothelial cells of the iris. Tight junctions within both layers restrict non-specific solute entry, thereby maintaining the transparency and chemical composition of the ocular fluids. Although the BAB exhibits strong structural integrity, its effectiveness in completely preventing drug permeation is not absolute (Chochaeva et al., 2023).

Mechanisms of Ocular Drug Delivery

Topical drug delivery to the eye primarily occurs via the corneal and non-corneal routes through the conjunctival sac (cul-de-sac). However, drugs administered by these methods are often rapidly eliminated by lacrimal fluid drainage, limiting their bioavailability. Among these pathways, the corneal route remains the principal mechanism for ophthalmic drug absorption. Nevertheless, due to the tight junctions and structural complexity of the corneal epithelium, this pathway is considered rate-limiting for drug uptake. An alternative absorption route involves trans-scleral delivery through the conjunctiva and sclera. This pathway, however, may be less efficient because local capillary networks can drain the drug into systemic circulation, thereby reducing intraocular drug levels. Despite this, drugs with low corneal permeability, such as prostaglandin $PGF_{2}\alpha$, gentamicin, and timolol maleate, have demonstrated the capacity to permeate via the conjunctival–scleral route to reach intraocular tissues (Chochaeva et al., 2023).

Ocular Delivery Pathways of Micelles

Micelles are self-assembled nanocarriers formed from amphiphilic copolymers or surfactants, typically measuring less than 50 nm in size. These surfactants possess both hydrophilic (polar) heads and hydrophobic (non-polar) tails, which arrange into spherical micelles in aqueous environments.

The polar head groups may be anionic, cationic, zwitterionic, or nonionic, while the tails generally consist of long hydrocarbon chains. Micellar systems have shown significant potential in enhancing the solubility, absorption, and bioavailability of both poorly water-soluble and highly water-soluble drugs (Binkhathlan et al., 2023). A key advantage of micelles is their prolonged circulation time, which delays systemic clearance and extends drug action. Additional benefits include straightforward preparation, tunable particle size, and the capacity for targeted delivery by incorporating mucoadhesive, thermosensitive, or pH-sensitive components. In ocular delivery, the hydrophilic corneal stroma, which accounts for approximately 85-90% of corneal thickness, acts as a rate-limiting barrier for hydrophobic drug penetration. Encapsulation of hydrophobic drugs within the lipophilic core of micelles overcomes this limitation, facilitating their passage through the stroma. Moreover, the nano-sized micelles can penetrate ocular tissues via both corneal and conjunctival-scleral pathways following topical administration. The extensive surface area of the conjunctiva and sclera enables lateral diffusion, allowing micelles to reach the posterior segment of the eye (Assiri et al., 2024). The hydrophilic corona of micelles promotes their transport through aqueous scleral pores while minimizing drug washout by conjunctival blood and lymphatic drainage. Upon reaching the posterior segment, micelles may be internalized by RPE cells via endocytosis, enabling the accumulation of therapeutic drug concentrations in deeper ocular tissues (Assiri et al., 2024).

Prepartion Methods of Micelles

Several formulation strategies are available for preparing micelles, each specifically tailored to accommodate various types of polymers, solvents, and active pharmaceutical ingredients.

The Direct Dissolution Method

This is the simplest technique for micelle preparation, suitable for highly water-soluble polymers. A polymer solution is prepared under continuous stirring, and when the polymer concentration exceeds the critical micelle concentration (CMC), the amphiphilic polymers spontaneously self-assemble into micelles in the aqueous medium (Kaur et al., 2021).

The Solvent Evaporation Method

In this method, the drug is first dissolved in an organic solvent and then added to a polymer solution. The mixture is stirred at 300 rpm for approximately four hours to facilitate the evaporation of the organic solvent. Following this, the formulation is ultracentrifuged to remove any unloaded drug and filtered to eliminate impurities. Finally, the micellar product is lyophilized to enhance its stability (Kaur et al., 2021).

The Dialysis Method

A polymer-drug solution is prepared using a water-miscible organic solvent such as dimethylformamide (DMF). Small volumes of water are added gradually under stirring to promote micelle formation. The mixture is then transferred to a dialysis bag and dialyzed against excess water for several hours to remove the organic solvent completely (Atanase, 2021).

The Oil-in-Water (O/W) Emulsion Method

Hydrophobic drugs and polymers are dissolved in a water-insoluble organic solvent (oil phase), which is then slowly added to an aqueous phase under stirring or homogenization. This creates an oil-in-water emulsion, wherein the hydrophobic components reorganize within the aqueous environment. Upon evaporation of the organic solvent, micelles are formed (Atanase, 2021).

The Film Hydration Method

Hydrophobic drugs and copolymers are dissolved in an organic solvent within a round-bottom flask and sonicated to ensure uniform dispersion. The solvent is then evaporated overnight under vacuum in a Petri dish, forming a thin film. This film is subsequently rehydrated with deionized water to generate a micellar solution, which is centrifuged to remove any free, unencapsulated drug (Thabet et al., 2022).

Structures of Micelles

Micelles are classified into 3 types based on their structures (Table 1) as surfactant micelles, polymeric micelles, and poly-ion complex micelles (Perumal et al., 2022).

Table 1. Different types of micelles and their properties

Property	Polyion Complex Micelles	Polymer Micelles	Surfactant Micelles
Formation Mechanism	Electrostatic interaction between oppositely charged polymers	Self-assembly of amphiphilic block copolymers	Self-assembly of surfactant molecules
Core Composition	Ionically crosslinked or electrostatic cores	Hydrophobic core	Hydrophobic region of surfactant molecules
Shell Composition	Hydrophilic polymers (e.g., PEG)	Hydrophilic polymer chains (e.g., PEG)	Hydrophilic head groups of surfactants
Encapsulation Efficiency	High for charged/hydrophilic drugs	High for hydrophobic drugs	Moderate for both hydrophilic and hydrophobic drugs
Stability in Tears	Sensitive to ionic strength and pH	Relatively stable	Sensitive to dilution or surfactant concentration
Drug Release Mechanism	Controlled by ionic dissociation or enzymatic degradation	Controlled by diffusion or degradation of the polymer matrix	Rapid release upon micelle breakdown
Surface Modification	Easily modified with targeting ligands	Modifiable with functional groups	Limited options for surface functionalization
Toxicity	Low if biodegradable polymers are used	Low to moderate depending on polymer choice	Higher risk of irritation from surfactants
Ease of Production	Moderate (requires careful control of charge ratios)	Moderate to high (scalable with polymer design)	High (simple mixing of surfactants)
Examples in Use	siRNA delivery for retinal diseases	Timolol for glaucoma, latanoprost for intraocular pressure	Antibiotics for bacterial infections

Surfactant Micelles

Surfactants are widely employed in ocular formulations due to their versatile physicochemical properties. Depending on their composition and function, they can serve as solubilizers, emulsifiers, wetting agents, stabilizers, antibacterial agents, charge inducers, gelling agents, or corneal permeability enhancers. In ocular micellar systems, surfactants play a crucial role by facilitating controlled drug release, enhancing aqueous solubility, and improving drug penetration into ocular tissues. By increasing corneal permeability, micelles enable drugs to bypass the corneal barrier and reach deeper ocular structures, thereby significantly enhancing ocular bioavailability (Zueva et al., 2025).

Polymeric Micelles

Polymeric micelles have attracted considerable attention in ocular drug delivery over the past two decades due to their structural stability and capacity to solubilize poorly water-soluble drugs. These micelles are nano-sized colloidal particles that spontaneously form from amphiphilic block copolymers, consisting of both hydrophilic and hydrophobic segments. The hydrophobic core encapsulates lipophilic drugs, while the hydrophilic corona provides stability and compatibility with aqueous environments. The size and performance of polymeric micelles are influenced by various factors, including the molecular weight of the block copolymer, aggregation number, physicochemical properties of both hydrophilic and hydrophobic segments, and the method of preparation. Additionally, polymeric micelles are being investigated as intelligent drug delivery systems, designed to respond to external stimuli such as pH or temperature changes. In ocular applications, they enhance drug bioavailability by facilitating paracellular and transcellular transport. Common polymers used in these micelles include polyethylene glycol (PEG), poly(glycolide) (PGA), poly(ε-caprolactone) (PCL), poly(lactide-co-glycolide) (PLGA), polypropylene oxide (PPO), and poly-L-lysine (Ludwig et al., 2021).

Polyionic Micelles

Polyionic micelles are formed through electrostatic interactions between polyionic block copolymers and oppositely charged molecules. Their structure, especially the charged corona, is stabilized by a combination of electrostatic forces and van der Waals interactions. These micelles offer several advantages, including simple synthesis, high drug loading capacity, enhanced structural stability, and prolonged circulation time in the bloodstream. Notably, polyionic micelles can be prepared entirely in aqueous media without the use of organic solvents, thereby eliminating the risk of toxicity from residual solvents. This characteristic makes them particularly suitable for ocular targeting, providing increased safety and efficiency in drug delivery (Zueva et al., 2025).

Ocular Diseases

Anterior Segment Disease (Table 2)

Cataract

Cataract refers to the clouding of the crystalline lens or its capsule, which impairs the passage of light to the retina. While cataracts are primarily age-related, congenital forms can affect infants, and trauma or disease may induce cataracts in adults. The pathological mechanisms include:

- Congenital cataracts: Abnormal lens development.
- Subcapsular cataracts: Fibrous changes in the lens epithelium.
- Cortical cataracts: Accumulation of moisture between cortical fibers.
- Nuclear cataracts: Pigment deposition, such as urochrome (Ting et al., 2021).

Conjunctivitis

Conjunctivitis is the inflammation of the conjunctiva, marked by dilation of conjunctival vessels, redness, edema, and often discharge.

- Infectious conjunctivitis: Most commonly caused by viruses, followed by bacterial infections.
- Non-infectious conjunctivitis: Includes allergic, toxic, immune-mediated, and malignant causes (Ting et al., 2021).

Anterior Uveitis

Anterior uveitis is inflammation affecting the iris (iritis), the ciliary body (cyclitis), or both (iridocyclitis). It is more common than posterior uveitis and generally poses a lower risk to vision. Symptoms include pain, photophobia, redness, blurred vision, and excessive tearing. Photophobia arises from ciliary muscle spasm, while blurred vision results from turbidity in the aqueous humor (Amador et al., 2022; Ting et al., 2021).

Dry Eye Syndrome (DES)

DES, also referred to as dysfunctional tear syndrome (DTS), keratoconjunctivitis sicca (KCS), or ocular surface disease (OSD), affects the preocular tear film. It commonly results from dysfunction within the nasolacrimal unit, which includes the lacrimal glands, eyelids, and corneal surface. DES is particularly prevalent among individuals over 48 years of age (Amador et al., 2022; Ting et al., 2021).

Glaucoma

Glaucoma is a group of progressive optic neuropathies characterized by irreversible vision loss. The two primary types are:

- Primary open-angle glaucoma (POAG): Commonly associated with older age, African descent, family history, and diabetes.
- Primary angle-closure glaucoma (PACG): More prevalent among females, Asians, and elderly populations (Amador et al., 2022; Ting et al., 2021).

Table 2. Micellar ocular drug delivery for anterior segment diseases

Disease	API	Polymer/surfactant	Study conclusion	Ref.
Dry Eye Syndrome	Cyclosporine A	mPEG-PLA	Corneal permeation is enhanced and it decreases CsA elimination from micelle.	(Li et al., 2022) (Mandal et al., 2017) (Li et al., 2022) (Assiri et al., 2024)
		Sympatens AS-200G	Drug intake is increased as compared with mPEG-PLA	
		PEG-PCL- g-PEI	Corneal penetration is enhanced both in-vivo and in-vitro	
	Rapamycin	PEG-DSPE	Lymphocytic infiltration is reduced	
Anterior Uveitis	Dextroamphetamine	PCL-PEG-PCL	Clinical symptoms of uveitis are reduced	(Alami-Milani et al., 2018)
Corneal Neovascularization	Bevacizumab	mPEG-PCL	Retention time on corneal surface is enhanced and inflammation is reduced	(Dastjerdi et al., 2010)
Corneal Wound healing	Ferulic acid	Pluronic F-68	Concentration at normal cornea is higher than damaged cornea	(Zueva et al., 2025)
Oxidative stress	Curcumin	PVCL-PVA-PEG	It promotes wound healing of corneal epithelium and enhanced its sensation	(Leng et al., 2020)
Ocular inflammation	Curcumin	PVCL-PVA-PEG	Corneal permeation is enhanced	(Sun et al., 2019) (Li et al., 2022) (Assiri et al., 2024) Assiri et al., 2024) (Binkhathlan et al., 2023) (Mandal et al., 2017) (Mandal et al., 2017)
		Rebaudioside A	In-vitro cell uptake and corneal permeation is enhanced and	
		Gellan gum	Corneal penetration is enhanced	
	Myricetin	PVCL-PVA-PEG	In-vivo anti-inflammatory activity and invitro anti-antioxidant activity is enhanced; cellular uptake is also enhanced	
		Kolliphor HS 15	Corneal permeation is enhanced	
	Triamcinolone	Hyaluronic acid	Retention time and corneal permeation is enhanced	
	Dextroamphetamine	PCL-PEG-PCL	Corneal permeation is enhanced	
	Diclofenac	Ginsenoside Rb1	Anti-inflammatory activity retained for longer time and Corneal permeation is enhanced	

Disease	API	Polymer/surfactant	Study conclusion	Ref.
	Neomycin- B	5-DUP3	It works better against bacterial growth and corneal permeation is enhanced	
	Tacrolimus	PEG-CCF	Inflammatory cells are reduced and thickness of corneal epithelial cells is restored	
	Sertaconazole nitrate	Solutol H15	In vivo and Ex vivo corneal permeation is increased by two folds	
	Lornoxicam	Tetronic 701	Ex-vivo penetration of the micelle is increased	
	Ofloxacin	Pluronic F-68/Pluronic F127	In vitro and in vivo retention time is increased	
	Gatifloxacin	Pluronic F127	Corneal permeation is enhanced	
	Fluorometholone	Soluplus	Corneal permeation is enhanced as compared to suspension form and physical mix	
	Terbinafine hydrochloride	Kolliphor HS 15	Corneal stroma permeation is enhanced	
	Celecoxib	Pluronic F127	Corneal permeation is enhanced compared to suspension and improved ocular bioavailability	
	Natamycin	Soluplus/Pluronic P103	Corneal permeation is enhanced	
	Itraconazole	B-cyclodextrin Poly(pseudo)rotaxanes	Ex-vivo and in-vivo corneal permeation is enhanced	
		Pluronic F127/Pluronic F68/Carbopol 934P	Ex vivo corneal permeation is increased five-fold compared with pure drug suspension and itraconazole eye drops	
Glaucoma	Methazolamide	mPEG-PCL	Intra ocular pressure is reduced	(Ghezzi et al., 2022)
	Latanoprost	mPEG-PLA	Retention time and bioavailability is increased	
	Melatonin	Pluronic F-127	Corneal permeability is enhanced	

Choroidal Melanoma

Uveal melanoma is a rare but serious malignant tumor arising from melanocytes in the uveal tract, which includes the choroid, ciliary body, and iris. It is the most common primary intraocular malignancy and is associated with oxidative DNA damage and genetic mutations, including adenine-to-cytosine transitions related to iris pigmentation (Soliman et al., 2023).

Posterior Segment Diseases (Table 3)

Retinitis Pigmentosa (RP)

Retinitis pigmentosa comprises a group of hereditary retinal degenerative disorders characterized by early night blindness and progressive loss of peripheral vision. Electroretinography (ERG) typically reveals a pronounced decline in rod photoreceptor function. The condition follows various inheritance patterns, including X-linked, autosomal dominant, and autosomal recessive, with symptoms commonly beginning in childhood (Wang et al., 2021).

Wet Age-Related Macular Degeneration (Wet AMD)

It is the leading cause of central vision loss in older adults in developed countries. The condition involves choroidal neovascularization (CNV), where abnormal blood vessels penetrate the Bruch's membrane, resulting in fluid leakage and damage to the macula (Wang et al., 2021).

Choroidal Neovascularization (CNV)

A serious complication of uveitis, CNV occurs when disruption between the Bruch's membrane and RPE triggers abnormal blood vessel growth. CNV can arise from both infectious and non-infectious forms of uveitis, with the latter often having an autoimmune origin (Wang et al., 2021).

Diabetic Macular Edema (DME)

The leading cause of visual impairment in diabetic patients, diabetic macular edema results from a breakdown in fluid regulation across the retina, causing accumulation of fluid within the retinal layers. Both extracellular and intracellular edema may develop, with intracellular swelling attributed to the buildup of lactate and sorbitol (Wang et al., 2021).

Diabetic Retinopathy (DR)
A chronic, progressive disease affecting diabetic patients, diabetic retinopathy manifests in two main stages:

- Non-proliferative diabetic retinopathy (NPDR): Characterized by microaneurysms, cotton wool spots, and hard exudates resulting from vascular leakage.
- Proliferative diabetic retinopathy (PDR): Marked by retinal ischemia and neovascularization driven by vascular endothelial growth factor (VEGF) (Tan and Wong, 2023).

Table 3. Micellar ocular drug delivery for posterior segment diseases

Disease	API	Polymer/ surfactant	Study conclusion	Ref.
Proliferative neo-vacuolization	Genistein	Hyaluronic acid	Decrease in neovascularization, increase corneal permeation	(Long et al., 2023)
Macular degeneration	Sirolimus	Cholesterol PEG	Scleral permeation is enhanced	(Suri et al., 2023)
	Aflibercept	PEG-PPG-PCL	Increased anti-angiogenic effects	
	Dextroamphetamine	Chitosan oligosaccharide stearic acid	Concentration is enhanced at retina, choroid and sclera	
Posterior Uveitis	Rapamycin	mPEG -PCL	Anti-inflammatory activity is increased	(Mofidfar et al., 2021)
	Dextroamphetamine	mPEG -PCL	Permeation at conjunctiva and sclera is enhanced	

Conclusion

Eye-related conditions affect a large portion of the global population and pose a significant threat to ocular health worldwide. The eye's complex anatomical and physiological barriers greatly hinder drug accumulation and reduce bioavailability, making effective drug delivery a major challenge. However, recent advancements in micelle-based ocular therapies offer promising solutions to these limitations. Micelles improve drug stability by protecting active compounds from degradation and enable efficient tissue penetration due to their small size and hydrophilic properties. These characteristics make micelles a valuable tool for targeted and sustained treatment of ocular disorders.

References

Ahmed, S., Amin, M. M., & Sayed, S. (2023). Ocular drug delivery: A comprehensive review. *AAPS PharmSciTech*, 24, 66.

Alami-Milani, M., Zakeri-Milani, P., Valizadeh, H., Salehi, R., & Jelvehgari, M. (2018). Preparation and evaluation of PCL-PEG-PCL micelles as potential nanocarriers for ocular delivery of dexamethasone. *Iranian Journal of Basic Medical Sciences*, 21(2), 153–164.

Amador, C., Shah, R., Ghiam, S., Kramerov, A. A., & Ljubimov, A. V. (2022). Gene therapy in the anterior eye segment. *Current Gene Therapy*, 22(2), 104–131.

Assiri, A. A., Glover, K., Mishra, D., Waite, D., Vora, L. K., & Thakur, R. R. S. (2024). Block copolymer micelles as ocular drug delivery systems. *Drug Discovery Today*, 29, 104098.

Atanase, L. I. (2021). Micellar drug delivery systems based on natural biopolymers. *Polymers*, 13(3), 477.

Binkhathlan, Z., Ali, R., Alomrani, A. H., Abul Kalam, M., Alshamsan, A., & Lavasanifar, A. (2023). Role of polymeric micelles in ocular drug delivery: An overview of decades of research. *Molecular Pharmaceutics*, 20(11), 5359–5382.

Chochaeva, T. I., Malinovskaia, D. G., Guchakova, A. A., Khaupsheva, M. I., Kokoeva, A. A., & Albekovna, T. A. (2023). Study of the permeability of blood-aqueous barrier with tetracycline group drugs in normal and pathological conditions. *Pharmacophore*, 14(6), 71–78.

Dastjerdi, M. H., Saban, D. R., Okanobo, A., Nallasamy, N., Sadrai, Z., Chauhan, S. K., Hajrasouliha, A. R., & Dana, R. (2010). Effects of topical and subconjunctival bevacizumab in high-risk corneal transplant survival. *Investigative Ophthalmology & Visual Science*, 51(5), 2411–2417.

Diwan, P., Jangde, R., Khunte, S., Bhardwaj, H., & Suresh, P. K. (2022). Ocular drug delivery system: Barrier for drug permeation, method to overcome barrier. *In Drug Development Life Cycle*. IntechOpen.

Ghezzi, M., Ferraboschi, I., Delledonne, A., Pescina, S., Padula, C., Santi, P., Sissa, C., Terenziani, F., & Nicoli, S. (2022). Cyclosporine-loaded micelles for ocular delivery: Investigating the penetration mechanisms. *Journal of Controlled Release*, 349, 744–755.

Kaur, J., Mishra, V., Singh, S. K., Gulati, M., Kapoor, B., Chellappan, D. K., Gupta, G., Dureja, H., Anand, K., Dua, K., Khatik, G. L., & Gowthamarajan, K. (2021). Harnessing amphiphilic polymeric micelles for diagnostic and therapeutic applications: Breakthroughs and bottlenecks. *Journal of Controlled Release*, 334, 64–95.

Leng, Q., Li, Y., Pang, X., Wang, B., Wu, Z., Lu, Y., Xiong, K., Zhao, L., Zhou, P., & Fu, S. (2020). Curcumin nanoparticles incorporated in PVA/collagen composite films promote wound healing. *Drug Delivery*, 27(1), 1676–1685.

Li, S., Chen, L., & Fu, Y. (2023). Nanotechnology-based ocular drug delivery systems: Recent advances and future prospects. *Journal of Nanobiotechnology*, 21, 232.

Li, S., Lu, Z., Huang, Y., Wang, Y., Jin, Q., Shentu, X., Ye, J., Ji, J., Yao, K., & Han, H. (2022). Anti-oxidative and anti-inflammatory micelles: Break the dry eye vicious cycle. *Advanced Science*, 9(28), 2200435.

Long, L., Ge, Z., Zhang, F., Dong, R., Yang, L., Chen, Z., Tang, S., & Wang, Y. (2023). Development of injectable hyaluronic acid-based hydrogels with antioxidant activity for the treatment of corneal neovascularization. *Chemical Engineering Journal*, 478, 147147.

Ludwig, M., Geisler, R., Prévost, S., & von Klitzing, R. (2021). Shape and structure formation of mixed nonionic–anionic surfactant micelles. *Molecules*, 26(14), 4136.

Mandal, A., Bisht, R., Rupenthal, I. D., & Mitra, A. K. (2017). Polymeric micelles for ocular drug delivery: From structural frameworks to recent preclinical studies. *Journal of Controlled Release*, 248, 96–116.

Mofidfar, M., Abdi, B., Ahadian, S., Mostafavi, E., Desai, T. A., Abbasi, F., Sun, Y., Manche, E. E., Ta, C. N., & Flowers, C. W. (2021). Drug delivery to the anterior segment of the eye: A review of current and future treatment strategies. *International Journal of Pharmaceutics*, 607, 120924.

Perumal, S., Atchudan, R., & Lee, W. (2022). A review of polymeric micelles and their applications. *Polymers*, 14(12), 2510.

Soliman, N., Mamdouh, D., & Elkordi, A. (2023). Choroidal melanoma: A mini review. *Medicines*, 10(1), 11.

Sun, F., Zheng, Z., Lan, J., Li, X., Li, M., Song, K., & Wu, X. (2019). New micelle myricetin formulation for ocular delivery: Improved stability, solubility, and ocular anti-inflammatory treatment. *Drug Delivery*, 26(1), 575–585.

Suri, R., Nag, T. C., Mehra, N., Neupane, Y. R., Shafi, S., Sharma, D., Sharma, K., Sultana, Y., & Kohli, K. (2023). Sirolimus loaded chitosan functionalized PLGA nanoparticles protect against sodium iodate-induced retinal degeneration. *Journal of Drug Delivery Science and Technology*, 82, 104369.

Tan, T.-E., & Wong, T. Y. (2023). Diabetic retinopathy: Looking forward to 2030. *Frontiers in Endocrinology*, 13, Article 1096073.

Thabet, Y., Elsabahy, M., & Eissa, N. G. (2022). Methods for preparation of niosomes: A focus on thin-film hydration method. *Methods*, 199, 9–15.

Thakur, S., Pal, R., Singh, A., Pandey, P., & Chanana, A. (2023). Recent approaches and various barriers involving in ocular drug delivery system. *World Journal of Pharmacy and Pharmaceutical Sciences*, 12(2), 472.

Ting, D. S. J., Foo, V. H., Yang, L. W. Y., Sia, J. T., Ang, M., Lin, H., Chodosh, J., Mehta, J. S., & Ting, D. S. W. (2021). Artificial intelligence for anterior segment diseases: Emerging applications in ophthalmology. *British Journal of Ophthalmology*, 105(2), 158–168.

Wang, R., Gao, Y., Liu, A., & Zhai, G. (2021). A review of nanocarrier-mediated drug delivery systems for posterior segment eye disease: Challenges analysis and recent advances. *Journal of Drug Targeting*, 29(7), 687–702.

Willoughby, C. E., Ponzin, D., Ferrari, S., Lobo, A., Landau, K., & Omidi, Y. (2010). Anatomy and physiology of the human eye: Effects of mucopolysaccharidoses disease on structure and function – A review. *Clinical & Experimental Ophthalmology*, 38(1), 2–11.

Wu, K. Y., Ashkar, S., Jain, S., Marchand, M., & Tran, S. D. (2023). Breaking barriers in eye treatment: Polymeric nano-based drug-delivery system for anterior segment diseases and glaucoma. *Polymers*, 15(6), 1373.

Zueva, O. S., Kazantseva, M. A., & Zuev, Y. F. (2025). Nanosized being of ionic surfactant micelles: An advanced view on micellization process. *Colloids and Interfaces*, 9(1), 15.

Chapter 10

Nanosuspensions in Ocular Drug Delivery

Swati[1,*]
Mansi[1]
Samantha[1]
Shivanki Joshi[2]
and Ram Babu Sharma[1]
[1]Himalayan Institute of Pharmacy, Kalaamb, Kala-Amb, Sirmour, Himachal Pradesh, India
[2]Guru Gobind Singh College of Pharmacy, Yamuna Nagar, Haryana, India

Abstract

Ocular drug delivery poses considerable challenges due to the eye's complex anatomical and physiological barriers, including the corneal epithelium, blood-retina barrier, and rapid nasolacrimal drainage. Traditional delivery systems, such as eye drops, often suffer from low bioavailability and short residence time, leading to suboptimal therapeutic outcomes. Nanosuspensions, colloidal dispersions of poorly water-soluble drugs in nanometer-sized particles, have emerged as a promising strategy to address these limitations. By reducing particle size, nanosuspensions enhance drug solubility, bioavailability, and ocular penetration, while also increasing retention time on the ocular surface. These systems enable targeted delivery to ocular tissues such as the cornea, retina, and vitreous humor, thereby improving therapeutic efficacy. Furthermore, surface modification of nanoparticles can enable controlled drug release and reduce systemic side effects. This chapter explores recent advancements, persistent challenges, and future

* Corresponding Author's Email: Swatipahwa129@gmail.com

In: Advances in Nanotechnology for Ocular Drug Delivery
Editors: Anjali Sharma, Ajmer Singh Grewal, Geeta Deswal et al.
ISBN: 979-8-89530-814-1
© 2025 Nova Science Publishers, Inc.

perspectives in nanosuspension-based ocular drug delivery, highlighting their potential to revolutionize the treatment of ocular diseases.

Keywords: nanosuspension, bioavailability, ocular penetration, ocular drug delivery, cornea, blood-retina barrier, nanotechnology

Abbreviations

AMD: Age-Related Macular Degeneration
API: Active Pharmaceutical Ingredient
CMC: Critical Micelle Concentration
DSC: Differential Scanning Calorimetry
FDA: Food and Drug Administration
GIT: Gastrointestinal Tract
HPH: High-Pressure Homogenization
HPLC: High Performance Liquid Chromatography
IOP: Intraocular Pressure
mPEG: Methoxy Polyethylene Glycol
NPs: Nanoparticles
PCL: Polycaprolactone
PEG: Polyethylene Glycol
PHEMA: Poly(hydroxyethyl methacrylate)
PLGA: Poly(lactic-co-glycolic acid)
ROS: Reactive Oxygen Species
RPE: Retinal Pigment Epithelium
SEM: Scanning Electron Microscopy
TPGS: D-alpha-Tocopheryl Polyethylene Glycol 1000 Succinate
VEGF: Vascular Endothelial Growth Factor
XRD: X-ray Diffraction

Introduction

Effective drug delivery to the eye remains a major challenge for pharmacologists and formulation scientists due to the eye's intricate anatomical and physiological barriers. Among available approaches, topical eye drops are most commonly used for treating anterior segment disorders owing to their ease of application and high patient compliance (Patel et al.,

2013; Gaudana et al., 2010). However, ocular drug delivery is hindered by several static barriers, such as the sclera, choroid, and retinal pigment epithelium (RPE), as well as dynamic factors including conjunctival and choroidal blood flow and lymphatic drainage from the episcleral and conjunctival tissues. To address limitations such as poor corneal permeability and short drug residence time, alternative delivery routes, such as intravitreal injections, periocular administration, and systemic delivery, have been explored. Additionally, various strategies have been developed to enhance precorneal retention and drug permeability, including the use of prodrugs, iontophoresis, cyclodextrins, and ion-pairing techniques. To further overcome ocular barriers, advanced drug delivery systems have been introduced. These include emulsions, suspensions, ointments, in situ forming gels, nano-micelles, nanoparticles, liposomes, dendrimers, nanosuspensions, drug-eluting contact lenses, and microneedle-based systems. Each of these technologies offers specific advantages in improving drug bioavailability and therapeutic efficacy in ophthalmic applications. In particular, nanoparticle-based systems have shown great promise due to their ability to reduce irritation, enhance bioavailability, and ensure compatibility with delicate ocular tissues, making them highly effective for the treatment of various eye diseases (Fathi-Karkan et al., 2024).

Nanosuspensions

Nanosuspensions are colloidal dispersions of drug nanoparticles, typically ranging from 100 to 500 nm in size, suspended in an aqueous medium. The nanoscale dimension of these particles markedly increases the surface area, thereby enhancing the dissolution rate and solubility of the drug, which in turn improves its bioavailability (Goel et al., 2019). This reduction in particle size is fundamental to the therapeutic efficacy of nanosuspensions. By significantly increasing the surface area, nanosuspensions promote faster drug dissolution and greater saturation solubility. However, the efficacy and safety of ocular nanosuspensions depend on several critical parameters, including particle size, surface charge (zeta potential), and overall colloidal stability (Pınar et al., 2023; El-Emam et al., 2020; Mu & Holm, 2018). Therefore, the selection of an appropriate preparation method is essential to achieving a stable, effective, and biocompatible ocular formulation.

Methods of Preparation of Ocular Nanosuspensions

Top-Down Approaches: Size Reduction Techniques

Top-down methods for preparing ocular nanosuspensions involve reducing the size of preformed drug particles to the nanoscale using mechanical or physical forces. These techniques are designed to break down larger particles into smaller ones, enhancing dissolution and bioavailability. The selection of an appropriate top-down method depends on various factors, including the physicochemical properties of the drug, the target particle size distribution, and the scalability and feasibility of the process (Pınar et al., 2023).

High-Pressure Homogenization

HPH is a widely used top-down technique for the preparation of nanosuspensions. In this process, a drug suspension is forced through a narrow orifice under extremely high pressure, typically ranging from 50 to 200 MPa (Pınar et al., 2023). The resulting high-pressure forces generate intense shear stress and cavitation, which fragment the drug particles into nanoscale dimensions (Gigliobianco et al., 2018). Several factors significantly influence the final particle size and distribution, including the number of passes through the homogenizer, the applied pressure, and the suspension temperature. Generally, increasing the number of passes leads to smaller particle sizes; however, excessive passes may result in drug degradation due to prolonged exposure to mechanical stress. Similarly, higher pressures tend to reduce particle size more effectively but may also cause degradation, particularly in sensitive drugs. Maintaining an appropriate temperature during the process is essential to prevent thermal degradation and preserve the stability of the suspension (Gigliobianco et al., 2018). HPH is known for its scalability and high efficiency, making it well-suited for large-scale nanosuspension production. However, due to the intense shear forces involved, careful optimization of process parameters is crucial to minimize drug degradation while achieving the desired particle size.

Wet Milling and Media Milling

Wet milling and media milling are mechanical size reduction techniques that employ milling media, typically small beads or particles, to grind drug particles within a liquid medium (Pınar et al., 2023; Gigliobianco et al., 2018). In wet milling, the milling media and drug suspension are agitated in a milling chamber, where the impact and shear forces generated by the moving media break down the drug particles. Media milling follows a similar principle but is often characterized by a more controlled and efficient process. Several parameters influence the final particle size and distribution, including the size and type of milling media, milling speed, and processing time. Smaller milling media tend to produce finer particles; however, prolonged milling must be carefully managed to prevent excessive particle degradation. Likewise, higher milling speeds can reduce particle size more effectively but may lead to increased equipment wear and energy consumption. Optimizing processing time is essential to achieve the desired particle size while minimizing drug degradation (Gigliobianco et al., 2018). Wet and media milling are generally applicable to a wide range of drug substances but may be less efficient than high-pressure homogenization (HPH) for certain compounds. Therefore, the careful selection of milling media and process parameters is critical to obtaining the desired nanosuspension characteristics while ensuring formulation stability and drug integrity.

Other Top-Down Methods

Several other top-down methods can be used to prepare ocular nanosuspensions, although they are less commonly employed than HPH, wet milling, and media milling. These include ultrasonication and microfluidization. Ultrasonication involves the use of high-frequency sound waves to generate cavitation and shear forces, which facilitate particle size reduction (Pınar et al., 2023). This technique is relatively simple and cost-effective but may offer lower efficiency compared to other methods and may not be suitable for all drug substances. Microfluidization, on the other hand, utilizes high-velocity fluid streams to break down drug particles through impact and shear forces. It provides excellent control over particle size and distribution but requires specialized equipment and is generally more expensive than other top-down approaches. The selection of a particular method depends on the physicochemical characteristics of the drug and the desired attributes of the final nanosuspension formulation.

Bottom-Up Approaches: Nanoparticle Assembly

Bottom-up methods for preparing ocular nanosuspensions involve the controlled assembly of drug molecules into nanoparticles from a solution. These techniques typically begin by dissolving the drug in a suitable solvent, followed by inducing precipitation or self-assembly to form nanoparticles. The selection of a specific bottom-up method depends on the physicochemical properties of the drug, the desired particle size and morphology, and any requirements for surface modifications (Pınar et al., 2023).

Anti-Solvent Precipitation

Anti-solvent precipitation is a simple and cost-effective bottom-up method for preparing nanosuspensions. In this technique, the drug is first dissolved in a suitable solvent, and then an anti-solvent, typically a liquid in which the drug has low solubility, is added to induce precipitation (Pınar et al., 2023; Gigliobianco et al., 2018). Rapid mixing of the solvent and anti-solvent promotes nucleation and the formation of drug nanoparticles. The choice of solvent and anti-solvent is critical for controlling particle size, morphology, and crystallinity. Factors such as mixing rate and temperature also play a significant role in determining the final characteristics of the nanoparticles. Faster mixing generally results in smaller particles, while elevated temperatures can lead to increased particle growth and aggregation. To prevent aggregation and enhance colloidal stability, stabilizers or surfactants are often added to the formulation (Wu et al., 2019). Although anti-solvent precipitation is a versatile method applicable to a wide range of drugs, precise control over particle size and distribution can be challenging, and uncontrolled aggregation remains a common issue.

Nanoprecipitation

Nanoprecipitation is a bottom-up technique that offers better control over nanoparticle nucleation and growth compared to anti-solvent precipitation (Pınar et al., 2023; Gigliobianco et al., 2018). Typically, the drug and a polymer, often a stabilizing agent, are dissolved together in a common solvent, which is then added dropwise into a non-solvent. The rapid change in solvent environment causes the drug to precipitate, while the polymer stabilizes the nanoparticles and prevents aggregation. Factors such as the type and concentration of polymer, choice of solvent and non-solvent, and mixing conditions significantly affect particle size, morphology, and surface characteristics. Nanoprecipitation can yield highly monodisperse nanoparticles with well-controlled properties; however, it requires specialized equipment and careful optimization of process parameters. This method is especially suitable for preparing nanoparticles of poorly water-soluble drugs.

Other Bottom-Up Methods

Several other bottom-up methods are available for preparing nanosuspensions, including emulsion-solvent evaporation and supercritical fluid technology. Emulsion-solvent evaporation involves dissolving the drug in an organic solvent, emulsifying this solution in an aqueous phase, and then evaporating the organic solvent to leave behind drug nanoparticles. This technique is suitable for preparing nanoparticles of both hydrophilic and hydrophobic drugs, although completely removing the organic solvent can be challenging. Supercritical fluid technology utilizes supercritical fluids, such as supercritical carbon dioxide, to dissolve and precipitate the drug, thereby forming nanoparticles (Pınar et al., 2023). This method provides excellent control over particle size and morphology but requires specialized equipment and is relatively costly. The choice of a bottom-up method depends on the specific formulation requirements, including the drug's properties, desired particle size and morphology, and any necessary surface modifications.

Hybrid Approaches: Combining Top-Down and Bottom-Up Strategies

The optimization of nanosuspension properties often benefits from combining top-down and bottom-up methods. Hybrid approaches capitalize on the advantages of both techniques to achieve superior control over particle size, morphology, and surface characteristics. For instance, a top-down method such as HPH can be used initially to reduce particle size, followed by a bottom-up technique like nanoprecipitation to further refine particle size and encapsulate the drug within a polymeric matrix for enhanced stability (Pınar et al., 2023). This combination can produce nanosuspensions with improved stability, higher drug loading, and controlled release properties. The choice of a hybrid approach depends on the specific formulation requirements and the characteristics of the drug substance. Careful optimization of process parameters is crucial to achieving the desired results. The synergistic effect of combining different preparation methods can yield nanosuspensions with enhanced performance compared to those prepared by a single method.

Characterization of Ocular Nanosuspensions

Rigorous characterization and quality control are essential in the development of ocular nanosuspensions to ensure the final product meets required safety and efficacy standards (Pınar et al., 2023; El-Emam et al., 2020; Mu & Holm, 2018). Various techniques are employed to assess the physical and chemical properties of nanosuspensions, including:

Particle Size Analysis

Methods such as dynamic light scattering (DLS) and laser diffraction are used to determine the nanoparticle size distribution. Particle size is a critical factor influencing drug dissolution rate, bioavailability, and potential toxicity.

Zeta Potential Measurement

Electrophoretic light scattering measures the surface charge of nanoparticles. A high absolute zeta potential value indicates good colloidal stability and reduces the risk of particle aggregation.

Morphological Analysis

Transmission electron microscopy (TEM) and scanning electron microscopy (SEM) provide visual insights into particle size, shape, and morphology. These techniques confirm the success of the preparation method and help identify any structural defects in the nanoparticles.

Drug Content Determination

High-performance liquid chromatography (HPLC) is commonly used to quantify the drug concentration in the nanosuspension, ensuring the desired drug loading is achieved.

Crystallinity Analysis

Techniques such as X-ray diffraction (XRD) and differential scanning calorimetry (DSC) assess the crystallinity of the drug within the nanosuspension. Crystallinity significantly affects the drug's dissolution rate and bioavailability.

Stability Studies

Accelerated stability testing evaluates the long-term stability of the nanosuspension under various storage conditions, ensuring the formulation maintains its quality and efficacy throughout its shelf life.

Advantages and Disadvantages of Nanosuspension

The application of nanosuspensions in ocular drug delivery offers numerous advantages, though it is important to consider potential limitations. The primary benefit arises from the nanoscale size of the drug particles, which significantly increases the surface area. This enhancement leads to improved dissolution rate and solubility, thereby boosting bioavailability and therapeutic efficacy, especially for poorly water-soluble drugs (Arora et al., 2018). Additionally, nanosuspension formulations can prolong the drug's residence time in the eye, reducing dosing frequency and improving patient compliance

(Liu et al., 2020; Qamar et al., 2019). This advantage is particularly valuable for patients requiring frequent administration of conventional eye drops. Moreover, the potential for targeted drug delivery may enhance therapeutic outcomes while minimizing systemic side effects (Akhter et al., 2022; Souto et al., 2019). However, several challenges must be acknowledged. A major concern is the tendency of nanosuspensions to aggregate and become unstable, which can decrease efficacy and cause adverse effects (Gigliobianco et al., 2018; Pınar et al., 2023). Careful selection of stabilizers and formulation optimization are essential to address these issues. Furthermore, the manufacturing process for nanosuspensions can be complex and costly, potentially limiting accessibility (Arora et al., 2018). Finally, limited clinical data for certain ocular applications restricts the widespread adoption of this technology (Aldeeb et al., 2024).

Applications of Nanosuspensions for Ocular Drug Delivery

Improvement in Bioavailability of Drug

Bioavailability refers to the percentage of a drug that reaches systemic circulation. Poor aqueous solubility and an incomplete dissolution profile are major factors contributing to the low bioavailability of lipophilic compounds. Nanosuspensions offer several advantages over traditional drug delivery forms, one of which is the enhancement of aqueous solubility and bioavailability (Yadollahi et al., 2015). Various preparation methods, such as high-pressure homogenization, media milling, precipitation, and melt emulsification, reduce drug particle size to the nanoscale. This reduction increases the surface area, thereby improving drug absorption from the gastrointestinal tract (Yadollahi et al., 2015). For example, the bioavailability of itraconazole, an antifungal drug, has been significantly improved through nanosuspension formulation.

Extended Release of Drugs

Various intraocular implants have been designed to provide localized, controlled drug release over extended periods. These devices deliver drugs directly to posterior ocular tissues and are implanted intravitreally through a minor surgical incision. Although implantation is an invasive procedure, it offers several advantages, including reduced systemic side effects and the ability to bypass the blood-retina barrier. Numerous implantable devices have been developed for ocular drug delivery, particularly for treating chronic vitreoretinal diseases. For example, the controlled-release intraocular implant of ganciclovir (Vitrasert®) is FDA-approved for managing acquired immune deficiency syndrome-associated cytomegalovirus retinitis. This implant provides long-term sustained drug release with minimal systemic toxicity and at a reduced cost (Patel et al., 2013).

Enhanced Stability

Chemically labile drugs can be protected by formulating them as nanosuspensions. In these formulations, stabilizer molecules surround the drug nanoparticles, shielding them from exposure to water and light. As a result, only the outer surface of the stabilizer is exposed and susceptible to degradation, thereby effectively minimizing the degradation of the drug itself. This makes nanosuspension an efficient technique to enhance the chemical stability of sensitive drugs. For example, the chemical stability of quercetin is significantly improved when prepared as a nanosuspension, which also reduces its toxicity (Wang et al., 2013; Wang et al., 2023).

Solubility and Dissolution Rate

Nanosuspensions provide a significant advantage over other formulations by simultaneously improving both the dissolution rate and saturation solubility of poorly soluble drugs. Different physiological media are required to accurately assess these two critical parameters. Evaluating saturation solubility and dissolution rate is essential for predicting the in vitro performance of the formulation. The reduction in particle size increases the dissolution rate, which in turn elevates the dissolution pressure, enhancing drug availability (Flach et al., 2018).

Improved Patient Compliance

Intravenous injections enable direct drug delivery to the posterior segment of the eye; however, frequent injections increase the risk of complications such as hemorrhage, endophthalmitis, retinal detachment, and patient discomfort (Jager et al., 2004). Controlled and sustained-release nanosuspensions can reduce the frequency of administration, which is especially important for managing chronic ocular diseases.

Protein Delivery

Nanoparticles have been widely used as carriers for delivering drugs, including anticancer agents and therapeutic proteins. Natural biomolecules, such as proteins, are attractive alternatives to synthetic polymers commonly used in nanoparticle formulations due to their inherent safety (Zhu et al., 2024). These protein-based nanoparticles offer significant advantages, including protection of proteins from degradation, extension of the half-life of protein drugs, responsive drug release, targeted delivery, and overall improvement in therapeutic efficacy.

Treatment of Ocular Infections

Glaucoma

Glaucoma is a progressive eye disease that can lead to significant vision loss and even blindness. Emerging drug delivery systems aim to enhance the bioavailability of therapeutic agents and improve patient compliance compared to traditional eye drops, while minimizing adverse effects. The primary goal in glaucoma management is to effectively reduce intraocular pressure (IOP) and maintain it over prolonged periods, ultimately preserving the patient's vision (Lavik et al., 2011).

Corneal Diseases

Nanosuspensions formulated with Compritol and stearic acid have shown significant potential as ocular drug delivery systems, owing to their small particle size, enhanced stability, and the use of physiologically compatible excipients. Additionally, PLGA nanospheres loaded with flurbiprofen have demonstrated increased drug accumulation within ocular tissues. Researchers have also developed nanoparticles containing cyclosporine A or labeled with fluorescein, employing Miglyol 840 and Precirol ATO 5 as liquid lipid components (Hashida et al., 2008).

Corneal Neovascularization

Corneal neovascularization is a vision-threatening condition often associated with abnormalities of the ocular surface. In experimental mouse studies, plasmids encoding Flt23k, an intracellular inhibitor of vascular endothelial growth factor (VEGF), were delivered using nanoparticles sized approximately 270.2 nm. Administered via subconjunctival injection, these nanoparticles effectively reduced VEGF activity, resulting in a significant decrease in leukocyte infiltration and macrophage migration. When combined with the steroid triamcinolone, the treatment substantially inhibited corneal neovascularization and lymphangiogenesis. Moreover, the survival rate of corneal grafts markedly improved with steroid use, increasing from about 20% without steroids to 91.6% over two months with steroids (Cho et al., 2012).

Age-Related Macular Degeneration (AMD)

siRNA-loaded PEGylated liposome-protamine-hyaluronic acid nanoparticles represent a promising therapeutic strategy for AMD. Additionally, the incorporation of cerium oxide nanoparticles in intravitreal injections has been shown to inhibit reactive oxygen species (ROS) and prevent the formation of neovascular lesions in both intraretinal and subretinal layers. This approach holds significant potential for treating neurodegenerative ocular conditions such as diabetic retinopathy and AMD (Zhou et al., 2013).

Different Formulations of Nanosuspensions

Polymer-Based Nanocarriers and Lipid-Based Nanocarriers

Polymeric nanoparticles, ranging from 1 to 1000 nm in size, are designed to carry active pharmaceutical compounds either entrapped within or adsorbed onto a polymeric core (Zielińska et al., 2020). Lipid-based nanocarriers, characterized by their non-spherical solid lipid cores and excellent biocompatibility, offer significant advantages in drug delivery, including enhanced penetration, increased drug retention and solubility, reduced toxicity, prolonged release, and targeted delivery (Han et al., 2023). Cyclosporine A, an anti-inflammatory and immunomodulatory agent, inhibits VEGF-induced endothelial cell migration and angiogenesis. To improve its solubility and bioavailability, micelles based on copolymers of methoxy polyethylene glycol (mPEG) and hexyl-substituted poly(lactides) (MPEG-hexPLA) have been developed.

Surfactant-Stabilized Nanosuspensions

Nanosuspensions are colloidal dispersions consisting of nano-sized drug crystals stabilized by suitable stabilizers such as povidone, polysorbate 80, synthetic and natural lipids, cellulosic derivatives, Soluplus®, poloxamer 188, poloxamer 407, sodium lauryl sulfate, polyvinyl alcohol, vitamin E TPGS, and polyethylene glycols. These stabilizers maintain the stability of high-energy nanocrystals produced by techniques like milling or high-pressure homogenization. The choice of stabilizer depends on the route of administration, and sometimes a combination of two or three stabilizers is used. These agents prevent particle agglomeration, thereby enhancing the physicochemical stability and in vivo performance of the nanosuspension (Truong et al., 2024).

Hydrogel-Based Nanosuspensions

Hydrogels are hydrophilic, cross-linked polymer networks capable of absorbing substantial amounts of water. When exposed to aqueous environments, they swell while maintaining their structural integrity. Both synthetic and natural polymers are commonly used to form hydrogels, including PEG, poly(N-isopropyl acrylamide), poly(hydroxyethyl methacrylate) (PHEMA), agarose, alginate, silk fibroin, chitosan, and hyaluronic acid. Their viscous nature contributes to prolonged ocular residence time, thereby enhancing the therapeutic efficacy of drug delivery systems. For example, hydroxypropyl methylcellulose (HPMC) hydrogels have been utilized as carriers for curcumin, a natural polyphenol recognized for its antioxidant, antitumor, antimicrobial, and anti-inflammatory activities (Sharma et al., 2016).

Nanocrystals

Drug nanocrystals are pure drug particles with sizes ranging from 10 to 1000 nm, stabilized by surfactants or polymeric stabilizers. The reduction in particle size markedly improves the solubility and dissolution rate of poorly water-soluble drugs (Hong et al., 2020).

Conclusion

Nanosuspensions represent a promising and innovative strategy for ocular drug delivery, offering significant advantages over conventional formulations. They improve the bioavailability, solubility, and stability of poorly water-soluble drugs, an ongoing challenge in ocular pharmacotherapy. Owing to their nanoscale particle size, nanosuspensions enhance penetration across corneal barriers, thereby increasing therapeutic efficacy and prolonging drug retention within ocular tissues. Moreover, they can be easily incorporated into various delivery systems such as eye drops, gels, and ocular inserts, providing versatility in administration. Their ability to sustain drug release while reducing systemic side effects makes them particularly suitable for managing chronic ocular conditions, including glaucoma, age-related macular degeneration, and dry eye disease. Despite these benefits, challenges such as production scalability, long-term stability, and formulation development to minimize ocular irritation and improve patient compliance remain critical areas for further research and optimization.

References

Akhter, M. H., Ahmad, I., Alshahrani, M. Y., Al-Harbi, A. I., Khalilullah, H., Afzal, O., Altamimi, A. S. A., Ullah, S. N. M. N., Ojha, A., & Karim, S. (2022). Drug delivery challenges and current progress in nanocarrier-based ocular therapeutic system. *Gels, 8*(2), 82.

Aldeeb, M., Wilar, G., Suhandi, C., Elamin, K., & Wathoni, N. (2024). Nanosuspension-based drug delivery systems for topical applications. *International Journal of Nanomedicine*, 19, 825–844.

Arora, D., Khurana, B., Rath, G., Nanda, S., & Goyal, A. (2018). Recent advances in nanosuspension technology for drug delivery. *Current Pharmaceutical Design,* 24(21), 2403–2415.

Cho, Y. K., Uehara, H., Young, J. R., Tyagi, P., Kompella, U. B., Zhang, X., et al. (2012). Flt23k nanoparticles offer additive benefit in graft survival and anti-angiogenic effects when combined with triamcinolone. *Investigative Ophthalmology & Visual Science, 53*(4), 2328–2336.

El-Emam, G., Girgis, G., El-Sokkary, M. A., Soliman, O. A. A., & Abd El Gawad, A. A. E. (2020). Ocular inserts of voriconazole-loaded proniosomal gels: Formulation, evaluation and microbiological studies. *International Journal of Nanomedicine*, 15, 7825-7840.

Evstropiev, S. K., Vasilyev, V. N., Nikonorov, N. V., Kolobkova, E. V., Volkova, N. A., & Boltenkov, I. A. (2018). Photoactive ZnO nanosuspension for intensification of organics contaminations decomposition. *Chemical Engineering and Processing - Process Intensification, 134*, 45–50.

Fathi-Karkan, S., Ramsheh, N. A., Arkaban, H., Narooie-Noori, F., Sargazi, S., Mirinejad, S., Roostaee, M., Sargazi, S., Barani, M., Shadman, S. M., Althomali, R. H., & Rahman, M. M. (2024). Nanosuspensions in ophthalmology: Overcoming challenges and enhancing drug delivery for eye diseases. *International Journal of Pharmaceutics, 658*, 124226.

Flach, F., Breitung-Faes, S., & Kwade, A. (2018). Model based process optimization of nanosuspension preparation via wet stirred media milling. *Powder Technology, 331*, 146–154.

Gaudana, R., Ananthula, H. K., Parenky, A., & Mitra, A. K. (2010). Ocular drug delivery. *AAPS Journal, 12*, 348–360.

Gigliobianco, M. R., Casadidio, C., Censi, R., & Di Martino, P. (2018). Nanocrystals of poorly soluble drugs: Drug bioavailability and physicochemical stability. *Pharmaceutics, 10*(3), 134.

Goel, S., Sachdeva, M., & Agarwal, V. (2019). Nanosuspension technology: Recent patents on drug delivery and their characterizations. *Recent Patents on Drug Delivery & Formulation,* 13(2), 91–104.

Han, H., Li, S., Xu, M., Zhong, Y., Fan, W., Xu, J., Zhou, T., Ji, J., Ye, J., & Yao, K. (2023). Polymer- and lipid-based nanocarriers for ocular drug delivery: Current status and future perspectives. *Advanced Drug Delivery Reviews, 196*, 114770.

Hashida, N., Ohguro, N., Yamazaki, N., Arakawa, Y., Oiki, E., Mashimo, H., et al. (2008). High-efficacy site-directed drug delivery system using sialyl-Lewis X conjugated liposome. *Experimental Eye Research, 86*(1), 138–149.

Hong, S., Choi, D. W., Kim, H. N., Park, C. G., Lee, W., & Park, H. H. (2020). Protein-based nanoparticles as drug delivery systems. *Pharmaceutics, 12*(7), 604.

Jager, R. D., Aiello, L. P., Patel, S. C., & Cunningham, E. T., Jr. (2004). Risks of intravitreous injection: A comprehensive review. *Retina, 24*(5), 676–698.

Lavik, E., Kuehn, M. H., & Kwon, Y. H. (2011). Novel drug delivery systems for glaucoma. *Eye, 25*(5), 578–586.

Liu, Y.-C., Lin, M. T.-Y., Ng, A., Wong, T., & Mehta, J. (2020). Nanotechnology for the treatment of allergic conjunctival diseases. *Pharmaceuticals, 13*(11), 351.

Mu, H., & Holm, R. (2018). Solid lipid nanocarriers in drug delivery: Characterization and design. *Expert Opinion on Drug Delivery,* 15(8), 771–785.

Patel, A., Cholkar, K., Agrahari, V., & Mitra, A. K. (2013). Ocular drug delivery systems: An overview. *World Journal of Pharmacology, 2*(2), 47–64.

Pınar, S. G., Oktay, A., Karakuş, A., & Çelebi, N. (2023). Formulation strategies of nanosuspensions for various administration routes. *Pharmaceutics, 15*(5), 1520.

Qamar, Z., Qizilbash, F. F., Iqubal, M. K., Ali, A., Narang, J., Ali, J., & Baboota, S. (2019). Nano-based drug delivery system: Recent strategies for the treatment of ocular disease and future perspective. *Recent Patents on Drug Delivery & Formulation,* 13(4), 246–254.

Sharma, O. P., Patel, V., & Mehta, T. (2016). Nanocrystal for ocular drug delivery: Hope or hype. *Drug Delivery and Translational Research,* 6, 399–413.

Souto, E. B., Dias-Ferreira, J., López-Machado, A., Ettcheto, M., Cano, A., Camins, A., Espina, M., García, M. L., & Sánchez-López, E. (2019). Advanced formulation approaches for ocular drug delivery: State-of-the-art and recent patents. *Pharmaceutics, 11*(9), 460.

Truong, D., Wu, K. Y., Nguyen, L., & Tran, S. D. (2024). Advancements in hydrogel technology for ocular drug delivery. *Explorations in Biomaterials-X, 1*, 331–352.

Wang, J., Xue, X., & Miao, X. (2023). Antioxidant effects of quercetin nanocrystals in nanosuspension against hydrogen peroxide-induced oxidative stress in a zebrafish model. *Pharmaceuticals, 16*(9), 1209.

Wang, Y., Zheng, Y., Zhang, L., Wang, Q., & Zhang, D. (2013). Stability of nanosuspensions in drug delivery. *Journal of Controlled Release, 172*(3), 1126–1141.

Wu, M., Feng, Z., Deng, Y., Zhong, C., Liu, Y., Liu, J., Zhao, X., & Fu, Y. (2019). Liquid antisolvent precipitation: An effective method for ocular targeting of lutein esters. *International Journal of Nanomedicine,* 14, 2667–2681.

Yadollahi, R., Vasilev, K., & Simovic, S. (2015). Nanosuspension technologies for delivery of poorly soluble drugs. *BioMed Research International, 2015*, 216375.

Zhou, H. Y., Hao, J. L., Wang, S., Zheng, Y., & Zhang, W. S. (2013). Nanoparticles in the ocular drug delivery. *International Journal of Ophthalmology, 6*(3), 390–396.

Zhu, C., Mu, J., & Liang, L. (2024). Nanocarriers for intracellular delivery of proteins in biomedical applications: Strategies and recent advances. *Journal of Nanobiotechnology, 22*, 688.

Zielińska, A., Carreiró, F., Oliveira, A. M., Neves, A., Pires, B., Venkatesh, D. N., Durazzo, A., Lucarini, M., Eder, P., Silva, A. M., Santini, A., & Souto, E. B. (2020). Polymeric nanoparticles: Production, characterization, toxicology and ecotoxicology. *Molecules, 25*(16), 3731.

Chapter 11

Nanoemulsions for Ocular Drug Delivery Systems

Parul Gupta[1,*]
Muskan[1]
Brijesh Duvey[2]
and Prerna Sharma[3]

[1]Guru Gobind Singh College of Pharmacy, Yamunanagar, Haryana, India
[2]Ch. Devi Lal College of Pharmacy, Jagadhri, Haryana, India
[3]Shri Krishna Ayush University, Kurukshetra, Haryana, India

Abstract

Nanoemulsions have emerged as a promising approach for ocular drug delivery, effectively addressing challenges posed by the eye's complex anatomy and protective barriers. These systems are formed by the interaction of surfactants, co-surfactants, oil, and water, producing droplet sizes typically ranging from 20 to 200 nm. Nanoemulsions enhance the stability, bioavailability, and solubility of both hydrophilic and hydrophobic drugs. Their prolonged retention time, improved corneal penetration, and controlled drug release make them particularly effective for treating ocular disorders such as glaucoma, dry eye syndrome, and uveitis. Generally biocompatible and non-irritating, nanoemulsions minimize systemic side effects and maintain patient comfort. Preparation techniques like ultrasonication and high-pressure homogenization ensure uniformity and scalability of formulations. Incorporating mucoadhesive agents and stimuli-responsive polymers further boosts efficacy by increasing ocular residence time and enabling

* Corresponding Author's Email: parultayal2000@gmail.com

In: Advances in Nanotechnology for Ocular Drug Delivery
Editors: Anjali Sharma, Ajmer Singh Grewal, Geeta Deswal et al.
ISBN: 979-8-89530-814-1
© 2025 Nova Science Publishers, Inc.

targeted drug delivery. Additionally, nanoemulsions facilitate reduced dosing frequency, thereby enhancing patient compliance. Nevertheless, challenges including long-term stability, cost-effective manufacturing, and regulatory approval remain significant obstacles to clinical translation. Future research should focus on overcoming these limitations and developing innovative formulations to fully harness the therapeutic potential of nanoemulsions.

Keywords: nanoemulsions, ocular drug delivery system, bioavailability, controlled release, corneal penetration, hydrophobic drugs, ophthalmic therapies

Abbreviations

API: Active Pharmaceutical Ingredient
BCS: Biopharmaceutics Classification System
CDR: Cumulative Drug Release
DED: Dry Eye Disease
DOPE: 1,2-Dioleoyl-sn-glycero-3-phosphoethanolamine
DOTAP: 1,2-Dioleoyl-3-trimethylammonium-propane
EMA: European Medicines Agency
FDA: Food and Drug Administration
NE: Nanoemulsion
NEs: Nanoemulsions
O/W: Oil in Water
PDI: Polydispersity Index
PDL: Poly(decalactone)
PEG: Polyethylene Glycol
PG: Propylene Glycol
POE: Polyoxyethylene
RH: Relative Humidity
SNEDDs: Self-Nanoemulsifying Drug Delivery Systems
TEM: Transmission Electron Microscopy
W/O: Water in Oil

Introduction

Topical administration of aqueous solutions containing active pharmaceutical ingredients (APIs) remains the most common method for ocular treatment; however, such formulations are rapidly cleared via the nasolacrimal drainage system (Putchakayala et al., 2025). Conventional ocular formulations, such as eye drops, have a limited residence time in the conjunctival sac, which restricts drug penetration into deeper ocular tissues to less than 5% of the administered dose. Some studies have reported ocular bioavailability as low as 1% of the instilled dose (Bisen et al., 2024). This rapid clearance is further aggravated by reflex lachrymation and blinking, particularly when medications are delivered in non-physiological volumes, such as eye drops or suspensions (Gawin et al., 2021). Moreover, factors like drug binding to tear film protcins, enzymatic degradation by tear fluid enzymes, and poor corneal permeability substantially reduce the bioavailability of topically applied ophthalmic drugs (Akhter et al., 2022). Despite these limitations, a significant proportion of ophthalmic products continue to be formulated as eye drops and suspensions. This persistence is primarily attributed to their low production costs, ease of patient compliance, and the convenience offered by noninvasive drug delivery. Nevertheless, the inherent drawbacks of conventional formulations have stimulated considerable interest in developing advanced ocular drug delivery systems. Novel nanocarrier-based platforms, such as nanomicelles, nanoemulsions, dendrimers, and nanosuspensions, are gaining increasing attention as promising alternatives to traditional formulations like emulsions, gels, suspensions, and ointments (Giri et al., 2024). Although only a limited number of nanocarrier-based products have reached the market to date, nanotechnology-driven formulations are widely regarded as one of the most promising approaches for enhancing ocular bioavailability and therapeutic efficacy (Giri et al., 2024).

Nanoemulsions for Ocular Drug Delivery

Nanoemulsions (NEs) are transparent, thermodynamically stable dispersions with droplet sizes typically ranging from 20 to 200 nm, although some studies report sizes up to 500 nm (Gawin et al., 2021). Ophthalmic oil-in-water (o/w) NEs comprise a continuous aqueous phase, a dispersed oil phase, and a mixture of surfactants and co-surfactants that effectively reduce interfacial tension between the immiscible phases. To ensure the safety and efficacy of

NEs for ocular applications, additional excipients such as preservatives, tonicity modifiers, buffering agents, viscosity enhancers, antioxidants, and solubilizers for APIs are incorporated, customized according to the target site. These auxiliary agents play a crucial role in maintaining and enhancing both the physicochemical properties (such as minimizing droplet size and improving colloidal stability) and the overall physical stability of the formulation (Kishore et al., 2025). NEs offer significant advantages in ocular drug delivery due to their excellent penetration capabilities and prolonged precorneal retention, both of which enhance ocular bioavailability (Fang et al., 2025). Surfactants and co-surfactants act as penetration enhancers by transiently disrupting tight junctions and mucosal barriers, facilitating drug transport across ocular tissues. Additionally, the submicron droplet size promotes cellular uptake via endocytosis in corneal epithelial cells, thereby improving drug absorption and therapeutic efficacy (Xie et al., 2023). Optimal therapeutic outcomes from topical ocular administration depend on proper instillation techniques, correct dosing frequency (often up to four times daily), and appropriate intervals between multiple ophthalmic treatments. However, patient adherence remains a significant challenge. Studies indicate that fewer than half of glaucoma patients consistently follow their prescribed dosing regimens. Similarly, poor compliance is common among patients undergoing treatment for dry eye syndrome, immune-mediated anterior ocular disorders, and various corneal infections in real-world clinical settings (Moore et al., 2023). NEs help overcome adherence challenges by enhancing drug absorption relative to conventional eye drops, allowing for reduced dosing frequency and simplified treatment regimens, which can significantly improve patient compliance (Chauhan et al., 2024). Unlike suspensions, gels, and ointments, which may provide variable dosing, NEs offer more uniform and consistent drug delivery. Hydrophobic APIs are particularly well-suited for incorporation into the lipid phase of NEs, improving their solubility and bioavailability (Choradiya et al., 2021). For example, an oil-in-water nanoemulsion containing 0.4% brinzolamide showed superior ocular tissue penetration and enhanced therapeutic efficacy in vivo compared to a commercially available 1% brinzolamide solution (Das et al., 2022). Similar enhanced drug delivery and therapeutic outcomes have been observed with nanoemulsions containing tacrolimus (Yang et al., 2022), moxifloxacin (Youseff et al., 2022), acetazolamide, and dorzolamide hydrochloride (Pardeshi et al., 2024), all demonstrating increased drug concentrations in the aqueous humor compared to conventional formulations. Several nanoemulsion-based ocular products are currently available on the market, such as Systane Complete, Durezol, Xelpros, Ikervis, Cyclokat, and Restasis (Khiev et al., 2021; Reimondez et al., 2015; Yeu et al., 2020; Choradiya & Patil, 2021). For instance, Vekacia, a cationic nanoemulsion containing 0.05% or 0.1% cyclosporin A, and Catioprost, containing 0.005% latanoprost, are approved treatments for vernal keratoconjunctivitis (Reimondez et al., 2015). In addition, several nanoemulsions without APIs have been approved for the treatment of conditions such as dry eye syndrome, keratitis, and glaucoma. By providing extensive surface contact between the drug carrier and the ocular surface, nanoemulsions significantly enhance corneal permeability, bioavailability, and therapeutic effectiveness. Their surface-active components improve interactions with the precorneal environment, facilitating better drug dispersion across the corneal surface. This results in a faster onset and prolonged duration of drug action on the corneal epithelium.

Benefits of NEs for Enhancing Ocular Surface Penetration and Bioavailability

NEs play a crucial role in the formulation of hydrophobic drugs, particularly within oil-in-water (o/w) systems. The dispersion of oil droplets markedly increases the surface area, thereby enhancing the penetration and bioavailability of hydrophobic compounds. Despite these advantages, delivering such drugs to the eye remains challenging due to the complex anatomy and physiology of the ocular environment. The lacrimal gland continuously secretes tear fluid at a rate of approximately 16% of the total tear volume per minute, which contributes to the rapid clearance of foreign substances from the ocular surface (Jansook et al., 2021). Two primary challenges hinder the ocular delivery of hydrophobic drugs. First, these drugs exhibit poor compatibility with conventional aqueous eye drop formulations. Second, the eye's efficient drainage mechanisms result in nearly 80% of the instilled eye drops being washed away within minutes, necessitating frequent administration to maintain sustained therapeutic effects. These limitations highlight the promise of NEs as an effective platform for ocular drug delivery. The Novasorb product line exemplifies the successful application of nanoemulsion technology to overcome these obstacles (Gupta, 2020). Unlike earlier solubility enhancement methods that often led to increased systemic toxicity and unfavorable pharmacokinetic profiles, NEs provide improved local bioavailability while maintaining a favorable safety profile (Naseema et al., 2021). A notable example is the development of a nanoemulsion-based formulation of latanoprost, a hydrophobic antiglaucoma agent. In this formulation, the commonly used surfactant benzalkonium chloride, known for its ocular toxicity with prolonged use, was replaced with potassium sorbate, a milder preservative. This substitution resulted in a well-tolerated nanoemulsion that provides a safer base for potential secondary formulations aimed at enhancing sustainability and improving patient compliance. Comparative studies demonstrated that the potassium sorbate-based nanoemulsion caused less ocular irritation and was better accepted by patients compared to traditional benzalkonium chloride-containing eye drops (Tau et al., 2022). The benefits and potential of nanoemulsion-based ocular drug delivery are visually summarized in Figure 1, which outlines the mechanisms and advantages of this advanced formulation approach.

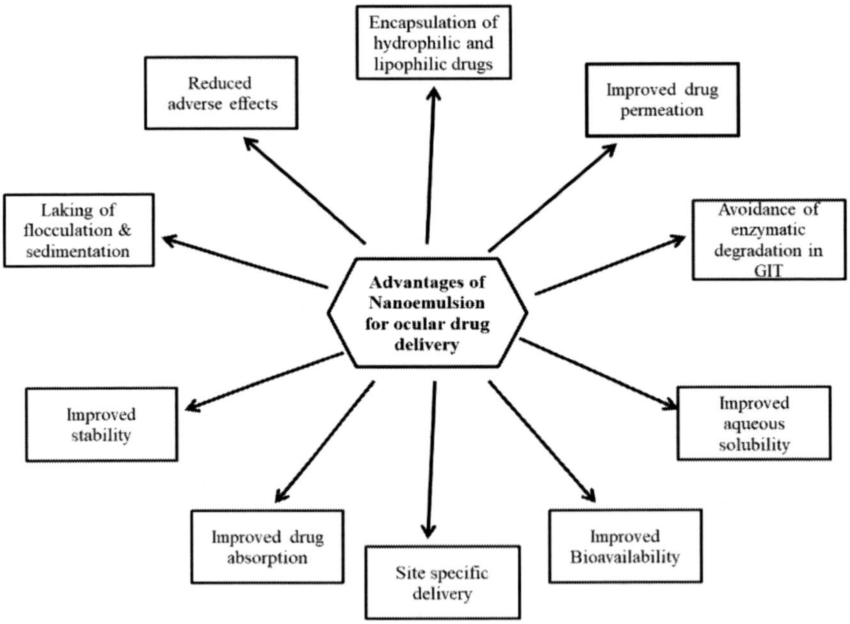

Figure 1. Advantages of nanoemulsion for ocular delivery of drug.

Components of NEs

Recent advancements in pharmaceutical science and technology have made a wide range of novel excipients increasingly accessible for developing innovative drug delivery systems. For a successful nanoemulsion formulation, the oil phase, surfactants, co-surfactants, and drug should ideally form a monophasic system at room temperature before being dispersed into the aqueous phase. However, excipients can interact with biological systems, other active or inert substances, or container-closure systems in both predictable and unpredictable ways. Therefore, it is essential to thoroughly evaluate their physicochemical properties, safety profiles, and compatibility. In ocular formulations, excipients must not only act as solvents to ensure drug solubility but also be non-irritating and non-toxic to the eye. Hence, the selection of oils, surfactants, and co-surfactants must be based on their solubilizing and emulsifying capacities as well as their ocular tolerability. Essential components of nanoemulsions include emulsifying agents, oil, and an aqueous phase. Commonly used oils in ophthalmic nanoemulsions include mineral oil, medium-chain triglycerides, and vegetable oils such as castor oil (Gupta et al., 2016).

Table 1. Various components used in ocular NEs (Gupta et al., 2016; Jaiswal et al., 2015)

Components	Examples	Properties
Surfactants	Miranol C2M, Poloxamer 407, Tween80, Tween40, and Tween20, brij 35, tyloxapol, span 20, soluphor p	Reduce the surface tension between the two phases in order to improve the oil and aqueous phase's miscibility.
Oil/ lipids	MCT, egg-lecithin, castor oil, peanut butter, Dermol®M5, Lipoid E-80, soy oil, Epikuron® 200, dermol M5, miglyol 812, triacetin, oleic acid, MCM, capmul, etanol G, ethyl oleate, isopropyl myristate	To improve medication loading and the solubility of lipophilic drugs
Cosurfactant	Cremophor rh40, cremophor el., PG, PEG 400, glycerin, transcutol-p (diethylene glycol monoethyl ether)	Reduce the amount of surfactant present for extremely low negative interfacial tension.
Cationic agent/ preservative	Benzoalkonium chloride, cetylpyridinium chloride benzoethonium chloride, cetrimide, cetalkonium chloride1, 1,2-di-(9Z-octadecenoyl), poly(llysine), stearylamine, polyethylenimine, oleylamine 1,2-snglycero-3-phosphoethanolamine (DOPE), and 1,2-di-(9Z-octadecenoyl)-3-trimethylammoniumpropane (DOTAP)	As preservatives, these chemicals determine the surface charge on the NE droplets.
Osmotic agents	Mannitol, sorbitol, propylene glycol, dextrose, and glycerin.	To keep osmolality stable

Emulsifying agents are essential for stabilizing nanoemulsions by preventing phase separation. These agents, typically surfactants, include compounds such as tyloxapol, polysorbates, poloxamers, vitamin E-TPGS, and cremophors. Ideal emulsifiers must be non-toxic, biocompatible, and compatible with both the formulation and ocular tissues. A summary of the various components commonly used in ocular nanoemulsion formulations is provided in Table 1.

Preparation Methods

Table 2. Method of preparation of NEs for ocular delivery

Method	Process	References
High pressure homogenization	This is the most common method, relying on multiple cycles and high pressure to achieve smaller droplet sizes. A macroemulsion is forced through a narrow orifice using high pressure, generating intense cavitation, turbulence, and shear forces. These combined effects produce NEs with droplet sizes as small as 1 nm.	(Rai et al., 2018)
Ultrasonication	Ultrasonic energy is applied, with varying power and duration, to create cavitation forces that break down macroemulsions into NEs. This technique helps achieve optimal droplet size.	(Dukovski et al., 2019)
Microfluidization	In this method, fluids are pumped through high-pressure microchannels (ranging from 500 to 20,000 psi) using a microfluidizer. The macroemulsion phases merge under the influence of shear, cavitation, and impact forces, forming NEs.	(Liu et al., 2015)
Phase inversion composition process	This technique involves changing the formulation's composition. For instance, water is added to an oil-surfactant mixture (for oil-in-water NEs) or oil is added to a water-surfactant system (for water-in-oil NEs), leading to phase inversion and nanoemulsion formation.	(Singh et al., 2017)
The temperature process of phase inversion	This process leverages the inverse relationship between temperature and the spontaneous curvature of surfactants. As the temperature increases, dehydration of polyoxyethylene (POE) groups in non-ionic surfactants occurs, increasing lipophilicity and altering curvature, thereby inducing nanoemulsion formation.	(Solans et al., 2012)
Self nanoemulsification method	This method involves the spontaneous formation of NEs when lipids and water are mixed in the presence of an adequate amount of surfactant or surfactant blend, without the need for external energy.	(Alexander et al., 2012)

Nanoemulsions can be prepared using either independent or combined approaches involving both high- and low-energy techniques. They are typically classified based on the energy input, phase inversion mechanisms, and self-emulsification processes (Singh et al., 2017; Rai et al., 2018; Liu et al., 2015; Solans et al., 2012; Alexander et al., 2012). Low-energy methods include phase inversion emulsification, which relies on changes in composition or temperature, catastrophic phase inversion occurring at the emulsion inversion point, and self-nanoemulsifying techniques. High-energy methods employ mechanical forces to reduce droplet size and stabilize emulsions, such as high-speed homogenization, microfluidization, ultrasonication, and high-pressure homogenization. Table 2 summarizes these techniques along with their key characteristics.

Research Based on NEs for Ocular Drug Delivery

Recent studies highlighting the application of nanoemulsions in ocular disease management are summarized in Table 3 (Fang et al., 2025; Tuncay et al., 2025; Srivastava et al., 2024; Mhboobian et al., 2024; Meshskar et al., 2024; Dehghani et al., 2024; Aboumanei et al., 2023; Jadhav et al., 2023).

Table 3. Various research based on NEs in ocular delivery

Drug	Ocular disease	Conclusion	References
Dexamethasone	Corneal inflammation	Pharmacokinetic studies revealed that DEX-NPH with 35% γ-cyclodextrin increased corneal bioavailability by 1.29 times compared to nanoemulsion, and by 4.09 times compared to the free drug solution.	(Fang et al., 2025)
Prednisolone derived corticosteroid	Ocular inflammation	After one year at 25°C and 40% RH, the formulation remained stable with particle size around 125.4 ± 41.20 nm and 99.29 ± 1.3% drug content. NE-containing PDC showed no observable toxic effects.	(Tuncay et al., 2025)
Brimonidine tartarate	Glaucoma	The optimized formulation showed 100.25% drug content and 92.46% CDR over 12 hours. Higuchi model (R^2 = 0.99) confirmed sustained drug release. The in vitro irritancy test confirmed ocular safety.	(Prajapati et al., 2024)

Table 3. (Continued)

Drug	Ocular disease	Conclusion	References
Thymoquinone	Glaucoma	TMQ-NE had a droplet size of 29.10 ± 0.87 nm, PDI of 0.096 ± 0.03, osmolarity of 302 ± 1.5 mOsm/kg, and pH of 6.2 ± 0.2. It demonstrated over 90% ex vivo permeability and 84.87 ± 0.18% in vitro drug release.	(Srivastava et al., 2024)
Ketoconazoe	Antifungal	The selected formulation had droplet sizes <20 nm with complete drug release throughout the day, indicating adequate physicochemical properties for ocular delivery.	(Mhboobian et al., 2024)
Nigella sativa. A	Dry eye	TEM analysis showed spherical nanoparticles around 10.4 nm. The nanoemulsion improved tear production in atropine-induced dry eye conditions.	(Meshskar et al., 2024)
Triamcinolone acetonide	Inflammation of posterior segment	TCA-loaded SNEDDs showed low toxicity and efficient ocular penetration, offering a potential topical alternative to intravitreal and intraconjunctival injections.	(Dehghani et al., 2024)
Tamoxifen	Photoreceptor degeneration	TAMNE-1 gel had a zeta potential of −27.86 ± 1.13 mV, particle size of 140.20 ± 1.50 nm, and PDI of 0.20 ± 0.00. It exhibited sustained release, improved biodistribution, and was non-irritating and safe for ocular use.	(Aboumanei et al., 2023)
Dexamethasone	Steroid responsive inflammation	The NE had a globule size of 181 ± 90 nm, zeta potential of -21.03 ± 1.68 mV, and viscosity of 19.99 cp. In vitro testing in simulated tear fluid showed sustained release up to 48 hours with excellent cell viability in cytotoxicity assays.	(Jadhav et al., 2023)

Applications of NEs for Various Ocular Disease

Dry Eye Disease

- Nanoemulsions enhance surface hydration and reduce tear evaporation, thereby improving dry eye symptoms more effectively than traditional eye drops (Ismail et al., 2020).

- They interact with hyperosmolar tear fluid by diluting and reducing lacrimal fluid osmolarity, while also stabilizing the tear film (Daull et al., 2020).
- Oil nanodroplets within nanoemulsions absorb nonpolar lipids, leading to an increased thickness of the tear film's lipid layer (Daull et al., 2020).
- Cationorm® (Santen Pharmaceutical, Osaka, Japan) is a cationic, API-free nanoemulsion approved in the EU for managing chronic dry eye disease (DED) (Watrobska et al., 2019).
- Restasis® (Allergan, Dublin, Ireland), a nanoemulsion containing cyclosporine A, is FDA-approved for dry eye treatment. A generic version was approved in 2022 through collaboration with Viatris (Watrobska et al., 2019).
- A nanoemulsion consisting of 75% water, 5% Nigella sativa seed extract, and 20% surfactant (in a 1:4 ratio) significantly increased tear production in an atropine-induced dry eye model (Meshkar et al., 2024).
- A novel nanoemulsion formulation containing 5.5% curcuminoids showed promising efficacy in animal models of dry eye disease (Hadipour et al., 2024).

Severe Keratitis

- Ikervis®, a cationic nanoemulsion containing cyclosporine A (1 mg/mL), was approved by the European Medicines Agency (EMA) in 2015 for the treatment of severe keratitis in adults (Watrobska et al., 2019).

Glaucoma

- Glaucoma treatment faces multiple challenges, including low corneal penetration, short ocular residence time, systemic absorption, poor bioavailability, unregulated drug release, limited drug retention, inadequate bioadhesion, and suboptimal patient compliance. Nanoemulsions offer promising solutions to overcome these limitations.
- Catioprost® is a preservative-free cationic nanoemulsion containing latanoprost, specifically designed to avoid the adverse effects associated with benzalkonium chloride, making it suitable for patients prone to dry eye disease (Watrobska et al., 2019).

- Novasorb®, another cationic nanoemulsion formulation of latanoprost for glaucoma, uses benzalkonium chloride to enhance mucoadhesion, but this may worsen dry eye symptoms in sensitive individuals (Daull et al., 2020).
- Quaternary ammonium compounds present in nanoemulsion formulations can mitigate preservative toxicity by lowering the free concentration of preservatives in the aqueous phase (Daull et al., 2020).

Conclusion

The emergence of NEs in ophthalmic drug delivery has attracted considerable interest due to their ability to improve drug solubility, bioavailability, and ocular retention time. Their small droplet size facilitates more effective penetration of ocular barriers, potentially leading to enhanced therapeutic outcomes. Furthermore, the sustained drug release provided by NEs can improve patient compliance by reducing dosing frequency. Their inherent biocompatibility and minimal irritation profile make them ideal candidates for ocular formulations. Despite these advantages, challenges such as formulation stability, surfactant-related toxicity, and scale-up limitations remain significant hurdles. To overcome these issues, ongoing research focuses on developing innovative stabilizers, safer surfactants, and advanced manufacturing technologies to enhance the safety, efficacy, and commercial viability of NEs. With continued progress in nanotechnology and pharmaceutical sciences, NEs hold substantial promise to revolutionize ocular drug delivery.

References

Aboumanei, M. H., & Mahmoud, A. F. (2023). Development of tamoxifen in situ gel nanoemulsion for ocular delivery in photoreceptor degeneration disorder: in vitro characterization, 131I-Radiolabeling, and in vivo biodistribution studies. *Journal of Pharmaceutical Innovation*, *18*(2), 369-380.

Akhter, M. H., Ahmad, I., Alshahrani, M. Y., Al-Harbi, A. I., Khalilullah, H., Afzal, O., & Karim, S. (2022). Drug delivery challenges and current progress in nanocarrier-based ocular therapeutic system. *Gels*, *8*(2), 82.

Alexander, A. (2012). A review on novel therapeutic strategies for the enhancement of solubility for hydrophobic drugs through lipid and surfactant-based self-micro emulsifying drug delivery system: A novel approach. *American Journal of Drug Discovery and Development, 2*, 143–183.

Bisen, A. C., Dubey, A., Agrawal, S., Biswas, A., Rawat, K. S., Srivastava, S., & Bhatta, R. S. (2024). Recent updates on ocular disease management with ophthalmic ointments. *Therapeutic Delivery*, 15(6), 1-8.

Chauhan, N., Gupta, N., & Kumar, A. (2024). A review on novel nanoemulgel for ocular drug delivery. *EPH-International Journal of Medical and Health Science*, *10*(1), 1-9.

Choradiya, B. R., & Patil, S. B. (2021). A comprehensive review on nanoemulsion as an ophthalmic drug delivery system. *Journal of Molecular Liquids*, *339*, 116751.

Das, B., Nayak, A. K., & Mallick, S. (2022). Lipid-based nanocarriers for ocular drug delivery: An updated review. *Journal of Drug Delivery Science and Technology*, *76*, 103780.

Daull, P., Amrane, M., Ismail, D., Georgiev, G., Cwiklik, L., Baudouin, C., Leonardi, A., Garhofer, G., & Garrigue, J. S. (2020). Cationic emulsion-based artificial tears as a mimic of functional healthy tear film for restoration of ocular surface homeostasis in dry eye disease. *Journal of Ocular Pharmacology and Therapeutics, 36*, 355–365.

Dehghani, M., Zahir-Jouzdani, F., Shahbaz, S., Andarzbakhsh, K., Dinarvand, S., Nasab, M. H. F., & Atyabi, F. (2024). Triamcinolone-loaded self nano-emulsifying drug delivery systems for ocular use: An alternative to invasive ocular surgeries and injections. *International Journal of Pharmaceutics*, *653*, 123840.

Fang, G., Zhao, R., Zhu, L., Wang, Q., Peng, S., Kang, L., & Tang, B. (2025). Nanoemulsion-based pseudopolyrotaxane hydrogel for enhanced corneal bioavailability and treatment of corneal inflammation. *Journal of Controlled Release*, *379*, 14-29.

Gawin-Mikołajewicz, A., Nartowski, K. P., Dyba, A. J., Gołkowska, A. M., Malec, K., & Karolewicz, B. (2021). Ophthalmic nanoemulsions: From composition to technological processes and quality control. *Molecular pharmaceutics*, *18*(10), 3719-3740.

Giri, B. R., Jakka, D., Sandoval, M. A., Kulkarni, V. R., & Bao, Q. (2024). Advancements in ocular therapy: a review of emerging drug delivery approaches and pharmaceutical technologies. *Pharmaceutics*, *16*(10), 1325.

Gupta, A. (2020). Nanoemulsions. In E. J. Chung, L. Leon, & C. Rinaldi (Eds.), *Nanoparticles for Biomedical Applications* (pp. 371–384). Elsevier.

Gupta, A., Eral, H. B., Hatton, T. A., & Doyle, P. S. (2016). Nanoemulsions: Formation, properties, and applications. *Soft Matter, 12*, 2826–2841.

Hadipour Jahromy, M., Qomi, M., Fazelipour, S., Sami, N., Faali, F., Karimi, M., & Adhami Moghadam, F. (2024). Evaluation of curcumin-based ophthalmic nanoemulsion on atropine-induced dry eye in mice. *Heliyon, 10*, e29009.

Ismail, A., Nasr, M., & Sammour, O. (2020). Nanoemulsion as a feasible and biocompatible carrier for ocular delivery of travoprost: Improved pharmacokinetic/ pharmacodynamic properties. *International Journal of Pharmaceutics, 583*, 119402.

Jadhav, A., Salave, S., Rana, D., & Benival, D. (2023). Development and in-vitro evaluation of dexamethasone enriched nanoemulsion for ophthalmic indication. *Drug Delivery Letters*, *13*(3), 196-212.

Jaiswal, M., Dudhe, R., & Sharma, P. K. (2015). Nanoemulsion: An advanced mode of drug delivery system. *3 Biotech, 5*, 123.

Jansook, P., Hnin, H. M., Loftsson, T., & Stefánsson, E. (2021). Cyclodextrin-based formulation of carbonic anhydrase inhibitors for ocular delivery – A review. *International Journal of Pharmaceutics, 606, 120955.*

Jurišić Dukovski, B., Bračko, A., Šare, M., Pepić, I., & Lovrić, J. (2019). *In vitro* evaluation of stearylamine cationic nanoemulsions for improved ocular drug delivery. *Acta Pharmaceutica, 69*, 621–634.

Khiev, D., Mohamed, Z. A., Vichare, R., Paulson, R., Bhatia, S., Mohapatra, S., Lobo, G. P., Valapala, M., Kerur, N., Passaglia, C. L., Mohapatra, S. S., & Biswal, M. R. (2021). Emerging nano-formulations and nanomedicines applications for ocular drug delivery. *Nanomaterials, 11*(1), 173.

Kishore, A., Jain, A., Asthana, N., Milan, R., Lakshmi, S. M., Gupta, M., & Kanoujia, J. (2025). Selection Criteria for Oils, Surfactants, and Co-Surfactants in Ocular Nanoemulsion Formulation: A Mini Review. *Current Pharmaceutical Design*, 2025, 10.2174/0113816128350573241202105210.

Liu, C.-H., Huang, Y.-C., Jhang, J.-W., Liu, Y.-H., & Wu, W.-C. (2015). Quercetin delivery to porcine cornea and sclera by solid lipid nanoparticles and nanoemulsion. *RSC Advances, 5*, 100923–100933.

Mahboobian, M. M., Azadi, N., & Pourmoslemi, S. (2024). Preparation of Ketoconazole-loaded Nanoemulsions for Ophthalmic Delivery: Characterization and In-vitro Antifungal Activity Evaluation. *Current Drug Therapy*, *19*(3), 327-335.

Meshksar, S., Jahromy, M. H., Qomi, M., Sami, N., & Faali, F. (2024). Formulation and evaluation of the effects of ophthalmic nanoemulsion of *Nigella sativa* seed extract on atropine-induced dry eye in mice. *Phytomedicine Plus, 4*, 100541.

Messmer, E. M. (2015). The pathophysiology, diagnosis, and treatment of dry eye disease. *Deutsches Ärzteblatt International, 112*, 71–81.

Naseema, A., Kovooru, L., Behera, A. K., Kumar, K. P. P., & Srivastava, P. (2021). A critical review of synthesis procedures, applications, and future potential of nanoemulsions. *Advances in Colloid and Interface Science, 287*, 102318.

Pardeshi, S. R., Gholap, A. D., Hatvate, N. T., Gharat, K. D., Naik, J. B., & Omri, A. (2024). Advances in dorzolamide hydrochloride delivery: harnessing nanotechnology for enhanced ocular drug delivery in glaucoma management. *Discover Nano*, *19*(1), 1-27.

Prajapati, B., Modi, C., Patel, U., & Kendre, P. (2024). Nanoemulsion based in-situ Gel for Ocular Delivery of Brimonidine Tartrate. *Current Drug Therapy*, *19*(3), 336-345.

Putchakayala, N. S. D., Morusu, K., Ramisetty, S., & Ravoru, N. (2025). Nanostructured Lipid Carriers: A Novel Platform in the Formulation of Targeted Drug Delivery Systems. *Current Nanomedicine*, *15*(1), 26-49.

Rai, V. K., Mishra, N., Yadav, K. S., & Yadav, N. P. (2018). Nanoemulsion as a pharmaceutical carrier for dermal and transdermal drug delivery: Formulation development, stability issues, basic considerations, and applications. *Journal of Controlled Release, 270*, 203–225.

Reimondez-Troitino, S., Csaba, N., Alonso, M. J., & de la Fuente, M. (2015). Nanotherapies for the treatment of ocular diseases. *European Journal of Pharmaceutics and Biopharmaceutics, 95*(Pt B), 279–293.

Sheppard, J., Shen Lee, B., & Periman, L. M. (2023). Dry eye disease: Identification and therapeutic strategies for primary care clinicians and clinical specialists. *Annals of Medicine, 55*, 241–252.

Singh, Y., Meher, J. G., Raval, K., Khan, F. A., Chaurasia, M., Jain, N. K., & Chourasia, M. K. (2017). Nanoemulsion: Concepts, development, and applications in drug delivery. *Journal of Controlled Release, 252*, 28–49.

Solans, C., & Solé, I. (2012). Nano-emulsions: Formation by low-energy methods. *Current Opinion in Colloid & Interface Science, 17*, 246–254.

Srivastava, V., Pardhi, E. R., Yadav, R., Singh, V., Khatri, D. K., & Mehra, N. K. (2024). QbD-driven thymoquinone laden nanoemulsion for glaucoma management: In vitro, ex vivo, and pre-clinical evaluation. *Journal of Drug Delivery Science and Technology*, *94*, 105493.

Tau, J., Passerini, M. S., del Papa, M., Aguilar, A., & Berra, A. (2022). A novel ophthalmic latanoprost 0.005% nanoemulsion: A cytotoxicity study. *Graefe's Archive for Clinical and Experimental Ophthalmology, 260*, 1941–1946.

Tuncay Tanriverdi, S., Gokce, E. H., Ozturk, N. Z. A., Turk, M., Entezari, B., Balci, A., & Ozer, O. (2025). Formulation, Characterization and in vitro Release of Topical Nanoemulsion Containing Prednisolone-Derived Corticosteroid. *Drug Development and Industrial Pharmacy*, 51(3), 180–192.

Walenga, R. L., Babiskin, A. H., Zhang, X., Absar, M., Zhao, L., & Lionberger, R. A. (2019). Impact of vehicle physicochemical properties on modeling-based predictions of cyclosporine ophthalmic emulsion bioavailability and tear film breakup time. *Journal of Pharmaceutical Sciences, 108*, 620–629.

Watrobska-Swietlikowska, D. (2019). Distribution of benzalkonium chloride into the aqueous phases of submicron dispersed systems: Emulsions, aqueous lecithin dispersion and nanospheres. *AAPS PharmSciTech, 21*, 7.

Xie, G., Lin, S., Wu, F., & Liu, J. (2023). Nanomaterial-based ophthalmic drug delivery. *Advanced drug delivery reviews*, 200, 115004.

Yang, J., Liang, Z., Lu, P., Song, F., Zhang, Z., Zhou, T., & Zhang, J. (2022). Development of a luliconazole nanoemulsion as a prospective ophthalmic delivery system for the treatment of fungal keratitis: in vitro and in vivo evaluation. *Pharmaceutics*, *14*(10), 2052.

Yeu, E., Silverstein, S., Guillon, M., Schulze, M.-M., Galarreta, D., Srinivasan, S., & Manoj, V. (2020). Efficacy and safety of phospholipid nanoemulsion-based ocular lubricant for the management of various subtypes of dry eye disease: A phase IV, multicenter trial. *Clinical Ophthalmology, 14*, 2561.

Youssef, A. A. A., Thakkar, R., Senapati, S., Joshi, P. H., Dudhipala, N., & Majumdar, S. (2022). Design of topical moxifloxacin mucoadhesive nanoemulsion for the management of ocular bacterial infections. *Pharmaceutics*, *14*(6), 1246.

Chapter 12

Inorganic Nanoparticles for Ocular Applications

Ketan Shah[1,*]
Yash Dudhwala[2]
Riya Mehta[2]
Devesh U. Kapoor[3]
and Reeti Shah[4]

[1]Department of Quality Assurance, Shree Naranjibhai Lalbhai Patel College of Pharmacy, Umrakh, Gujarat, India
[2]Department of Pharmaceutics, Shree Naranjibhai Lalbhai Patel College of Pharmacy, Umrakh, Gujarat, India
[3]Dr. Dayaram Patel Pharmacy College, Bardoli, Gujarat, India
[4]Department of Pharmaceutical Sciences, University of Surrey, Guildford, England, United Kingdom

Abstract

Inorganic nanoparticles have emerged as powerful platforms in ocular drug delivery, offering innovative solutions to overcome longstanding challenges such as poor bioavailability, limited drug retention, and inadequate targeting within the eye. The unique anatomical and physiological barriers of the ocular region pose significant obstacles to conventional therapies for diseases like glaucoma, age-related macular degeneration, diabetic retinopathy, and other retinal disorders. Inorganic nanoparticles, including gold, silica, magnetic, and ceramic-based systems, provide distinct advantages over traditional delivery methods.

* Corresponding Author's Email: ketanshah4281@gmail.com

In: Advances in Nanotechnology for Ocular Drug Delivery
Editors: Anjali Sharma, Ajmer Singh Grewal, Geeta Deswal et al.
ISBN: 979-8-89530-814-1
© 2025 Nova Science Publishers, Inc.

Their precise engineering enables enhanced cellular uptake, controlled drug release, and targeted delivery to specific ocular tissues such as the cornea and retina. Advances in surface functionalization, such as ligand conjugation and polymeric coatings, have further improved their ability to cross the blood-retinal barrier, increase ocular bioavailability, and sustain drug release, thereby reducing dosing frequency and enhancing patient compliance. Despite these promising features, challenges persist, especially in ensuring biocompatibility, long-term stability, and safe clearance from ocular tissues. This chapter also reviews current preclinical and clinical studies while discussing ongoing challenges and future prospects for integrating inorganic nanoparticles into personalized ocular therapies to improve clinical outcomes.

Keywords: inorganic nanoparticles, ocular drug delivery, targeted drug delivery, blood-retinal barrier, personalized ocular medicine

Abbreviations

AMD: Age-Related Macular Degeneration
BAB: Blood-Aqueous Barrier
BRB: Blood Retinal Barrier
EPR: Enhanced Permeability and Retention
INPs: Inorganic Nanoparticles
MNP: Magnetic Nanoparticle
MSNs: Mesoporous Silica Nanoparticles
NPs: Nanoparticles
QD: Quantum Dots
RPE: Retinal Pigment Epithelial

Introduction

Ocular drug delivery faces significant challenges due to the eye's complex anatomical and physiological barriers, which restrict effective therapeutic penetration. The human eye is protected by several defense mechanisms, including the corneal epithelium and specialized barriers such as the blood-aqueous barrier (BAB) and the blood-retinal barrier (BRB) (Bachu et al., 2018). These protective structures limit drug absorption, often resulting in reduced drug availability and rapid clearance from the ocular surface.

Conventional methods for delivering medication to the eye include topical application of eye drops on the ocular surface and intraocular injections. Among these, eye drops are the most commonly used due to their ease of administration and patient compliance (Patel et al., 2013). Traditional ocular drug delivery methods are not very effective, as they typically exhibit bioavailability of less than 5%. This low efficiency is largely due to rapid tear turnover and frequent blinking, which quickly wash away the medication. Systemic administration, such as injecting medication into the bloodstream, requires high doses to achieve therapeutic levels in the inner eye tissues, increasing the risk of systemic side effects. Intravitreal injections provide a more direct approach by delivering drugs to the posterior segment of the eye; however, they are invasive and carry risks such as endophthalmitis and retinal detachment. Furthermore, these injections must be administered by healthcare professionals, limiting their convenience. Given these challenges, nanotechnology-based approaches have emerged as promising alternatives. Inorganic nanoparticles, including gold (Au), silica (Si), and quantum dots (QD), offer advantages over conventional formulations due to their exceptional stability, tunable physical and chemical properties, and enhanced drug-carrying capacity (Gaudana et al., 2010). Inorganic nanoparticles offer promising strategies to overcome ocular barriers such as the cornea, sclera, and retina, which typically limit drug bioavailability. These nanosystems can penetrate deeper into ocular tissues more effectively and can be engineered with surface ligands, such as hyaluronic acid or antibodies, to enable receptor-mediated targeted drug delivery (Janagam et al., 2017; Elumalai et al., 2024). Magnetic nanoparticles enable targeted drug delivery by using external magnetic fields to direct and concentrate the particles at specific ocular sites. Compared to polymeric carriers, inorganic nanoparticles such as gold and silica exhibit greater stability and resistance to enzymatic degradation, ensuring prolonged shelf-life and maintained structural integrity (Dykman et al., 2024). Additionally, their multifunctional properties facilitate theranostic applications that combine therapeutic delivery with real-time imaging, making them exceptionally well-suited for advanced ocular therapies.

Anatomical and Physiological Barriers of the Eye

The eye is a highly specialized organ featuring complex anatomical and physiological barriers that significantly impede efficient drug permeation, posing major challenges for ocular drug delivery. These barriers are categorized as pre-corneal, corneal, conjunctival, and intraocular, all of which markedly reduce drug bioavailability and necessitate novel strategies such as the use of inorganic nanoparticles for sustained and targeted therapy (Khazaeli et al., 2025). Among the dynamic barriers, the tear film, composed of mucin, aqueous, and lipid layers, rapidly clears topically applied drugs through tear turnover (~16% per minute) and blink reflex (~5-10 seconds), thereby minimizing ocular residence time (Kulkarni et al., 2025). The cornea, consisting of five layers, acts primarily as a lipophilic barrier allowing permeation of small lipophilic molecules, while hydrophilic drugs larger than 500 Da are restricted by the tight junctions of the epithelium. Additionally, the hydrophilic stroma limits lipophilic drug passage, and the endothelium restricts the movement of large molecules (Mao et al., 2024). Beneath the cornea, the conjunctival vasculature facilitates systemic absorption, reducing the local drug concentration. More profound intraocular barriers include BAB, formed by tight junctions in the ciliary body and iris epithelium, which restricts hydrophilic and macromolecular drug penetration into the anterior chamber. Furthermore, BRB, along with efflux transporters such as P-glycoprotein and MRP-1, actively export drugs, significantly lowering bioavailability in the posterior segment (Kulkarni et al., 2025).

Types of Inorganic Nanoparticles for Ocular Drug Delivery

Drug delivery to the eye is significantly challenged by anatomical and physiological barriers, including the corneal epithelium, BRB, and rapid tear drainage, all of which limit drug absorption and retention (Hedaya, 2024). Figure 1 illustrates the principal pathways utilized for introducing drugs into the ocular cavity.

AuNPs in Ocular Drug Delivery

AuNPs are increasingly recognized as efficient carriers in ophthalmic formulations due to their excellent biocompatibility, chemical stability, and high functionalization versatility. Their size-dependent optical properties make them especially suitable for targeted drug delivery, photothermal therapy, and imaging applications in ocular diseases (Ahmed et al., 2024). AuNPs can be functionalized with therapeutic agents, peptides, or antibodies to enhance their specificity for targeted ocular tissues.

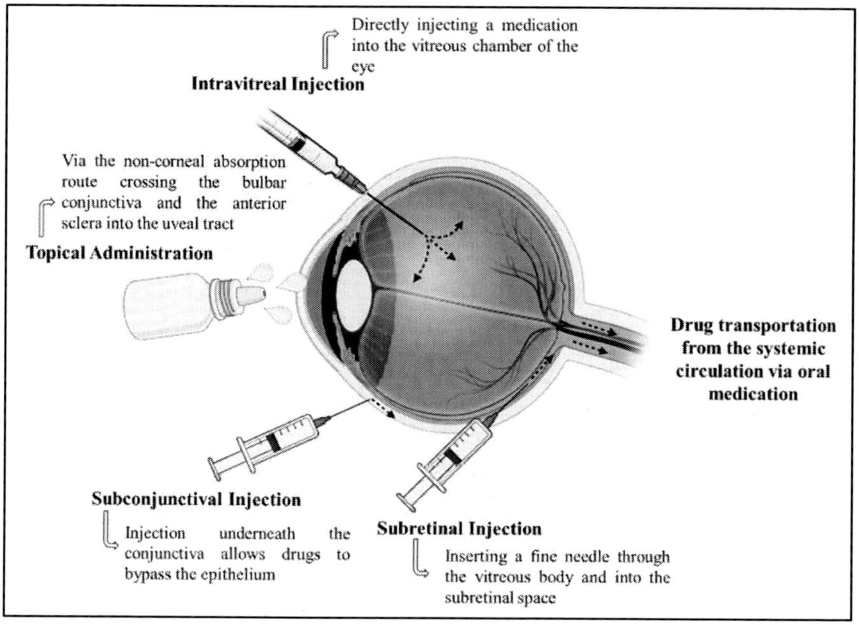

Figure 1. Strategic pathways for ocular drug delivery: overcoming anatomical barriers.

Silica NPs: Versatile Carriers for Ocular Applications

Mesoporous silica NPs (MSNs) and other forms of silica NPs (SiNPs) have garnered significant interest in ocular drug delivery due to their high drug-loading capacity, tunable porosity, and controlled release properties. Their inherent biodegradability and modifiable surface characteristics make them especially suitable for sustained drug delivery, gene therapy, and bioimaging applications in ophthalmology (Ahmed et al., 2024). Notably, MSNs have demonstrated enhanced ability to traverse ocular barriers, particularly when functionalized with mucoadhesive polymers or targeting ligands.

Magnetic NPs for Targeted Ocular Therapy

Iron oxide magnetic NPs (MNPs) have opened new avenues in targeted ophthalmic drug delivery. Their superparamagnetic properties allow the application of an external magnetic field to direct drug-loaded nanoparticles to specific intraocular sites, enabling localized and sustained drug release. This approach is particularly advantageous in the treatment of posterior segment disorders, such as retinal diseases, where conventional systemic therapies face limitations due to the restrictive nature of the BRB (Zhang et al., 2024).

Ceramic-Based NPs: Biocompatible and Stable Drug Carriers

NPs, particularly those derived from calcium phosphate, zirconia, and titanium oxide, are gaining increasing attention in ophthalmic drug delivery due to their excellent biocompatibility, inherent biodegradability, and high stability under physiological conditions. These properties make ceramic NPs well-suited for sustained and controlled drug release. They have demonstrated considerable potential in the treatment of chronic ocular conditions such as uveitis and dry eye syndrome, where prolonged therapeutic action is essential (Wadzer et al., 2024).

Mechanisms of Drug Delivery Using Inorganic Nanoparticles

Inorganic NPs present promising avenues for advancing ocular drug delivery by effectively overcoming key physiological barriers such as the corneal epithelium, conjunctival clearance, and the BRB. By leveraging their diverse physicochemical properties and surface modifiability, inorganic NPs can be engineered to enhance ocular bioavailability, provide controlled and sustained drug release, and optimize therapeutic efficacy (Narayana et al., 2024).

Endocytosis-Mediated Uptake

NPs deliver drugs to ocular tissues through several cellular internalization pathways. One primary mechanism is phagocytosis, in which larger NPs, such as mesoporous silica or gold nanospheres, are engulfed by retinal pigment epithelial (RPE) cells, enabling targeted drug release at the site of action (Paul & Sharma, 2020). Another major route is clathrin- and caveolae-mediated endocytosis, which facilitates the uptake of smaller NPs (10-100 nm) by corneal and retinal cells. These pathways are particularly advantageous because they often bypass lysosomal degradation, thereby enhancing intracellular drug delivery (Paul & Sharma, 2020). Additionally, macropinocytosis, a non-specific uptake mechanism, allows for the internalization of a larger volume of NPs into the aqueous humor, thereby increasing drug accumulation in the anterior segment of the eye (Mohammadpour & Ghandehari, 2022).

Stimuli-Responsive Drug Release

One of the key advantages of inorganic NPs lies in their ability to facilitate controlled and stimuli-responsive drug release, which ensures sustained therapeutic efficacy while reducing systemic toxicity (Pham et al., 2020). Several stimuli-responsive mechanisms have been explored in ocular drug delivery:

pH-Responsive Systems

Certain silica and gold NPs are engineered to undergo structural transformations in response to acidic microenvironments, such as those found in inflamed ocular tissues. These pH-sensitive systems enable the targeted and timely release of therapeutic agents specifically at sites of inflammation.

Light-Activated Release

Inorganic NPs can be functionalized to respond to specific wavelengths of light, triggering drug release upon irradiation. This approach allows for spatial and temporal control over drug delivery, minimizing off-target effects (Abdo et al., 2020).

Targeted Drug Delivery Mechanisms

Inorganic NPs can be precisely engineered to preferentially target diseased ocular tissues, thereby enhancing local drug concentrations while reducing off-target effects. Targeted delivery strategies often rely on ligand-mediated mechanisms, in which NPs are surface-functionalized with targeting moieties such as hyaluronic acid, transferrin, or folic acid. These ligands exhibit high binding affinity to specific receptors overexpressed on ocular cells, enabling selective uptake and improved therapeutic efficacy (Sultana et al., 2013; Tietjen et al., 2018).

Sustained and Controlled Drug Release

Prolonged drug release significantly decreases dosing frequency, thereby improving both patient compliance and therapeutic outcomes. MSNs are particularly well-suited for achieving sustained and controlled drug delivery. Owing to their large surface area, tunable pore size, and versatile surface functionalization, MSNs enable high drug loading and gradual, extended release, often sustaining therapeutic levels for several weeks (Cohen et al., 2000).

Formulation Strategies and Recent Advancements in Inorganic Nanoparticles for Ocular Drug Delivery

The synthesis of inorganic NPs for ocular drug delivery requires precise engineering to surmount anatomical barriers, improve bioavailability, and facilitate controlled drug release. Numerous formulation strategies have been devised to optimize the physicochemical characteristics of NPs, thereby enhancing their therapeutic efficacy in ocular applications (Hatem et al., 2024). Table 1 summarizes various surface functionalization techniques and their corresponding effects on ocular drug delivery performance.

Table 1. Surface functionalization strategies and their effect on ocular drug delivery

Strategy	Material that Can be Used	Effect on Ocular Drug Delivery	Example	Reference
Polymer	Chitosan, Hyaluronic acid, PEG, PLGA	Enhances biocompatibility, extends circulation half-life, and enhances muco-adhesion, resulting in enhanced drug residence on the ocular surface.	Chitosan-coated gold NPs for improved ocular drug delivery	(Dandamudi et al., 2021)
Peptide	Cell-penetrating peptides (CPPs), RGD peptide	Allows receptor-mediated uptake, facilitates corneal penetration. RGD-functionalized NPs have been designed for targeted gene delivery to the retina with enhanced specificity and retinal tissue uptake.	RGD-modified gold NPs for retinal therapy	(Zhang et al., 2021)

Strategy	Material that Can be Used	Effect on Ocular Drug Delivery	Example	Reference
Aptamer	Nucleic acid aptamers (e.g., Anti-VEGF aptamers)	Offers high specificity binding to ocular disease markers like VEGF, preventing pathological angiogenesis in diseases like age-related macular degeneration (AMD).	Anti-VEGF-Aptamer Modified C-Dots for Topical Treatment of Ocular Vascular Disorders	(Shoval et al., 2019)
DNA	DNA oligonucleotides, DNA nanostructures	Enables gene therapy strategies through the delivery of genetic material to target ocular cells. DNA-functionalized NPs can enable the repair or modification of genetic factors leading to ocular pathology.	DNA NPs for ophthalmic drug delivery	(Willem de Vries et al., 2018)
Protein	Albumin, Transferrin, Antibodies (e.g., Anti-VEGF)	Improve targeting and uptake by ocular tissues. Transferrin-conjugated.	Transferrin, RGD peptide and dual-targeted NPs	(Singh et al., 2009)

Surface Functionalization and Targeting Approaches

Surface modification plays a crucial role in enhancing nanoparticle stability, improving ocular tissue penetration, and enabling site-specific drug delivery. Functionalization with biopolymers such as chitosan and hyaluronic acid significantly increases mucoadhesion, thereby prolonging the residence time on the ocular surface (Ly et al., 2024). Additionally, ligand conjugation facilitates selective targeting of specific ocular tissues, such as the retina or corneal epithelium, which enhances therapeutic efficacy while minimizing systemic side effects (Fathi-Karkan et al., 2024).

Encapsulation and Controlled Release Mechanisms

Encapsulation methodologies, including drug loading via mesoporous silica and conjugation with gold NPs, facilitate sustained and controlled drug release. Advanced strategies, such as pH-sensitive and enzyme-responsive coatings, have been developed to trigger drug release selectively within specific ocular compartments (Yadav et al., 2024). Moreover, magnetic NPs offer an additional level of precision by enabling targeted drug localization through the application of an external magnetic field.

Hybrid Nanoparticle Systems

The integration of inorganic NPs with biodegradable polymers significantly enhances biocompatibility and alleviates toxicity concerns. Hybrid formulations, such as gold-silica core-shell structures and lipid-coated inorganic NPs, offer tunable drug release profiles alongside improved ocular penetration. Recent advancements include the use of quantum dot-based nanoparticles for imaging-guided therapy, enabling real-time monitoring of drug delivery efficiency and distribution within ocular tissues (Zhou et al., 2013).

Nanoparticle Delivery Systems: Eye Drops, Gels, and Implants

Traditional eye drops suffer from rapid clearance due to blinking and tear drainage. Incorporating inorganic NPs into mucoadhesive gels or in situ-forming hydrogels markedly improves ocular retention time. Additionally, NP-loaded intraocular implants offer sustained drug delivery, reducing dosing frequency for chronic ocular conditions (Onugwu et al., 2023).

Recent Advancements and Preclinical Evaluations

Recent studies have demonstrated the potential of SiNPs and AuNPs for targeted delivery of anti-VEGF therapeutics in treating AMD. MNPs have shown promising applications in gene therapy for retinal diseases. Advances in nanofabrication techniques, particularly microfluidic synthesis, have improved batch-to-batch reproducibility, thereby accelerating the translation of these nanotechnologies into clinical use (Kuno & Fujii, 2011).

Table 2. Preclinical investigations of inorganic NPs for ocular drug delivery

Nanoparticle Type	Surface Functionalization	Findings	Research Source
AuNPs	Chitosan coating	Enhanced corneal penetration and extended drug delivery in rabbit models.	(Yang et al., 2022)
AgNPs	Unmodified	Shown antifungal and anti-inflammatory activities; however, in animal models, it also had ocular surface toxicity.	(Khiev et al., 2021)
Iron oxide NPs	PEGylation	Increased biocompatibility and retinal targeting described in rodent models.	(Unnikrishnan et al., 2023)
Zirconium-based metal-organic frameworks (UiO-67)	Polyurethane matrix embedding	Demonstrated long-term release of ocular medications for 14 days in vitro, which indicates the potential for long-term ocular drug delivery.	(Gandara-Loe et al., 2020)
AuNPs	RGD peptide functionalization	Targeted delivery to retinal cells with increased uptake in animal models.	(Yang et al., 2022)

Therapeutic Applications of Inorganic Nanoparticles in Ocular Drug Delivery

The use of NPs has shown promising potential in treating ocular conditions by improving drug delivery, thereby enhancing therapeutic effectiveness and ensuring optimal drug action, key factors for better clinical outcomes and overall treatment success (Meza-Rios et al., 2020).

Glaucoma Treatment

Glaucoma, caused by elevated IOP, results in optic nerve damage and vision loss. Conventional treatments often suffer from limited corneal penetration and short drug retention times. Inorganic NPs such as gold and silica NPs offer controlled drug release and enhanced permeability. For instance, timolol maleate conjugated with AuNPs demonstrated a more sustained reduction in IOP compared to conventional eye drop formulations (Kambhampati & Kannan, 2013; Li et al., 2025).

Age-Related Macular Degeneration (AMD)

Treatment for AMD typically involves repeated intravitreal injections of anti-VEGF agents, which carry risks of retinal damage and infection. Silica NPs loaded with bevacizumab enable sustained, long-term intraocular drug delivery, reducing the need for frequent injections. Additionally, anti-angiogenic peptide-tagged AuNPs exhibit significant anti-neovascularization effects, demonstrating considerable therapeutic potential (Kumbhar et al., 2024).

Diabetic Retinopathy (DR)

DR is driven by oxidative stress and inflammation of the retinal vasculature. Cerium oxide and selenium NPs offer antioxidant protection, limiting retinal damage. Moreover, dexamethasone-loaded mesoporous silver NPs enhance anti-inflammatory effects and enable targeted delivery to the retina, thereby minimizing systemic side effects (Ramadan et al., 2024; Elbedwehy et al., 2024).

Corneal Diseases and Wound Healing

Silver and zinc oxide NPs exhibit broad-spectrum antimicrobial activity, making them effective in treating keratitis and corneal ulcers. Growth factor-loaded silica NPs and hydroxyapatite-based scaffolds promote epithelial regeneration, thereby enhancing corneal wound healing and infection control (Elbedwehy et al., 2024).

Retinal Disorders and Gene Therapy

Genetic retinal disorders face significant delivery challenges due to the BRB. Quantum dot NPs and magnetic NPs serve as non-viral gene vectors, enabling effective transfection. siRNA-conjugated, functionalized superparamagnetic iron oxide NPs (SPIONs) and AuNPs have demonstrated efficient gene silencing and targeted delivery, offering safer and more precise alternatives to viral vectors (Haase et al., 2024; Elbedwehy et al., 2024).

Ocular Infections and Anti-Inflammatory Therapies

Inorganic NPs such as silver, TiO_2, and cerium oxide exhibit potent antimicrobial and anti-inflammatory properties. When formulated into hydrogels or eye drops, they offer sustained ocular retention, enhance therapeutic efficacy, and reduce bacterial resistance and oxidative damage (Mohamed, 2024).

Future Perspectives

The future of inorganic NPs in ocular therapy lies in enhancing their biocompatibility, targeting specificity, and clinical efficacy. Advanced surface engineering techniques such as PEGylation and biomimetic coatings are increasingly employed to reduce toxicity and improve ocular compatibility. Simultaneously, personalized nanomedicine guided by molecular and genetic profiling holds promise to tailor drug-loaded nanocarriers to individual disease phenotypes, particularly in complex conditions like AMD and DR (Patel & Geed, 2024). Artificial intelligence (AI) is emerging as a transformative tool for optimizing NP design, surface properties, and drug release kinetics, while enabling improved *in vivo* tracking and personalized dosing strategies (Peng et al., 2024). Research into smart, biodegradable inorganic NPs and hybrid systems offers controlled degradation and stimuli-responsive release triggered by local pH, oxidative stress, or enzymatic activity (Aldeeb et al., 2024; Ma et al., 2024). Successful clinical translation of these technologies will require multidisciplinary collaboration to address regulatory, safety, and scalability challenges. Encouraging outcomes from ongoing clinical trials involving gold- and silica-based NPs indicate their potential to revolutionize ocular drug delivery by offering safer, more effective, and patient-specific therapies (Mulkutkar et al., 2024). Rigorous adherence to regulatory guidelines, clinical trial protocols, and comprehensive safety assessments remains essential for their successful incorporation into ophthalmic treatments (Rojekar et al., 2024; Sahu et al., 2024).

Conclusion

Inorganic NPs offer a revolutionary solution for ocular drug delivery, addressing major concerns such as low bioavailability, rapid drug elimination, and limited tissue penetration. Their precise engineering facilitates controlled drug release, enhanced corneal permeability, and targeted retinal delivery, making them highly effective in diseases like glaucoma, AMD, and diabetic retinopathy. Despite their promise, many challenges remain, including biocompatibility, toxicity, and regulatory hurdles. Although surface modifications have improved efficacy and safety, scalability and clinical translation continue to pose significant challenges. Standardization through regulation and large-scale, long-term safety studies are essential for clinical application. Future developments depend on advanced surface functionalization, AI-aided NP design, and personalized medical approaches to enhance therapeutic precision. Collaboration among researchers, clinicians, and regulatory bodies will be crucial to overcoming these challenges. With continued research, inorganic NPs hold great promise for revolutionizing ocular therapeutics, paving the way for more efficient and patient-friendly therapies.

References

Abdo, G. G., Zagho, M. M., & Khalil, A. (2020). Recent advances in stimuli-responsive drug release and targeting concepts using mesoporous silica nanoparticles. *Emergent Materials*, 3(3), 407–425.

Ahmed, B., Jaiswal, S., Naryal, S., Shah, R. M., Alany, R. G., & Kaur, I. P. (2024). In situ gelling systems for ocular drug delivery. *Journal of Controlled Release*, 371, 67–84.

Aldeeb, M. M. E., Wilar, G., Suhandi, C., Elamin, K. M., & Wathoni, N. (2024). Nanosuspension-based drug delivery systems for topical applications. *International Journal of Nanomedicine*, 19, 825–844.

Al-Thani, A. N., Jan, A. G., Abbas, M., Geetha, M., & Sadasivuni, K. K. (2024). Nanoparticles in cancer theragnostic and drug delivery: A comprehensive review. *Life Sciences*, 352, 122899.

Bachu, R. D., Chowdhury, P., Al-Saedi, Z. H. F., Karla, P. K., & Boddu, S. H. S. (2018). Ocular drug delivery barriers-Role of nanocarriers in the treatment of anterior segment ocular diseases. *Pharmaceutics*, 10(1), 28.

Cohen, H., Levy, R. J., Gao, J., Fishbein, I., Kousaev, V., Sosnowski, S., Slomkowski, S., & Golomb, G. (2000). Sustained delivery and expression of DNA encapsulated in polymeric nanoparticles. *Gene Therapy*, 7(22), 1896–1905.

Dandamudi, M., McLoughlin, P., Behl, G., Rani, S., Coffey, L., Chauhan, A., Kent, D., & Fitzhenry, L. (2021). Chitosan-coated PLGA nanoparticles encapsulating triamcinolone acetonide as a potential candidate for sustained ocular drug delivery. *Pharmaceutics*, 13(10), 1590.

De Vries, J. W., Schnichels, S., Hurst, J., Strudel, L., Gruszka, A., Kwak, M., Bartz-Schmidt, K., Spitzer, M. S., & Herrmann, A. (2017). DNA nanoparticles for ophthalmic drug delivery. *Biomaterials*, 157, 98–106.

Janagam, D. R., Wu, L., & Lowe, T. L. (2017). Nanoparticles for drug delivery to the anterior segment of the eye. *Advanced Drug Delivery Reviews*, 122, 31–64.

Dykman, L., Khlebtsov, B., & Khlebtsov, N. (2024). Drug delivery using gold nanoparticles. *Advanced Drug Delivery Reviews*, 216, 115481.

Elbedwehy, A. M., Wu, J., Na, H. K., Baek, A., Jung, H., Kwon, I. H., Lee, S. W., Kim, J. H., & Lee, T. G. (2024). ROS-responsive charge reversal mesoporous silica nanoparticles as promising drug delivery system for neovascular retinal diseases. *Journal of Controlled Release*, 373, 224–239.

Ramadan, A. E. H., Elsayed, M. M. A., Elsayed, A., Fouad, M. A., Mohamed, M. S., Lee, S., Mahmoud, R. A., Sabry, S. A., Ghoneim, M. M., Hassan, A. H. E., Abd Elkarim, R. A., Belal, A., & El-Shenawy, A. A. (2024). Development and optimization of vildagliptin solid lipid nanoparticles loaded ocuserts for controlled ocular delivery: A promising approach towards treating diabetic retinopathy. *International Journal of Pharmaceutics: X*, 7, 100232.

Gaudana, R., Ananthula, H. K., Parenky, A., & Mitra, A. K. (2010). Development of an ophthalmic nanoemulsion topical formulation for corneal wound healing. *Journal of Ocular Pharmacology and Therapeutics*, 12(4), 348–360.

Haase, A., Miroschnikov, N., Klein, S., Doege, A., Dünker, N., Meenen, D., Junker, A., Göpferich, A., Apaolaza, P., & Busch, M. (2024). New retinoblastoma (RB) drug delivery approaches: anti-tumor effect of atrial natriuretic peptide (ANP)-conjugated hyaluronic-acid-coated gold nanoparticles for intraocular treatment of chemoresistant RB. *Molecular Oncology*, 18, 832-849.

Gandara-Loe, J., Souza, B. E., Missyul, A., Giraldo, G., Tan, J. C., & Silvestre-Albero, J. (2020). MOF-based polymeric nanocomposite films as potential materials for drug delivery devices in ocular therapeutics. *ACS Applied Materials & Interfaces*, 12(27), 30189–30197.

Elumalai, K., Srinivasan, S., & Shanmugam, A. (2024). Review of the efficacy of nanoparticle-based drug delivery systems for cancer treatment. *Biomedical Technology*, 5, 109–122.

Khazaeli, P., Tabatabaei, M. S., Sayed-Tabatabaei, L., & Estabragh, M. A. R. (2025). Ocular novel drug delivery systems. In *Novel Drug Delivery Systems: Fundamentals and Applications*, 22(23), pp. 393–418. Wiley.

Khiev, D., Mohamed, Z. A., Vichare, R., Paulson, R., Bhatia, S., Mohapatra, S., Lobo, G. P., Valapala, M., Kerur, N., Passaglia, C. L., Mohapatra, S. S., & Biswal, M. R. (2021). Emerging nano-formulations and nanomedicines applications for ocular drug delivery. *Nanomaterials*, 11(1), 173.

Kulkarni, N. S., Josowitz, A., James, R., Liu, Y., Rayaprolu, B., Sagdullaev, B., Shameem, M. (2025). Latest trends & strategies in ocular drug delivery. *Methods*, 235, 100-117.

Kumbhar, P., Kolekar, K., Vishwas, S., Shetti, P., Kumbar, V., & Pinto, T. J. A. (2023). Treatment avenues for age-related macular degeneration: Breakthroughs and bottlenecks. *Ageing Research Reviews*, 98, 102322.

Kuno, N., & Fujii, S. (2011). Recent advances in ocular drug delivery systems. *Polymers*, 3(1), 193-221.

Ly, P. D., Ly, K. N., Phan, H. L., Nguyen, H. H. T., Duong, V. A., & Nguyen, H. V. (2024). Recent advances in surface decoration of nanoparticles in drug delivery. *Frontiers in Nanotechnology*, 6, 1456939.

Ma, Y., Zhang, Z., Yu, Y., Wang, X., Liu, S., Sun, B., & Zhao, J. J. N. T. (2024). Nanomaterials in the diagnosis and treatment of ophthalmic diseases. *Nanomaterials*, 54, 102117.

Mao, B., Tang, B., Yu, S., Ying, J., Wu, J., Lan, L., Wang, Y., Zan, X., Zheng, Q., & Li, J. (2024). A promising strategy for ocular noninvasive protein delivery: The case in treating corneal neovascularization. *Acta Biomaterialia*, 128, 349–361.

Meza-Rios, A., Navarro-Partida, J., Armendariz-Borunda, J., & Santos, A. (2020). Therapies based on nanoparticles for eye drug delivery. *Ophthalmology and Therapy*, 9(3), 1–14.

Mohamed, A. A. (2024). Nanoparticles for ocular drug delivery. *Al-Zahraa Journal for Health and Medical Sciences*, 2(2), 71–83.

Mohammadpour, R., & Ghandehari, H. (2022). Mechanisms of immune response to inorganic nanoparticles and their degradation products. *Advanced Drug Delivery Reviews*, 180, 114022.

Mulkutkar, M., Damani, M., & Sawarkar, S. (2024). Polymeric microneedles for the eye: An overview of advances and ocular applications for minimally invasive drug delivery. *European Journal of Pharmaceutics and Biopharmaceutics*, 114, 209.

Narayana, S., Gowda, B. H. J., Hani, U., Shimu, S. S., & Paul, K. (2024). Inorganic nanoparticle-based treatment approaches for colorectal cancer: Recent advancements and challenges. *Journal of Nanobiotechnology*, 22(1), 427.

Onugwu, A. L., Nwagwu, C. S., Onugwu, O. S., & Echezona, A. C. (2023). Nanotechnology-based drug delivery systems for the treatment of anterior segment eye diseases. *Journal of Controlled Release*, 354, 465–488.

Patel, A., Cholkar, K., Agrahari, V., & Mitra, A. K. (2013). Ocular drug delivery systems: An overview. *World Journal of Pharmacology*, 2(2), 47–59.

Patel, P., & Geed, S. R. (2024). Recent advancements in the application of nanomaterial in modern drug delivery and future perspective. In *Biogenic Nanomaterials for Environmental Sustainability: Principles, Practices, and Opportunities* (pp. 319–351). Elsevier.

Paul, W., & Sharma, C. P. (2020). Inorganic nanoparticles for targeted drug delivery. In *Biointegration of Medical Implant Materials* (pp. 333-373). Elsevier.

Peng, S., Wang, W., Zhang, R., Wu, C., Pan, X., & Huang, Z. (2024). Nano-formulations for pulmonary delivery: Past, present, and future perspectives. *Pharmaceutics*, 16(2), 161.

Pham, S. H., Choi, Y., & Choi, J. (2020). Stimuli-responsive nanomaterials for application in antitumor therapy and drug delivery. *Pharmaceutics*, 12(7), 630.

Rojekar, S., Parit, S., Gholap, A. D., Manchare, A., Nangare, S. N., & Hatvate, N. (2024). Revolutionizing eye care: Exploring the potential of microneedle drug delivery. *Pharmaceutics*, 16(11), 1398.

Fathi-Karkan, S., Ramsheh, N. A., Arkaban, H., Narooie-Noori, F., Sargazi, S., & Mirinejad, S. (2024). Nanosuspensions in ophthalmology: Overcoming challenges and enhancing drug delivery for eye diseases. *International Journal of Pharmaceutics*, 658, 124226.

Hatem, S., Mohammed, D., & Ezzat, N. (2024). Nanotechnology-based Strategies Overcoming the challenges of Retinoblastoma: A Comprehensive overview and future Perspectives. *Future Journal of Pharmaceutical Sciences*, 10(1), 537–554.

Sahu, D., Rath, G., & Gupta, G. (2024). Addressing ocular drug delivery challenges with solid nanofiber variants and supramolecular nanofibrous gel composite. *Journal of Drug Delivery Science and Technology*, 105, 476.

Shoval, A., Markus, A., Zhou, Z., Liu, X., Cazelles, R., Willner, I., & Mandel, Y. (2019). Anti-VEGF-aptamer modified C-DOTs—a hybrid nanocomposite for topical treatment of ocular vascular disorders. *Nano-Micro Small*, 15(40), 1902776.

Singh, S. R., Grossniklaus, H. E., Kang, S. J., Edelhauser, H. F., Ambati, B. K., & Kompella, U. B. (2009). Intravenous transferrin, RGD peptide, and dual-targeted nanoparticles enhance anti-VEGF intraceptor gene delivery to laser-induced CNV. *Gene Therapy*, 16(5), 645–659.

Kambhampati, S. P., & Kannan, R. M. (2013). Dendrimer nanoparticles for ocular drug delivery. *Journal of Ocular Pharmacology and Therapeutics*, 29(2), 151–165.

Sultana, S., Khan, M. R., Kumar, M., Kumar, S., & Ali, M. (2013). Nanoparticles-mediated drug delivery approaches for cancer targeting: A review. *Journal of Drug Targeting*, 21(2), 107–125.

Tietjen, G. T., Bracaglia, L. G., Saltzman, W. M., & Pober, J. S. (2018). Focus on fundamentals: Achieving effective nanoparticle targeting. *Trends in Molecular Medicine*, 24(7), 598–606.

Unnikrishnan, G., Joy, A., Megha, M., Kolanthai, E., & Senthilkumar, M. (2023). Exploration of inorganic nanoparticles for revolutionary drug delivery applications: A critical review. *Discover Nano*, 18(1), 157.

Vieira, I. R. S., Tessaro, L., & Conte-Junior, C. A. (2024). Gold nanoparticles for treatment of infectious diseases. In *Gold Nanoparticles for Drug Delivery* (pp. 277–303). Springer.

Wadzer, Y., Htwe, Y. Z. N., & Mamat, H. (2024). A review of ceramic nanofluids and their applications. *Journal of Dispersion Science and Technology*, 2024, 1–24.

Yadav, R. H., Kenchegowda, M., Angolkar, M., TS, M., & Osmani, R. A. M. (2024). A review of silk fibroin-based drug delivery systems and their applications. *European Polymer Journal*, 113, 286.

Yang, C., Yang, J., Lu, A., Gong, J., Yang, Y., Lin, X., Li, M., & Xu, H. (2022). Nanoparticles in ocular applications and their potential toxicity. *Frontiers in Molecular Biosciences*, 9, 931759.

Zhang, H., Mao, Y., Nie, Z., Li, Q., Wang, M., Cai, C., & Hao, W. (2024). Iron oxide nanoparticles engineered macrophage-derived exosomes for targeted pathological angiogenesis therapy. *ACS Nano*, 18(10), 7644–7655.

Zhang, J., Zhu, J., Zhao, L., Mao, K., Gu, Q., Li, D., Zhao, J., & Wu, X. (2021). RGD-modified multifunctional nanoparticles encapsulating salvianolic acid A for targeted treatment of choroidal neovascularization. *Journal of Nanobiotechnology*, 19, 1–16.

Zhou, H. Y., Hao, J. L., Wang, S., Zheng, Y., & Zhang, W. S. (2013). Nanoparticles in ocular drug delivery. *International Journal of Ophthalmology*, 6(3), 390.

Chapter 13

Nanotechnology in the Management of Retinal Disorders

Gaurav Mude[1,3]
Mohammad Tauqeer Sheikh[1,*]
Umesh Telrandhe[2]
Shantilal Singune[3]
and Sanjay Nagdev[4]
[1]K. C. Bajaj College of Pharmacy and Research, Jaripatka, Nagpur, Maharashtra, India
[2]Datta Meghe College of Pharmacy, DMIHER (DU), Sawangi, Wardha, Maharashtra, India
[3]Institute of Pharmaceutical Sciences, SAGE University, Indore, Madhya Pradesh, India
[4]Shri. Prakashchand Jain College of Pharmacy and Research, Palaskheda, Jamner, Maharashtra, India

Abstract

Nanotechnology is transforming ophthalmology by introducing innovative approaches for the detection, evaluation, and treatment of retinal disorders. This chapter examines how nanotechnology is addressing major challenges in managing retinal diseases such as diabetic retinopathy, age-related macular degeneration, and inherited retinal dystrophies. The unique properties of nanomaterials, including their nanoscale size, modifiable surfaces, and high drug-loading capacity, enable the development of advanced drug delivery systems that offer targeted, sustained, and efficient therapeutic effects with minimal systemic side effects. Additionally, nanotechnology-enhanced diagnostic

---

[*] Corresponding Author's Email: mohdtauqeer06@gmail.com

In: Advances in Nanotechnology for Ocular Drug Delivery
Editors: Anjali Sharma, Ajmer Singh Grewal, Geeta Deswal et al.
ISBN: 979-8-89530-814-1
© 2025 Nova Science Publishers, Inc.

tools, such as nanoparticle-based imaging techniques, provide improved sensitivity and precision for the early detection and monitoring of retinal conditions. The chapter also explores emerging medical applications of nanotechnology, including nanostructured retinal implants, nanomaterial-supported cell therapies, and gene delivery systems utilizing nanocarriers. These cutting-edge strategies aim to restore vision and prevent further retinal degeneration. Despite encouraging preclinical and early clinical outcomes, several challenges persist, including regulatory barriers, concerns regarding biocompatibility and toxicity, and the complexities involved in large-scale manufacturing. Nevertheless, the integration of nanotechnology into retinal therapeutics offers substantial promise for improving patient outcomes and advancing the field of precision medicine.

Keywords: nanotechnology, retinal disorders, targeted drug delivery, diagnostic imaging, retinal health, vision restoration

Abbreviations

AAV: Adeno-Associated Virus
AMD: Age-Related Macular Degeneration
BRB: Blood-Retinal Barrier
CNT: Carbon Nanotube
CRISPR: Clustered Regularly Interspaced Short Palindromic Repeats
DME: Diabetic Macular Edema
DR: Diabetic Retinopathy
iPSC: Induced Pluripotent Stem Cell
LCA: Leber Congenital Amaurosis
NPDR: Non-Proliferative Diabetic Retinopathy
OCT: Optical Coherence Tomography
PCL: Polycaprolactone
PDR: Proliferative Diabetic Retinopathy
PLA: Polylactic Acid
PLGA: Poly(lactic-co-glycolic acid)
RP: Retinitis Pigmentosa
RPE: Retinal Pigment Epithelium
VEGF: Vascular Endothelial Growth Factor

Introduction

Retinal diseases are a leading global cause of vision impairment and blindness, representing a significant public health concern. The retina, a light-sensitive tissue located at the back of the eye, is essential for converting light into neural signals that are transmitted to the brain. Among the most common and debilitating retinal disorders are age-related macular degeneration (AMD), diabetic retinopathy (DR), and inherited retinal dystrophies such as retinitis pigmentosa (RP). AMD, which primarily affects the aging population, exists in two main forms: dry (non-neovascular) and wet (neovascular). The dry form, accounting for 85–90% of cases, is marked by drusen accumulation and geographic atrophy, while the wet form involves rapid vision loss due to choroidal neovascularization (Wong et al., 2014). Genetic predisposition, particularly polymorphisms in the CFH and ARMS2 genes, along with oxidative stress and chronic inflammation, are major contributors to the pathogenesis of AMD (Fritsche et al., 2016). DR progresses from non-proliferative DR (NPDR) to proliferative DR (PDR) and can lead to diabetic macular edema (DME). It is characterized by microaneurysms, retinal hemorrhages, and pathological neovascularization, resulting from prolonged hyperglycemia-induced damage to the retinal vasculature (Yau et al., 2012). Since DR is often asymptomatic in its early stages, routine screening of individuals with diabetes is crucial for timely diagnosis and intervention. Despite advancements in ophthalmic diagnostics and therapeutic strategies, significant challenges remain, particularly in the early detection and long-term management of retinal diseases. Many of these conditions are asymptomatic at onset, leading to delayed diagnosis and potentially irreversible vision loss. Although current diagnostic modalities such as fundoscopy, optical coherence tomography (OCT), and fluorescein angiography are effective in identifying structural and vascular abnormalities, they may fail to detect early functional or molecular changes (Spaide et al., 2018). However, recent advancements in imaging technologies have greatly improved the early detection and monitoring of retinal diseases, enabling the identification of subclinical edema, neovascular changes, and ischemic regions, key indicators that facilitate timely and effective intervention (De Carlo et al., 2015).

Basics of Nanotechnology

Definition of Nanotechnology

The term "nano" is derived from the Greek word nanos, meaning "dwarf," and refers to a scale of one-billionth of a meter. At this nanoscale, materials exhibit unique behaviors that differ significantly from their bulk counterparts.

Core Principles of Nanotechnology

Quantum Effects
At the nanoscale, electron behavior changes material properties, influencing electrical, magnetic, and optical characteristics (Salata, 2004).

Bottom-Up vs. Top-Down Approaches
Nanostructures can be built atom-by-atom or molecule-by-molecule (bottom-up), or produced by reducing larger structures through methods such as etching or milling (top-down) (Buzea et al., 2007).

Properties of Nanomaterials Relevant to Biomedical Applications

Nanomaterials possess several features that make them particularly suitable for biomedical use:

Large Surface Area
A key characteristic of nanomaterials is their high surface area-to-volume ratio, which improves their ability to carry therapeutic agents. This enhances drug loading and facilitates efficient interactions with biological targets (Farokhzad & Langer, 2009).

Size and Penetrability
Due to their nanoscale dimensions, nanomaterials can penetrate biological barriers such as cell membranes, the blood-brain barrier, and tumor vasculature, making them ideal for molecular imaging and targeted drug delivery (Sun et al., 2009).

Overview of Types of Nanomaterials

Nanomaterials are classified based on their shape, composition, and intended applications. Common categories include nanofibers, nanostructures, and nanoparticles.

Nanoparticles

Nanoparticles are solid colloidal particles ranging from 1 to 100 nanometers in size and are widely used in various biomedical fields.

- Lipid-Based Nanoparticles: This group includes liposomes and solid lipid nanoparticles, valued for their biocompatibility and use in drug and gene delivery systems (Torchilin, 2005).
- Polymeric Nanoparticles: Made from biodegradable polymers such as PLGA (poly(lactic-co-glycolic acid)), these nanoparticles enable controlled and sustained drug release (Danhier et al., 2012).

Nanotechnology in Retinal Drug Delivery

Retinal abnormalities, such as DR, AMD, and retinal vein occlusion, are among the leading causes of irreversible blindness worldwide. Conventional ocular drug delivery methods, including topical (eye drops), systemic (oral or intravenous), and intraocular (intravitreal injections) routes, face considerable limitations, especially in effectively targeting the posterior segment of the eye.

Topical Administration

Topical formulations such as eye drops are non-invasive and convenient; however, their effectiveness in treating posterior segment diseases is limited. The corneal and conjunctival barriers significantly restrict drug penetration, with less than 5% of the administered dose typically reaching intraocular tissues (Gaudana et al., 2010).

Systemic Delivery

Oral and intravenous routes exhibit poor ocular bioavailability due to the restrictive nature of the blood-retinal barrier (BRB). Achieving therapeutic concentrations in retinal tissues often requires high systemic doses, thereby increasing the risk of systemic toxicity (Macha et al., 2001).

Nanocarriers for Targeted and Sustained Drug Delivery

Nanotechnology has facilitated the development of nanocarriers, engineered nanoscale materials capable of encapsulating therapeutic agents to enhance targeted delivery to retinal tissues while enabling sustained drug release.

Liposomes

Liposomes are spherical vesicles composed of phospholipid bilayers that can be functionalized with targeting ligands. Liposomal formulations of triamcinolone acetonide have demonstrated prolonged retention in the vitreous, thereby reducing the need for frequent intravitreal injections (Kim et al., 2009).

Polymeric Nanoparticles

Constructed from biodegradable polymers such as PLGA, polymeric nanoparticles offer controlled and sustained drug release. PLGA nanoparticles loaded with anti-VEGF agents have shown extended inhibition of neovascularization in AMD models (Koo et al., 2011).

Case Studies on Nanotechnology-Based Drug Delivery Systems

Case Study 1: Liposomal Delivery of Triamcinolone Acetonide

In a preclinical study, Kim et al. (2009) investigated the intravitreal administration of liposomal triamcinolone acetonide in a rabbit model. The findings demonstrated prolonged intraocular drug retention and a significant reduction in retinal inflammation, highlighting the potential of liposomes for sustained corticosteroid delivery.

Case Study 2: Anti-VEGF PLGA Nanoparticles for AMD
In a preclinical study, Kim et al. (2009) investigated the intravitreal administration of liposomal triamcinolone acetonide in a rabbit model. The findings demonstrated prolonged intraocular drug retention and a significant reduction in retinal inflammation, highlighting the potential of liposomes for sustained corticosteroid delivery.

Nanoparticle-Enhanced Imaging for Retinal Diagnostics

Nanoparticles provide a promising approach to enhance the efficacy of diagnostic tools in ophthalmology by enabling contrast enhancement, molecular targeting, and real-time monitoring capabilities.

Role of Nanoparticles in Improving Imaging Techniques

Nanoparticles exhibit highly tunable surface properties that allow functionalization with ligands, antibodies, or aptamers specifically designed to bind biomarkers associated with retinal diseases. When conjugated with targeting molecules, gold nanoparticles (AuNPs) can selectively highlight diseased retinal tissues or pathological changes, significantly improving detection sensitivity beyond conventional dyes and contrast agents (Huang et al., 2006).

Applications in Early Detection and Disease Monitoring

Engineered nanoparticles interact with specific biological structures or processes, providing targeted and enhanced imaging contrast. This precision facilitates acquisition of high-resolution, high-sensitivity images, enabling identification of pathological features at the earliest stages of disease progression (Chen et al., 2016). Early diagnosis is critical in conditions such as AMD and DR, where timely intervention can prevent irreversible vision loss. A notable example is the integration of nanoparticles with optical coherence tomography (OCT). Contrast agents like gold nanorods and nanoshells markedly enhance OCT signal intensity due to their superior light-scattering properties compared to native biological tissues. Studies in animal models have demonstrated that nanoparticle-assisted OCT improves visualization of retinal architecture and pathological lesions, highlighting their diagnostic potential (Qin et al., 2014).

Gene Therapy Using Nanotechnology

Gene therapy represents a transformative approach for treating inherited retinal diseases (IRDs), such as RP and Leber congenital amaurosis (LCA). Nanotechnology offers a powerful non-viral delivery platform that overcomes many limitations of traditional viral vectors, including immunogenicity, restricted cargo capacity, and challenges in repeated administration.

Nanocarriers for Gene Delivery to Retinal Cells

Nanocarriers can be precisely engineered to traverse ocular barriers, resist enzymatic degradation, and release therapeutic payloads in a controlled manner. For example, intravitreal administration of lipid-based nanoparticles has demonstrated efficient transfection of RPE cells with therapeutic genes, while minimizing inflammation and cytotoxicity (Han et al., 2012). These features underscore their potential for safe and effective ocular gene delivery.

Advances in Nanotechnology-Enhanced CRISPR/Cas Systems

The advent of CRISPR/Cas technology has revolutionized gene editing; however, targeted and efficient delivery of Cas9 proteins and guide RNAs into retinal cells remains challenging. Nanoparticle-based delivery systems have successfully facilitated gene knockout and correction in preclinical models of IRDs. Notably, Zhang et al. (2020) demonstrated that nanocarriers could deliver CRISPR/Cas components to retinal cells, resulting in successful correction of CEP290 gene mutations associated with LCA.

Potential for Treating Inherited Retinal Diseases

Nanotechnology-enabled gene delivery offers several advantages over viral vector-based methods:

Reduced Immunogenicity

Nanocarriers elicit minimal immune responses, addressing a major limitation of viral gene therapy.

Repeatable and Scalable Dosing

Unlike adeno-associated viruses (AAVs), which are limited by payload size and immune memory that restricts repeat dosing, nanoparticle-based systems allow for adjustable and repeated administration.

Nanotechnology in Cell-Based Therapies

Cell-based therapies, particularly those utilizing stem cells, have emerged as promising strategies for treating retinal degenerative diseases such as AMD and RP. Nanotechnology plays a crucial role in enhancing these therapies by supporting stem cell delivery, integration, and tissue regeneration.

Use of Nanomaterials in Stem Cell Delivery

Nanomaterials significantly improve stem cell-based approaches by enhancing cell survival, guiding targeted delivery, and promoting site-specific differentiation. They also enable real-time monitoring of transplanted cells, allowing researchers to track cell migration, viability, and integration within host retinal tissue (Bulte et al., 2004). This capability is essential for optimizing therapeutic outcomes and understanding stem cell behavior in vivo.

Nanostructured Scaffolds for Retinal Tissue Regeneration

Electrospun nanofibers fabricated from biodegradable polymers such as polycaprolactone (PCL), polylactic acid (PLA), and PLGA provide a highly porous structure with a large surface area, which promotes cellular attachment and proliferation. When functionalized with extracellular matrix proteins like laminin or fibronectin, these nanoscaffolds markedly enhance the adhesion, viability, and differentiation of RPE cells and photoreceptors, key components for restoring retinal structure and function (Hinton et al., 2017).

Examples of Preclinical and Clinical Progress

Promising results have been reported in preclinical studies. For example, PLGA-based nanoscaffolds seeded with human induced pluripotent stem cell (iPSC)-derived RPE cells have demonstrated successful integration into host tissue, leading to structural and functional retinal restoration in animal models of AMD (Lu et al., 2018).

Nanostructured Retinal Implants

Retinal implants represent a significant advancement in efforts to restore visual function in individuals affected by severe photoreceptor degeneration, such as in RP or late-stage AMD. By integrating nanoscale technologies, these devices aim to replicate or stimulate the natural phototransduction processes lost due to disease.

Development of Nanoscale Retinal Prosthetics

Recent innovations have focused on retinal prosthetics utilizing advanced nanomaterials, including silicon nanowires, carbon nanotubes (CNTs), and quantum dots, to enhance both the spatial and temporal resolution of visual signal processing. These materials offer high electrical conductivity, mechanical flexibility, and biocompatibility, enabling more precise interaction with retinal neurons (Chen et al., 2020).

Applications in Restoring Vision for Degenerative Disorders

Preclinical studies have demonstrated that carbon nanotube-based retinal interfaces can establish functional synaptic connections with retinal ganglion cells. These interactions restore downstream visual signaling, showing promise for partial recovery of visual perception (Chen et al., 2020). Such approaches provide potential alternatives to more invasive or biologically limited treatments.

Challenges and Future Perspectives

Despite promising progress, several challenges must be overcome to enable widespread clinical use of nanoscale retinal prosthetics:

Long-Term Biocompatibility and Immune Response
Prolonged implantation of nanomaterials in the eye requires thorough assessment of biocompatibility. Chronic inflammation, gliosis, and mechanical mismatch with delicate retinal tissue can impair implant function and patient safety.

Resolution and Signal Fidelity
While nanoscale materials improve electrode density and spatial resolution, converting these advances into meaningful, high-fidelity visual perception remains challenging. Disease progression, variability in retinal cell survival, and individual anatomical differences further complicate this translation.

Biocompatibility and Toxicity of Nanomaterials

While nanotechnology holds immense potential to revolutionize the treatment of retinal diseases, addressing the biocompatibility and toxicity of nanomaterials remains a critical challenge before widespread clinical adoption can be realized.

Evaluation of Safety Concerns

Nanomaterials exhibit unique physicochemical properties, such as ultra-small size, high surface area-to-volume ratio, and increased chemical reactivity, that enable novel interactions with biological systems. However, these same features may also lead to unpredictable and potentially harmful biological effects. For instance, nanoparticles can induce cytotoxicity, generate reactive oxygen species (ROS) causing oxidative stress, trigger inflammatory responses, or even cause genotoxicity through interactions with DNA (Soenen et al., 2011). As summarized in Table 1, several nanomaterials commonly used in ophthalmology offer significant therapeutic benefits but also pose potential toxicological risks. These effects depend on factors including material composition, shape, surface charge, coating, dosage, and duration of exposure. Consequently, thorough in vitro and in vivo safety evaluations are essential for every new nanomaterial proposed for ocular applications.

Table 1. Exploring the potential toxicity of common nanomaterials in ocular applications

Nanomaterial	Benefits	Potential Toxicity
Gold nanoparticles	Biocompatible, easy functionalization	Oxidative stress at high doses (Khlebtsov et al., 2011)
Quantum dots	Strong fluorescence for imaging	Heavy metal leakage (e.g., cadmium), cytotoxicity (Derfus et al., 2004)
Carbon nanotubes	Electrical conductivity, scaffold use	Inflammatory response, poor degradation (Pulskamp et al., 2007)
Silver nanoparticles	Antimicrobial, imaging	DNA damage, mitochondrial dysfunction (Kim et al., 2009)
Lipid nanoparticles	Efficient drug carriers	Generally safe but can cause immune responses (Paliwal et al., 2012)

Strategies for Minimizing Toxicity

To mitigate concerns related to nanotoxicity, several design and engineering strategies have been developed to enhance the safety and biocompatibility of nanomaterials used in retinal therapies.

Biodegradable Materials

Employing biodegradable polymers such as PLGA and chitosan allows nanoparticles to degrade into non-toxic, metabolizable by-products. This approach prevents the accumulation of residual particles in ocular tissues and reduces the risk of long-term cytotoxicity.

Targeted Delivery Systems

Functionalizing nanoparticles with ligands, peptides, or monoclonal antibodies enables precise targeting of diseased retinal cells. This targeted delivery not only improves therapeutic efficacy but also significantly minimizes off-target effects and systemic exposure, thereby enhancing the overall safety profile of the treatment.

Regulatory Considerations for Clinical Applications

As nanomedicine advances, regulatory agencies have begun establishing specific guidelines to ensure the safety, efficacy, and quality of nano-enabled therapeutics. Key institutions such as the U.S. Food and Drug Administration (FDA) and the European Medicines Agency (EMA) require the following:

Comprehensive Long-Term Biocompatibility Studies
To assess the chronic effects of nanomaterials within ocular environments.

Adherence to Good Manufacturing Practices (GMP)
To guarantee consistency, quality control, and reproducibility throughout production processes.

Compliance with ISO 10993 Standards
Especially relevant for biomedical devices, including ocular implants and drug delivery systems, focusing on biological evaluation and risk assessment (U.S. Food and Drug Administration, 2014).

Personalized Medicine and Nanotechnology

Personalized medicine aims to tailor therapeutic interventions to an individual's unique genetic, molecular, and phenotypic characteristics, with nanotechnology playing a crucial role in enabling this approach. By functionalizing nanomaterials with ligands or antibodies that specifically recognize disease-associated biomarkers, therapies can be selectively delivered to affected retinal cells, thereby enhancing efficacy and minimizing systemic exposure. For example, mutation-targeted delivery systems using nanoparticles loaded with RNA interference (RNAi) or CRISPR components have shown promise in correcting genetic defects such as those causing Leber's congenital amaurosis due to RPE65 mutations (Wang et al., 2019). Furthermore, patient-derived induced pluripotent stem cells (iPSCs) can be differentiated into RPE cells or photoreceptors and re-implanted using nanostructured scaffolds, offering a personalized regenerative strategy tailored to the individual's disease profile.

Preclinical and Clinical Studies

The expanding potential of nanotechnology in retinal therapies is increasingly validated by a growing body of preclinical and clinical studies demonstrating promising outcomes.

Summary of Key Preclinical Breakthroughs

Numerous animal model investigations have demonstrated the efficacy of nanoparticle-based treatments for retinal disorders such as AMD, diabetic retinopathy, and inherited retinal dystrophies. These studies underscore the ability of various nanocarrier systems to deliver therapeutic agents with precision and sustained effectiveness (Table 2). For instance, in mouse models of AMD, PLGA nanoparticles encapsulating anti-VEGF agents have successfully achieved prolonged suppression of pathological neovascularization (Kim et al., 2009).

Table 2. Notable preclinical studies in nanotechnology for retinal disorders

Study Model	Nanocarrier Type	Target Disease	Key Outcome
Rat (laser-induced CNV)	PLGA-anti-VEGF	AMD	Suppressed neovascularization
Mouse (rd1 model)	Lipid nanoparticles	RP	Restored rhodopsin expression
Rabbit	Chitosan nanoparticles	Gene therapy delivery	High retinal uptake, low toxicity

Overview of Ongoing and Completed Clinical Trials

Although challenges such as long-term safety and regulatory approval persist, clinical trials are actively underway and have reported encouraging outcomes. Notable examples include:

NCT05254075

A trial investigating nanocarrier-based delivery systems for retinal diseases, which has shown promising preliminary results.

NCT04049294

A study evaluating a contact lens-based biosensor for intraocular pressure monitoring in glaucoma patients, demonstrating a strong correlation between biosensor readings and standard tonometry, along with improved patient adherence (Mansouri et al., 2012).

Success Stories and Lessons Learned

While biodegradable nanocarriers generally exhibit favorable safety profiles, concerns remain about the potential accumulation of nanomaterials in ocular tissues. This highlights the critical need to design nanocarriers that ensure efficient clearance and biodegradability to minimize long-term risks (Fadeel et al., 2017).

Future Directions and Opportunities

Retinal diseases remain a leading cause of global vision loss, yet nanotechnology holds transformative potential for diagnosis, drug delivery, and vision restoration. Future advances in retinal healthcare are expected to be driven by enhanced precision, integration, and personalization. A particularly promising area is nanoscale neuroregeneration, where innovations such as self-assembling peptide nanostructures that mimic the extracellular matrix and graphene-based nanomaterials that support photoreceptor renewal show significant potential for treating conditions like AMD and RP (Holmes et al., 2002). Progress in this field depends heavily on interdisciplinary collaboration among materials scientists, ophthalmologists, nanorobotic engineers, AI specialists, and regulatory agencies. This collaborative ecosystem is exemplified by initiatives like the NIH/NIBIB in the United States, which supports translational nanotechnologies for ocular gene therapies. Looking ahead to 2035, retinal care is anticipated to incorporate smart nanomedicine, real-time diagnostics, and AI-enhanced decision-making, resulting in a more integrated and personalized model of care. Emerging nanotechnologies poised to shape clinical practice include CRISPR-loaded lipid nanoparticles for precise, non-viral gene editing in diseases such as LCA; quantum dot-enhanced OCT for early detection of diabetic microaneurysms; nano-robotic carriers enabling targeted, minimally invasive drug delivery; smart contact lenses embedded with nanoparticles for at-home monitoring of intraocular pressure in glaucoma; and AI-powered nano-analytics platforms for individualized treatment modeling.

Conclusion

Nanotechnology is revolutionizing the diagnosis, treatment, and restoration of vision by overcoming major challenges in managing retinal diseases. Conditions such as inherited retinal dystrophies, DR, and AMD often evade early detection and effective treatment through traditional methods. Nanotechnology offers targeted, sustained, and personalized therapies that improve treatment outcomes while minimizing systemic side effects. A key advantage is its ability to bypass anatomical barriers like the blood-retinal barrier, which conventional drug delivery methods such as systemic injections or eye drops often fail to overcome effectively. Various nanocarriers, including lipid-based systems, polymeric nanoparticles, and liposomes, enable precise and prolonged delivery of therapeutics to the retina. Clinical studies with liposomal triamcinolone and PLGA-based anti-VEGF nanoparticles have demonstrated extended therapeutic effects with reduced injection frequency and systemic exposure. Nanoparticles also enhance diagnostic imaging techniques such as OCT by improving contrast and enabling molecular targeting. Nanostructures like gold nanoparticles facilitate earlier detection of retinal diseases, allowing timely intervention and better management of chronic conditions like DR and AMD through real-time monitoring of disease progression. Gene therapy has advanced significantly, with nanoparticle-based delivery systems overcoming the immunogenicity and cargo limitations of viral vectors, enabling safer and more reliable treatment of inherited retinal diseases such as retinitis pigmentosa and Leber congenital amaurosis. Recent innovations include CRISPR-based systems and lipid nanoparticle delivery of therapeutic genes, which have shown promise in correcting genetic defects in preclinical models. Furthermore, nanotechnology supports stem cell therapies by enhancing cell survival, integration, and differentiation. Nanostructured scaffolds made from materials such as PLGA and PCL mimic the extracellular matrix and facilitate the growth of photoreceptors and retinal pigment epithelial cells, promoting tissue regeneration and functional recovery in retinal degeneration models. Despite these advancements, significant challenges remain to be addressed for broader clinical translation.

References

Abràmoff, M. D., Lavin, P. T., Birch, M., Shah, N., & Folk, J. C. (2018). Pivotal trial of an autonomous AI-based diagnostic system for detection of diabetic retinopathy in primary care offices. *NPJ digital medicine*, *1*(1), 39.

Bulte, J. W., & Kraitchman, D. L. (2004). Iron oxide MR contrast agents for molecular and cellular imaging. *NMR in Biomedicine*, *17*(7), 484-499.

Buzea, C., Pacheco, I. I., & Robbie, K. (2007). Nanomaterials and nanoparticles: sources and toxicity. *Biointerphases*, *2*(4), MR17-MR71.

Chen, J., Liu, Y., Peng, D., Wang, Y., Zhang, H., Li, X., & Zhao, Q. (2020). Nanostructured interfaces for neural stimulation. *Advanced Functional Materials, 30*(15), 1909035.

Chen, Y., Zhang, Y., Hu, Y., Zhou, T., Zhao, X., Wang, L., & Li, J. (2016). VEGF-targeted multifunctional nanoparticles for imaging and treatment of ocular neovascularization. *ACS Nano, 10*(1), 1236–1248.

Danhier, F., Ansorena, E., Silva, J. M., Coco, R., Le Breton, A., & Préat, V. (2012). PLGA-based nanoparticles: an overview of biomedical applications. *Journal of controlled release*, *161*(2), 505-522.

De Carlo, T. E., Romano, A., Waheed, N. K., & Duker, J. S. (2015). A review of optical coherence tomography angiography (OCTA). *International Journal of Retina and Vitreous, 1*, 5.

Derfus, A. M., Chan, W. C., & Bhatia, S. N. (2004). Probing the cytotoxicity of semiconductor quantum dots. *Nano Letters, 4*(1), 11–18.

Fadeel, B., Pietroiusti, A., & Shvedova, A. A. (2017). *Adverse Effects of Engineered Nanomaterials: Exposure, Toxicology, and Impact on Human Health*. Academic Press.

Farokhzad, O. C., & Langer, R. (2009). Impact of nanotechnology on drug delivery. *ACS Nano, 3*(1), 16–20.

Fritsche, L. G., Igl, W., Bailey, J. N. C., Grassmann, F., Sengupta, S., Bragg-Gresham, J. L., Burdon, K. P., Hebbring, S. J., Wen, C., Gorski, M., Kim, I. K., Cho, D., Zack, D., Souied, E. H., Scholl, H. P. N., Issa, P. C., Guymer, R. H., Chew, E. Y., Klein, M. L., & Zhang, K. (2016). A large genome-wide association study of age-related macular degeneration highlights contributions of rare and common variants. *Nature Genetics, 48*(2), 134–143.

Gao, W., & Wang, J. (2014). The dawn of synthetic micro/nanomotors. *Annual Review of Biomedical Engineering, 16*, 321–350.

Gaudana, R., Ananthula, H. K., Parenky, A., & Mitra, A. K. (2010). Ocular drug delivery. *AAPS Journal, 12*(3), 348–360.

U.S. Food and Drug Administration. (2014). Considering Whether an FDA-Regulated Product Involves the Application of Nanotechnology. https://www.fda.gov/

regulatory-information/search-fda-guidance-documents/considering-whether-fda-regulated-product-involves-application-nanotechnology.

Chapter 14

Emerging Nanotechnologies in Glaucoma Therapy

Navin Chandra Pant[1]
Himani Bajaj[2,*]
Vaseem Fateh[3]
and Arun Kumar Singh[1]
[1]Faculty of Pharmaceutical Sciences, Amrapali University, Haldwani, Uttarakhand, India
[2]AVIPS, Shobhit University, Gangoh, Saharanpur, Uttar Pradesh, India
[3]Burman Group of Institutions Roorkee, Haridwar, Uttarakhand, India

Abstract

Glaucoma is a leading cause of irreversible blindness worldwide and is characterized by a chronic, progressive loss of retinal ganglion cells. Despite treatments primarily focused on lowering intraocular pressure (IOP), many patients continue to experience vision loss, partly due to the poor bioavailability of current drugs within the eye caused by natural protective barriers that limit absorption. To enhance therapeutic efficacy, various drug delivery systems, including inorganic particles, polymers, hydrogels, and contact lens-based methods, have been explored for their ability to carry and release medications more effectively. Nanotechnology offers promising advancements in this area by utilizing small, biodegradable, and biocompatible nanoparticles to deliver IOP-lowering drugs and antifibrotic agents with sustained release profiles. This controlled delivery reduces dosing frequency, protects vulnerable ocular cells, and improves patient adherence. Furthermore, nanotechnology enables customization of treatment modalities, such as

[*] Corresponding Author's Email: himanibajaj@gmail.com

In: Advances in Nanotechnology for Ocular Drug Delivery
Editors: Anjali Sharma, Ajmer Singh Grewal, Geeta Deswal et al.
ISBN: 979-8-89530-814-1
© 2025 Nova Science Publishers, Inc.

eye drops, ocular inserts, or drug-loaded contact lenses, tailored to individual patient needs. Although significant progress has been made in slowing glaucoma progression, challenges remain in repairing or regenerating damaged tissues like the trabecular meshwork and retinal ganglion cell layer. Nevertheless, nanotechnology holds strong potential to overcome existing treatment limitations and facilitate more effective, long-term management of glaucoma.

Keywords: nanocarrier, ophthalmic drug delivery system, glaucoma, drug bioavailability, nanoparticles, ocular barriers, nanotechnology

Abbreviations

CSF: Cerebrospinal Fluid
IOP: Intraocular Pressure
LUVs: Large Unilamellar Vesicles
MLVs: Multilameller Vesicles
NPs: Nanoparticles
PNPs: Polymeric nanoparticles
SLNs: Solid Lipid Nanoparticles
SPNs: Spherical Polymeric Nanoparticles

Introduction

Glaucoma is a complex ocular disease that can result in irreversible blindness if not properly managed. It is primarily characterized by elevated intraocular pressure (IOP), which serves as a key indicator of the condition. In the United States, glaucoma ranks as the second leading cause of permanent blindness, especially among the elderly. The disease is broadly classified into primary and secondary types, with further subcategories including open-angle and angle-closure glaucoma. Among these, primary open-angle glaucoma is the most prevalent form affecting adults (Asrani et al., 2024). However, other forms, including angle-closure glaucoma and secondary open- or angle-closure glaucoma, also occur. Glaucoma is a progressive optic neuropathy characterized by degeneration of retinal ganglion cells and cupping of the optic disc (Pei et al., 2024). It affects approximately 3 to 5% of individuals aged 40 to 80 years and remains the leading cause of irreversible blindness worldwide (Iqbal et al., 2024). Its prevalence is rising in low-income countries with limited healthcare resources and is projected to affect over 112 million people by 2040 (Yadav et al., 2024). Early detection is crucial to slowing disease progression; however, diagnosis is often delayed because visual field loss typically remains asymptomatic until the later stages (Gautam et al., 2024). Several risk factors are associated with glaucoma, including elevated intraocular pressure, advancing age, non-Caucasian ethnicity, and a positive family history (Matuwana et al., 2024). Once diagnosed, individuals with glaucoma generally require lifelong medical management (Pardeshi et al., 2024).

Pathophysiology of Glaucoma

The exact cause of glaucoma remains unclear; however, a well-established association exists between elevated IOP and the degeneration of retinal ganglion cells (Klézlová et al., 2024). IOP is regulated by two primary processes: the production of aqueous humor by the ciliary body and its drainage through two pathways, the trabecular meshwork and the uveoscleral outflow. Although distinct, these systems work together to maintain balanced IOP. In open-angle glaucoma, increased resistance within the trabecular meshwork impedes fluid outflow, whereas in angle-closure glaucoma, the drainage angle becomes physically blocked (Roberti et al., 2020). Elevated IOP exerts mechanical stress on the posterior segment of the eye, particularly targeting the lamina cribrosa and adjacent tissues (Wang & Johnson, 2022). The lamina cribrosa, a perforated structure within the sclera through which optic nerve axons exit the eye, is considered the eye's weakest point under pressure (Schuster et al., 2020). Increased IOP can compress and deform this delicate region, potentially disrupting the retrograde axonal transport of vital neurotrophic factors from the brain, especially from the lateral geniculate nucleus to the retinal ganglion cells. This disruption may cause axonal injury and contribute significantly to glaucomatous optic nerve damage. Experimental studies in monkeys and cats have demonstrated that ocular hypertension disrupts both anterograde and retrograde axonal transport at the lamina cribrosa. This disruption occurs early in the disease process and results in the accumulation of vesicles, as well as disorganization of microtubules and neurofilaments within the prelaminar and postlaminar regions of the optic nerve. Comparable structural alterations have been observed in post-mortem optic nerves of human eyes affected by glaucoma. Additionally, mitochondrial dysfunction in retinal ganglion cells and astrocytes, caused by pressure-induced metabolic stress, may compromise their ability to meet high energy demands, increasing their susceptibility to injury. Notably, some individuals develop glaucomatous optic neuropathy despite maintaining normal IOP. One hypothesis proposes that low cerebrospinal fluid (CSF) pressure in the subarachnoid space surrounding the optic nerve creates a substantial pressure gradient across the lamina cribrosa, which may contribute to optic nerve damage (Pircher et al., 2025). Additional factors involved in glaucoma pathogenesis include immune system dysregulation, excitotoxicity, oxidative stress, and impaired microcirculation. Furthermore, primary neurodegeneration of retinal ganglion cells may trigger secondary degeneration of neurons in other parts of the central visual pathway by altering the cellular environment and increasing the vulnerability of these neurons to damage.

Nanocarrier-Based Formulations for Ocular Drug Delivery

Nanocarriers are specialized drug delivery systems characterized by their nanoscale size, typically ranging from 10 to 1000 nanometers, and distinct surface charges. These surface charges enhance nanocarrier stability in colloidal solutions, enable surface functionalization, and help maintain structural integrity at target sites. For instance, particles with similar charges repel each other, reducing aggregation risk. In ocular drug delivery, the surface charges of eye tissues are crucial: both the conjunctiva and cornea have negatively charged surfaces that strongly attract positively charged (cationic) nanocarriers. This interaction improves corneal retention and enables more effective topical drug delivery to the anterior segment. Conversely, negatively charged (anionic) nanocarriers tend to diffuse more efficiently to the retina when delivered via intravitreal injection, enhancing drug delivery to the posterior segment. The small size and tailored surface charge of nanocarriers are key to their ability to bypass ocular barriers and deliver therapeutics precisely to target tissues.

Liposomes

Liposomes, first introduced by Alec Bangham in 1965, are bilayer vesicles composed of phospholipids, typically ranging in size from 0.08 to 10.00

micrometers. Based on size and lamellarity, they are classified into three types: small unilamellar vesicles (SUVs, 10-100 nm), large unilamellar vesicles (LUVs, 100-300 nm), and multilamellar vesicles (MLVs, greater than 500 nm) (Batur et al., 2024). Liposomes have been widely used to deliver various therapeutic agents, including enzymes, vitamins, and drugs with anticancer, antifungal, antiviral, or anesthetic properties. Their structural similarity to biological cell membranes enhances their effectiveness as drug carriers (Natarajan et al., 2012). Liposomes are considered safe and can be functionalized by attaching molecules to their lipid bilayer, enabling targeted drug delivery. Due to their compatibility with ocular tissues and ability to carry both hydrophilic and lipophilic drugs, liposomes represent a promising system for ocular drug delivery. Studies have demonstrated that liposomal formulations effectively deliver drugs to both the anterior and posterior segments of the eye. They improve drug bioavailability and biocompatibility while facilitating controlled and sustained drug release (Zhai et al., 2021). The polar aqueous core and lipophilic bilayer of liposomes enable them to encapsulate and protect both water-soluble and fat-soluble drugs from enzymatic degradation, thereby enhancing drug stability. Although topical liposomal applications have demonstrated limited ocular penetration, studies indicate that subconjunctival delivery can partially sustain intraocular pressure reduction (González-Cela-Casamayor et al., 2022). However, the precise mechanisms governing liposome circulation and cellular uptake within the subconjunctival space remain unclear. Consequently, successful development of sustained-release ocular drug carriers requires formulations with high drug loading capacity that also maintain optical transparency to prevent visual disturbance.

Nanoparticles

Solid lipid nanoparticles (SLNs) are colloidal particles dispersed within a liquid medium, characterized by their nanoscale size, large surface area, and high drug encapsulation efficiency, making them well-suited for drug delivery applications. They are composed of biodegradable and biocompatible lipids such as fatty acids, waxes, and fats (Baig et al., 2024). To improve formulation stability, surfactants, stabilizers, and cosurfactants are incorporated. Nanostructured lipid carriers (NLCs) are an advanced form of SLNs created by blending solid and liquid lipids. Together, these lipid nanoparticles (LNPs), a class that includes SLNs and NLCs, offer an alternative to traditional nanoemulsions by replacing the oily phase with solid lipids. A limitation of SLNs is drug expulsion during lipid crystallization, which reduces drug loading capacity; this issue is addressed by NLCs, where liquid lipids enhance drug loading efficiency and storage stability (Gohil et al., 2025). Polymeric nanoparticles (PNPs) provide another effective strategy for ocular drug delivery challenges (Katiyar et al., 2014). These submicron particles are coated with biodegradable and biocompatible polymers, enabling encapsulation of diverse therapeutic agents such as small molecules, proteins, and genes for targeted, sustained delivery to ocular tissues (Kwon et al., 2020). PNPs show promise in treating various eye diseases including glaucoma, cataracts, age-related macular degeneration, dry eye syndrome, and diabetic retinopathy. Polymers like chitosan, polyethylene glycol (PEG), and poly(lactic-co-glycolic acid) (PLGA) are particularly valuable in ophthalmic use due to their safety, biodegradability, and their capacity to enhance drug absorption, prolong retention time, and reduce systemic side effects (Sánchez-López et al., 2018).

Nanomicelles

Nanomicelles are a promising and efficient drug delivery system known for their simplicity, safety, and high effectiveness. These self-assembling colloidal dispersions range from 10 to 100 nm in size and consist of a hydrophobic core surrounded by a hydrophilic shell (Gohar et al., 2024). Similar to liposomes, dendrimers, and other nanoparticles, nanomicelles enable controlled and targeted release of therapeutic agents over extended periods. Their advantages include excellent biocompatibility, high efficacy, and the capacity to encapsulate poorly water-soluble drugs. In ocular drug delivery, nanomicelles improve drug penetration through eye tissues while minimizing irritation and protecting drugs from enzymatic degradation. Their small size, straightforward manufacturing, and superior drug-loading capacity make nanomicelles an ideal platform for delivering drugs to the eye (Gohar et al., 2024).

Niosomes

Niosomes are nanocarriers that form through the self-assembly of nonionic surfactants in aqueous solutions, resulting in closed bilayer vesicles. Initially developed for cosmetic applications, niosomes were soon identified as promising candidates for drug delivery and have since been widely studied in pharmaceutical research (Durak et al., 2020). Researchers and pharmaceutical companies are actively exploring niosomes as bi- or multilayer nanoparticles due to their ability to encapsulate and release drugs or proteins in a controlled manner. Key areas of research include optimizing drug and protein loading, release kinetics, and effective encapsulation techniques. Niosomes have gained particular interest for their potential to improve the bioavailability and therapeutic efficacy of medications used to treat chronic inflammation and various types of cancer. For example, sustained release of BSA-loaded niosomes has been demonstrated. Additionally, the cholesterol content within niosome formulations can be adjusted to control drug release profiles. Niosomes offer several advantages as drug delivery systems, including low production cost, solution and storage stability, and the ability to carry both hydrophilic and hydrophobic drugs. They are also biocompatible, biodegradable, and structurally flexible. Due to these properties, niosomes show strong potential as nanocarriers in the treatment of ocular diseases (Shadab et al., 2025).

Dendrimers

Dendrimers are a relatively new class of polymeric compounds characterized by a unique, highly branched nanostructure, which has attracted significant attention for their potential in delivering drugs and imaging agents (Wang et al., 2023). These nanosized, "tree-like" polymers offer several advantages, including enhanced water solubility, bioavailability, and biocompatibility, making them well suited for various drug delivery applications. In ocular drug delivery, dendrimers have shown promise due to their customizable size, structural flexibility, and multiple modifiable surface groups that enable targeted delivery. They are capable of transporting medications to both the surface and deeper regions of the eye, such as the choroid and retina, potentially avoiding the need for invasive procedures (Yavuz et al., 2013). Dendrimers can either be chemically conjugated to drug molecules or used to encapsulate them. A notable example is polyamidoamine (PAMAM) dendrimers, which interact effectively with the corneal surface due to their cationic surface charge, spherical shape, and nanoscale size, thereby enhancing precorneal retention. Further advancements have introduced dendrimer hydrogels (DHs), three-dimensional polymer networks formed by cross-linking dendrimers. These hydrogels can be engineered to optimize the balance between fluidity and viscosity, making them ideal for topical ocular drug formulations (Pitha et al., 2023). Micro- and nano-sized dendrimer hydrogels (micro-/nano-gels) are especially attractive for ophthalmic applications. Their large surface area, adhesive properties, and ability to provide sustained drug release make them effective carriers for anti-glaucoma medications and other ocular therapies (Kambhampati & Kannan, 2013).

Nanofibers

Nanofibers have attracted considerable attention in ocular drug delivery and tissue engineering due to their unique properties, including a high surface-area-to-volume ratio, porous architecture, mechanical strength, flexibility, and structural similarity to the extracellular matrix (Cegielska et al., 2022). To enhance their interaction with the biological environment and to tailor drug release profiles, various formulation and processing modifications have been explored (Josyula et al., 2023). Among these methods, electrospinning remains the most widely used and scalable technique for producing nanofibers. Further improvements in scaffold functionality have been achieved through hybrid systems, such as incorporating nanoparticles into nanofibers, or employing advanced electrospinning techniques like coaxial electrospinning. Electrospun nanofibers create a matrix-like framework that supports cell regeneration and can deliver proteins, antibiotics, and anti-inflammatory agents directly to ocular tissues. Additionally, integrating nanofiber technology into medical devices, including contact lenses, opens innovative possibilities for practical ocular applications (Sakpal et al., 2022). Through nanofiber-based systems, researchers have made significant progress in addressing challenges such as poor drug bioavailability and tissue incompatibility in ocular treatments.

Nanosuspensions

Hydrophobic pharmaceuticals, characterized by poor solubility in both aqueous and organic solvents, often face limited bioavailability. Nanosuspensions provide a promising approach to overcome this limitation. The emergence of nanosuspension-based drug delivery has effectively addressed solubility challenges associated with poorly water-soluble drugs, thereby enhancing absorption and bioavailability (Aldeeb et al., 2024). These systems are attracting growing attention due to their straightforward formulation processes and broad applicability, especially as topical and mucosal delivery methods continue to advance. Extensive research has highlighted the potential of nanosuspensions for various topical applications, including ophthalmic, pulmonary, and dermatological uses. Within this field, nanocrystals have emerged as a notable innovation, particularly for cutaneous delivery. Although research on skin permeability and drug accumulation via nanosuspensions remains limited, mainly because these systems were originally developed for oral administration, preliminary results suggest that nanosuspensions may significantly improve cutaneous bioavailability, especially for active pharmaceutical ingredients with low to moderate solubility. Enhanced saturation solubility, dissolution rate, and skin adhesion afforded by nanocrystals contribute to their effectiveness in dermal drug delivery. With their submicron particle size and unique physicochemical characteristics, nanosuspensions offer a versatile platform for drugs with low solubility in both water and lipids. Current advancements in nanosuspension technology underscore their applicability across multiple administration routes, including pulmonary, intravenous, transdermal, and oral delivery. Additionally, recent progress focuses on converting liquid oral nanosuspensions into more stable solid dosage forms such as powders, pellets, granules, tablets, and capsules, addressing important stability challenges (Jacob et al., 2025).

Nano and Microemulsions

Nanoemulsions are among the most promising platforms for enhancing ocular drug delivery, especially for hydrophobic pharmaceutical compounds. Successful development of ocular nanoemulsions requires a comprehensive understanding of both pharmaceutical and technological factors, particularly the choice of excipients and formulation methods (Srivastava et al., 2024). Oil-in-water nanoemulsions are particularly advantageous for hydrophobic drugs because the dispersed oil droplets substantially increase the surface area available for drug dissolution and improve membrane penetration, thereby enhancing overall bioavailability. However, formulating effective ocular nanoemulsions remains challenging due to the eye's complex anatomy and physiology. The ocular surface is highly sensitive and protected by robust defense mechanisms, including rapid clearance of foreign substances through blinking and tear secretion. For example, the lacrimal glands produce tear fluid at a rate accounting for approximately 16% of the total tear volume per minute under normal physiological conditions, which contributes to rapid elimination of instilled drugs. These factors create two major challenges: limited residence time and reduced drug absorption. Low-energy emulsification methods often show variable effectiveness depending on the formulation composition, making the process less robust for some applications. Conversely, high-energy emulsification techniques are increasingly favored due to their scalability for industrial production. A critical formulation challenge is ensuring ocular tolerability, especially concerning preservative choice. Benzalkonium chloride (BAK), although widely used, is known to cause ocular surface toxicity with long-term use. To address this, potassium sorbate has been selected as a milder alternative. Comparative studies between potassium sorbate-preserved nanoemulsions and conventional BAK-containing ophthalmic solutions have shown that the former are less irritating and more biocompatible with the ocular surface (Gautam et al., 2024).

Conclusion

Glaucoma affects individuals of all ages and remains a leading cause of irreversible blindness worldwide. A major challenge in glaucoma management is patient non-compliance, often linked to the limited bioavailability of conventional topical eye drops. This challenge is especially critical given that glaucoma is a chronic condition requiring consistent, lifelong treatment with frequent dosing. Nanomedicine offers promising solutions to these limitations by enhancing drug bioavailability, prolonging therapeutic action, reducing side effects, and improving patient adherence. Nanocarrier-based therapies have demonstrated potential in overcoming several pharmacological barriers, including poor penetration of the blood-retinal and blood-brain barriers, unfavorable pharmacokinetics, and short in vivo half-lives of drugs. Through targeted and sustained drug delivery, nanocarriers can maintain therapeutic efficacy while lowering required dosages. Moreover, they can simultaneously carry both hydrophilic and hydrophobic drugs, either via chemical conjugation or encapsulation within nanoscale cavities, enabling combination therapies in a single formulation. Although preclinical studies on neuroprotective nanocarriers have shown encouraging outcomes, clinical research remains in the early stages. Further clinical trials are essential to establish their safety and therapeutic effectiveness. While not yet a standard glaucoma treatment, neuroprotective nanocarriers hold significant promise as a future strategy for more effective and patient-friendly glaucoma care.

References

Aldeeb, M. M. E., Wilar, G., Suhandi, C., Elamin, K. M., & Wathoni, N. (2024). Nanosuspension-based drug delivery systems for topical applications. *International journal of nanomedicine*, *19*(2024), 825-844.

Asrani, S. G., McGlumphy, E. J., Al-Aswad, L. A., Chaya, C. J., Lin, S., Musch, D. C., Pitha, I., Robin, A. L., Wirostko, B., & Johnson, T. V. (2024). The relationship between intraocular pressure and glaucoma: An evolving concept. *Progress in Retinal and Eye Research*, *103*, 101303.

Baig, M. S., Karade, S. K., Ahmad, A., Khan, M. A., Haque, A., Webster, T. J., Faiyazuddin, M., & Al-Qahtani, N. H. (2024). Lipid-based nanoparticles: innovations in ocular drug delivery. *Frontiers in Molecular Biosciences*, *11*, 1421959.

Batur, E., Özdemir, S., Durgun, M. E., & Özsoy, Y. (2024). Vesicular Drug Delivery Systems: Promising Approaches in Ocular Drug Delivery. *Pharmaceuticals*, *17*(4), 511.

Cegielska, O., Sierakowski, M., Sajkiewicz, P., Lorenz, K., & Kogermann, K. (2022). Mucoadhesive brinzolamide-loaded nanofibers for alternative glaucoma treatment. *European Journal of Pharmaceutics and Biopharmaceutics*, *180*, 48-62.

Durak, S., Esmaeili Rad, M., Alp Yetisgin, A., Eda Sutova, H., Kutlu, O., Cetinel, S., & Zarrabi, A. (2020). Niosomal drug delivery systems for ocular disease—Recent advances and future prospects. *Nanomaterials*, *10*(6), 1191.

Gautam, D., Talwan, P., Chaurasia, H., Kumar, S., & Singh, R. (2024). Nanotechnology Carriers for the Management, Electrochemical Detection and Diagnosis of Glaucoma. *Electrocatalytic Materials*, 527-559.

Gohar, S., Iqbal, Z., Nasir, F., Khattak, M. A., e Maryam, G., Pervez, S., Alasmari, F., Neau, S. H., Zainab, S. R., & Ali, A. T. (2024). Self-assembled latanoprost loaded soluplus nanomicelles as an ophthalmic drug delivery system for the management of glaucoma. *Scientific Reports*, *14*(1), 27051.

Gohil, U., Chandarana, C., Prajapati, P., & Prajapati, B. (2025). Polymeric Nanoparticles in Ophthalmology: A Comprehensive Review of Therapeutic Applications. *BioNanoScience*, *15*(1), 1-19.

González-Cela-Casamayor, M. A., López-Cano, J. J., Bravo-Osuna, I., Andrés-Guerrero, V., Vicario-de-la-Torre, M., Guzmán-Navarro, M., Benítez-del-Castillo, J. M., Herrero-Vanrell, R., & Molina-Martínez, I. T. (2022). Novel osmoprotective DOPC-DMPC liposomes loaded with antihypertensive drugs as potential strategy for glaucoma treatment. *Pharmaceutics*, *14*(7), 1405.

Iqbal, H., Razzaq, A., Zhou, D., Lou, J., Xiao, R., Lin, F., & Liang, Y. (2024). Nanomedicine in glaucoma treatment; Current challenges and future perspectives. *Materials Today Bio*, 101229.

Jacob, S., Kather, F. S., Boddu, S. H., Attimarad, M., & Nair, A. B. (2025). Nanosuspension Innovations: Expanding Horizons in Drug Delivery Techniques. *Pharmaceutics*, *17*(1), 136.

Josyula, A., Mozzer, A., Szeto, J., Ha, Y., Richmond, N., Chung, S. W., Rompicharla, S. V. K., Narayan, J., Ramesh, S., & Hanes, J. (2023). Nanofiber-based glaucoma drainage implant improves surgical outcomes by modulating fibroblast behavior. *Bioengineering & Translational Medicine*, *8*(3), e10487.

Kambhampati, S. P., & Kannan, R. M. (2013). Dendrimer nanoparticles for ocular drug delivery. *Journal of Ocular Pharmacology and Therapeutics*, *29*(2), 151-165.

Katiyar, S., Pandit, J., Mondal, R. S., Mishra, A. K., Chuttani, K., Aqil, M., Ali, A., & Sultana, Y. (2014). In situ gelling dorzolamide loaded chitosan nanoparticles for the treatment of glaucoma. *Carbohydrate polymers*, *102*, 117-124.

Klézlová, A., Bulíř, P., Klápšťová, A., Netuková, M., Šenková, K., Horáková, J., & Studený, P. (2024). Novel Biomaterials in Glaucoma Treatment. *Biomedicines*, *12*(4), 813.

Kwon, S., Kim, S. H., Khang, D., & Lee, J. Y. (2020). Potential therapeutic usage of nanomedicine for glaucoma treatment. *International journal of nanomedicine*, 5745-5765.

Matuwana, D., Hong, E., Huang, S., Xu, X., Jang, G., Xiao, R., Rao, S., & Wang, Q. (2024). Near-infrared activated liposomes for neuroprotection in glaucoma. *Journal of Materials Chemistry B*, *12*(42), 10902-10914.

Natarajan, J. V., Ang, M., Darwitan, A., Chattopadhyay, S., Wong, T. T., & Venkatraman, S. S. (2012). Nanomedicine for glaucoma: liposomes provide sustained release of latanoprost in the eye. *International journal of nanomedicine*, *7*, 123-131.

Pardeshi, S. R., Gholap, A. D., Hatvate, N. T., Gharat, K. D., Naik, J. B., & Omri, A. (2024). Advances in dorzolamide hydrochloride delivery: harnessing nanotechnology for enhanced ocular drug delivery in glaucoma management. *Discover Nano*, *19*(1), 1-27.

Pei, K., Georgi, M., Hill, D., Lam, C. F. J., Wei, W., & Cordeiro, M. F. (2024). Neuroprotective Nanocarriers in Glaucoma. *Pharmaceuticals*, *17*(9), 1190.

Pircher, A., Berberat, J., Remonda, L., Roberts, C. J., Neutzner, A., & Killer, H. E. (2025). Incidence of optic nerve kinking in a cohort of patients with Normal tension glaucoma. *Eye*, 1-6.

Pitha, I., Kambhampati, S., Sharma, A., Sharma, R., McCrea, L., Mozzer, A., & Kannan, R. M. (2023). Targeted microglial attenuation through Dendrimer–drug conjugates improves Glaucoma neuroprotection. *Biomacromolecules*, *24*(3), 1355-1365.

Roberti, G., Oddone, F., Agnifili, L., Katsanos, A., Michelessi, M., Mastropasqua, L., Quaranta, L., Riva, I., Tanga, L., & Manni, G. (2020). Steroid-induced glaucoma: Epidemiology, pathophysiology, and clinical management. *Survey of Ophthalmology*, *65*(4), 458-472.

Sakpal, D., Gharat, S., & Momin, M. (2022). Recent advancements in polymeric nanofibers for ophthalmic drug delivery and ophthalmic tissue engineering. *Biomaterials Advances*, *141*, 213124.

Sánchez-López, E., Egea, M. A., Davis, B. M., Guo, L., Espina, M., Silva, A. M., Calpena, A. C., Souto, E. M. B., Ravindran, N., & Ettcheto, M. (2018). Memantine-loaded

PEGylated biodegradable nanoparticles for the treatment of glaucoma. *Small*, *14*(2), 1701808.

Schuster, A. K., Erb, C., Hoffmann, E. M., Dietlein, T., & Pfeiffer, N. (2020). The diagnosis and treatment of glaucoma. *Deutsches Ärzteblatt International*, *117*(13), 225.

Shadab, S., Alam, M. A., Sharma, P. K., & Paliwal, D. (2025). Overcoming the Therapeutic Limitation of Ocular Drug Delivery with the Help of Novel Drug Carriers. *Current Nanomaterials*, *10*(1), 34-42.

Srivastava, V., Pardhi, E. R., Yadav, R., Singh, V., Khatri, D. K., & Mehra, N. K. (2024). QbD-driven thymoquinone laden nanoemulsion for glaucoma management: In vitro, ex vivo, and pre-clinical evaluation. *Journal of Drug Delivery Science and Technology*, *94*, 105493.

Wang, J., Li, B., Kompella, U. B., & Yang, H. (2023). Dendrimer and dendrimer gel-derived drug delivery systems: Breaking bottlenecks of topical administration of glaucoma medications. *MedComm–Biomaterials and Applications*, *2*(1), e30.

Wang, Q., & Johnson, T. V. (2022). Wang, Q., & Johnson, T. V. (2022). Pathophysiology of neovascular glaucoma. In M. Qiu (Ed.), *Neovascular Glaucoma: Current Concepts in Diagnosis and Treatment* (pp. 11-21). Springer.

Yadav, K. S., Tripathi, S., Vyas, J., Singhal, D., Joshi, G., & Soni, G. (2024). Advances in drug-loaded contact lenses for glaucoma: Materials, evaluation parameters, and novel drug delivery strategies with modified nanoparticles. *Journal of Drug Delivery Science and Technology*, 99, 105949.

Yavuz, B., Bozdağ Pehlivan, S., & Ünlü, N. (2013). Dendrimeric systems and their applications in ocular drug delivery. *The Scientific World Journal*, *2013*(1), 732340.

Zhai, Z., Cheng, Y., & Hong, J. (2021). Nanomedicines for the treatment of glaucoma: current status and future perspectives. *Acta Biomaterialia*, *125*, 41-56.

Chapter 15

Advances in Nanocarrier-Based Ocular Drug Delivery for Glaucoma Management

Poonam Talwan[1]
Nitasha Chauhan[2]
Avichal Kumar[3]
Sanjay Kumar[4]
and Darsh Gautam[5,*]
[1]Department of Pharmaceutical Chemistry, Gautam College of Pharmacy, Hamirpur, Himachal Pradesh, India
[2]Department of Pharmaceutical Sciences and Technology, Maharaja Ranjit Singh Punjab Technical University (MRSPTU), Bathinda, Punjab, India
[3]Department of Pharmaceutics, KLE College of Pharmacy, Bengaluru, Karnataka, India
[4]Department of Pharmacognosy, Gautam College of Pharmacy, Hamirpur, Himachal Pradesh, India
[5]Department of Pharmaceutics, Gautam College of Pharmacy, Hamirpur, Himachal Pradesh, India

Abstract

Glaucoma is a leading cause of irreversible blindness worldwide, primarily resulting from the progressive degeneration of the optic nerve. Conventional treatments face significant challenges, including low ocular bioavailability, frequent dosing schedules, systemic side effects, and poor patient adherence. These limitations underscore the urgent need for more effective, sustained, and patient-friendly therapeutic strategies. Nanocarrier-based drug delivery systems have emerged as a promising

---

[*] Corresponding Author's Email: drsgautam86@gmail.com

In: Advances in Nanotechnology for Ocular Drug Delivery
Editors: Anjali Sharma, Ajmer Singh Grewal, Geeta Deswal et al.
ISBN: 979-8-89530-814-1
© 2025 Nova Science Publishers, Inc.

approach, offering enhanced drug solubility, improved ocular penetration, prolonged residence time, and controlled release profiles. This chapter examines the evolving role of nanocarriers in glaucoma management, beginning with an overview of key anatomical and physiological barriers such as the corneal epithelium, tear film, and blood-aqueous barrier that impede efficient drug delivery to ocular tissues. It further explores how nanotechnology can overcome these obstacles through particle size optimization, surface engineering, and advanced drug encapsulation techniques. Special attention is given to smart delivery systems, including stimuli-responsive nanocarriers that release drugs in response to environmental cues like pH changes, and targeted nanocarriers designed for specific ocular tissues to maximize therapeutic efficacy. The chapter also discusses the potential of combination therapies to achieve synergistic effects and improve clinical outcomes. Finally, critical considerations regarding safety, toxicity, and regulatory challenges are addressed, emphasizing the pathway toward clinical translation. Overall, nanotechnology holds great promise for transforming glaucoma treatment by enhancing drug delivery and patient outcomes.

Keywords: glaucoma, nanocarriers, sustained release, stimuli-responsive nanoparticles, targeted ocular therapy, intraocular pressure, ocular bioavailability

Abbreviations

AI: Artificial intelligence
ATMPs: Advanced therapy medicinal products
GUV: Giant unilamellar vesicles
IOP: Elevated intraocular pressure
LUV: Large unilamellar vesicles
MIGS: Minimally invasive glaucoma surgeries
ML: Machine learning
MMPs: Matrix metalloproteinases
NTG: Normal tension glaucoma
PAMAM: Poly(amidoamine)
PCL: Polycaprolactone
PEG: Polyethylene glycol
PLA: Polylactic acid
PLGA: Poly(lactic-co-glycolic acid)

PMMP: Poly(methylmethacrylate)
PNIPAM: Polymers like poly(N-isopropylacrylamide)
PNPs: Polymeric nanoparticles
POAG: Primary open-angle glaucoma
RGCs: Retinal ganglion cells
SLT: Selective laser trabeculoplasty
SUV: Small unilamellar vesicles

Introduction

Glaucoma is a chronic and progressively worsening ocular condition characterized by the degeneration of retinal ganglion cells (RGCs), thinning of the retinal nerve fiber layer, and structural deterioration of the optic nerve head, ultimately leading to visual impairment (Zhai et al., 2021). Due to its gradual onset and the brain's ability to compensate for lost visual input, diagnosis is often delayed until the disease has advanced significantly. Elevated intraocular pressure (IOP) is a major contributor to disease progression, closely associated with optic nerve damage and accelerated RGC degeneration (Quigley, 1999). Often called the "silent thief of sight," glaucoma is among the leading causes of irreversible blindness globally, affecting over 95 million people, with approximately 10 million experiencing blindness in one eye (Jayaram et al., 2023; Vijaya et al., 2008; Tham et al., 2014). Glaucoma is broadly classified into primary and secondary types, each further subdivided into open-angle and angle-closure forms (Sommer et al., 1991). Primary glaucoma, which includes primary open-angle glaucoma (POAG) and normal tension glaucoma (NTG), arises without identifiable causes (Ciulla et al., 2017). POAG typically involves elevated IOP, whereas NTG results in optic nerve damage despite normal IOP levels. Secondary glaucoma occurs due to known causes such as trauma or inflammation (Leske, 1983). Angle-closure glaucoma encompasses both primary (acute or chronic) and secondary forms. Acute primary angle-closure glaucoma leads to sudden spikes in IOP, while chronic forms develop gradually. Although less common, angle-closure glaucoma carries a higher risk of blindness. Glaucoma also varies by age of onset. Early-onset glaucoma, occurring before age 40, is often linked to Mendelian genetic mutations that are rare but highly penetrant. In contrast, adult-onset glaucoma typically appears after age 40 and involves complex inheritance patterns with more common but milder genetic variants (Harasymowycz et al., 2016). Management of glaucoma presents significant challenges, including maintaining consistent intraocular pressure (IOP) control, enhancing patient adherence to treatment regimens, and minimizing adverse side effects. While conventional pharmacological therapies remain the mainstay, they often suffer from limited efficacy, poor ocular bioavailability, and frequent dosing requirements. Emerging treatments such as selective laser trabeculoplasty (SLT) and minimally invasive glaucoma surgeries (MIGS) provide promising alternatives with improved safety profiles and efficacy. In parallel, advanced drug-eluting devices and nanotechnology-based delivery systems are gaining attention for their potential to optimize drug bioavailability, prolong therapeutic action, and reduce treatment burden. This chapter delves into the latest nanocarrier-based innovations aimed at overcoming current limitations and improving glaucoma management outcomes.

Physiological and Anatomical Barriers in Ocular Drug Delivery

The human eye captures and processes light through interconnected layers and compartments that ensure clear vision (Ng & Oliver, 2018). The outer layer consists of the cornea, which focuses light, and the sclera, which maintains the shape of the eye and protects its internal structures. The uvea includes the iris, ciliary body, and choroid; it controls light entry, produces aqueous humor, and nourishes the retina. The retina is the innermost layer and contains rods for vision in dim light and cones for color and detail. The lens focuses light onto the retina with support from the aqueous and vitreous humor, which also help maintain the eye's shape. The optic nerve transmits visual signals to the brain (Figure 1).

Glaucoma drug delivery encounters multiple barriers that reduce drug bioavailability and ocular retention. Tear drainage, blinking, and conjunctival absorption limit the effectiveness of topical medications (Ghate and Edelhauser, 2008). The cornea's lipophilic epithelium and hydrophilic stroma restrict drug permeability while the blood-aqueous and blood-retinal barriers limit systemic and posterior segment drug access (Gunda et al., 2008). Efflux transporters such as P-glycoprotein and enzymatic degradation further decrease drug levels. Rapid clearance often necessitates frequent dosing which can lead to poor patient compliance. To address these challenges advanced delivery systems including sustained release implants, nanocarriers, drug eluting devices, and prodrugs are being developed to improve drug delivery, retention, and therapeutic outcomes in glaucoma treatment (Akhter et al., 2022; Zeng et al., 2023).

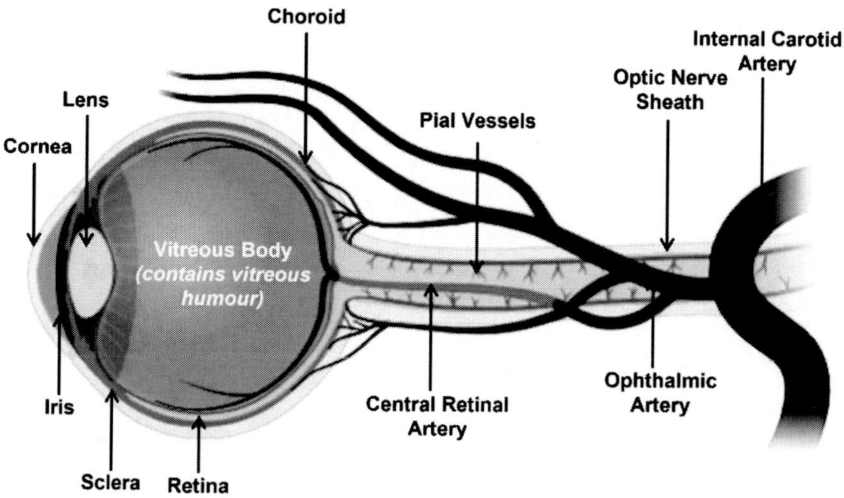

Figure 1. A comprehensive depiction of the human eye's anatomy.

Nanocarrier-Based Approaches for Glaucoma Therapy

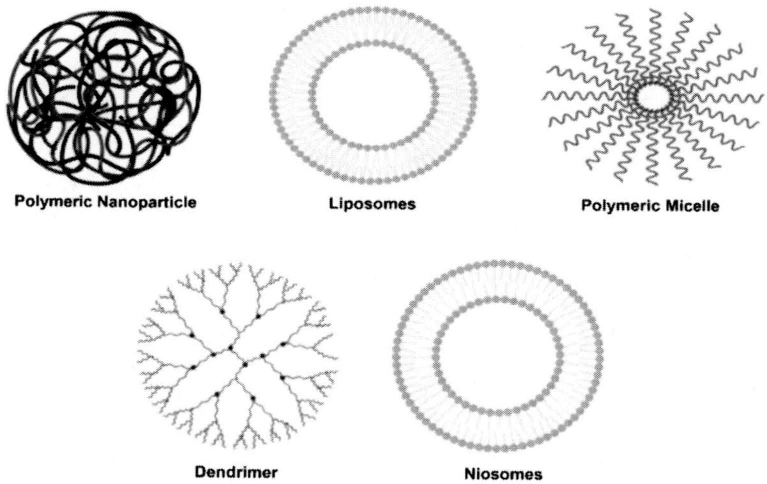

Figure 2. Nano-based delivery platforms designed to improve glaucoma therapy through ocular administration.

Nanocarrier-based strategies are emerging as promising alternatives to improve drug delivery in glaucoma treatment. These advanced systems, including liposomes, polymeric nanoparticles, dendrimers, niosomes, and micelles, are designed to overcome ocular barriers, enhance drug retention, and increase bioavailability (Akhter et al., 2022). By enabling sustained release and reducing the frequency of administration, they improve patient compliance and offer potential benefits for both intraocular pressure regulation and neuroprotection. Figure 2 presents an overview of various nanocarriers investigated for ocular drug delivery in glaucoma management.

Liposomes

Liposomes, first discovered by Alec Bangham in 1965, are spherical vesicles made up of phospholipid bilayers that enclose an aqueous core (Zeng et al., 2023; Syama et al., 2022). Their ability to carry both water-soluble and fat-soluble drugs has made them a vital tool in drug delivery, as seen with the USFDA-approved formulation Doxil, alongside a significant rise in research with over 10,000 publications in the past decade (Baryakova et al., 2023; Dillinger et al., 2018). Liposomes can be single- or multilamellar and are commonly classified by size into small unilamellar vesicles (SUV), large unilamellar vesicles (LUV), and giant unilamellar vesicles (GUV) (Lipowsky, 2019). Their biocompatibility, ease of degradation, and surface properties make them highly effective for drug delivery. In ophthalmology, liposomes are valued for providing controlled release and targeted drug delivery, enhancing the treatment efficacy for various ocular diseases (Mishra et al., 2011).

Polymeric Nanoparticles (PNPs)

Nanoparticles offer an innovative strategy to improve drug delivery across the eye's protective barriers, a critical advantage in treating diseases like glaucoma where effective tissue penetration and retention are major therapeutic challenges (Suri et al., 2020). These nanoscale carriers can encapsulate drugs by either embedding them within a solid core, as seen in nanospheres, or by enclosing them within a polymeric or lipid-based shell, as in nanocapsules (Reis et al., 2006). Various materials, including synthetic and natural polymers, are used to develop these systems (Panda et al., 2023;

Niculescu & Grumezescu, 2021). Their small size and modifiable surface properties grant nanoparticles strong bioadhesive abilities and extended retention on the ocular surface, thereby improving drug absorption. Additionally, surface functionalization with mucoadhesive agents or targeting ligands enhances their capacity to deliver drugs deeper into ocular tissues while providing controlled release. These technological advances highlight the significant potential of nanoparticles to overcome the limitations of traditional ocular therapies, offering more effective treatment options for glaucoma and other eye disorders.

Biodegradable PNPs

Biodegradable polymeric nanoparticles are increasingly recognized as effective carriers for ocular drug delivery in glaucoma treatment due to their biocompatibility and capacity for sustained drug release. These nanoparticles are commonly formulated using polymers such as poly(lactic-co-glycolic acid) (PLGA), polylactic acid (PLA), and polycaprolactone (PCL) (Figure 3) (Soppimath et al., 2001; Sánchez-López et al., 2018). PLGA is particularly favored because it safely degrades into lactic and glycolic acids, which are naturally metabolized by the body. These biodegradable carriers can encapsulate a wide range of drugs, both hydrophobic and hydrophilic, enabling gradual release that supports stable intraocular pressure control in glaucoma patients (Tsung et al., 2023). Additionally, they protect drugs from enzymatic degradation within the eye, thereby improving drug availability. Recent advancements include surface modification with hydrophilic agents like polyethylene glycol (PEG), which enhances retention in the tear film and facilitates corneal penetration (Prajapati et al., 2024). Targeting ligands such as hyaluronic acid and chitosan are also used to increase mucoadhesion and improve drug delivery to the eye's posterior segment.

Non-Biodegradable PNPs

Non-biodegradable polymeric nanoparticles, such as those fabricated from poly(methylmethacrylate) (PMMA), polystyrene, or ethyl cellulose (Figure 3), provide sustained drug release without undergoing metabolic degradation, making them well-suited for long-term glaucoma management (Singh et al., 2024). Their rigid structure supports extended drug retention, particularly for hydrophobic drugs. However, concerns remain regarding ocular accumulation and biocompatibility. To mitigate these issues, hybrid systems featuring biocompatible surface coatings are being developed. Further advancements in mucoadhesive and stimuli-responsive coatings that respond to pH or temperature changes enhance targeting and retention within intraocular tissues (Sapino et al., 2019; Jung et al., 2013). These innovations contribute to improved delivery efficiency and better patient compliance.

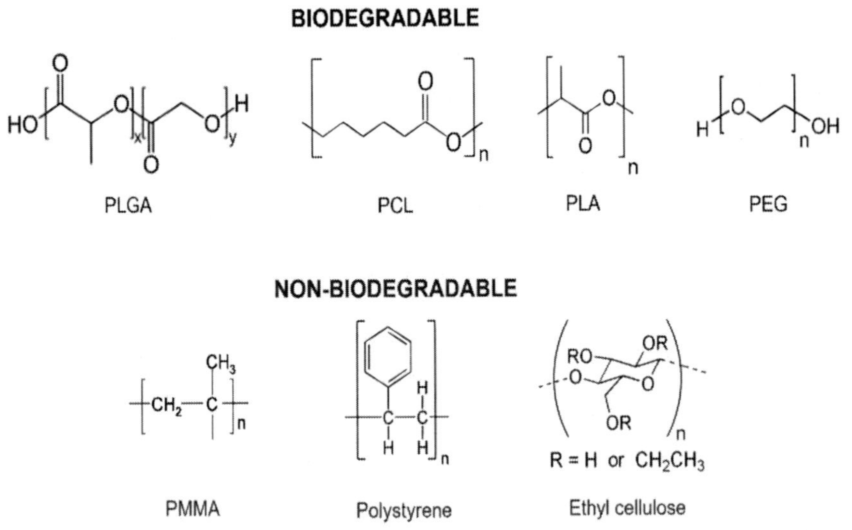

Figure 3. The chemical structures of biodegradable and non-biodegradable polymers utilized in the formulation of ocular nanoparticulate systems.

Dendrimers

The term "dendrimers" originates from the Greek words "dendron," meaning tree, and "meros," meaning part. These highly branched, tree-like macromolecules were first introduced by Vogtle in 1978 (Parat & Felder-Flesch, 2016). Structurally, dendrimers have a globular or ellipsoidal shape composed of a central core, multiple branching layers, and surface functional groups. They are water-soluble and contain internal cavities capable of encapsulating drugs, which makes them well-suited for drug delivery applications (Holden et al., 2012). In ocular drug delivery, dendrimers provide advantages over other water-soluble polymers by enhancing drug retention and solubility, particularly for hydrophilic drugs (Spataro et al., 2010; Khare et al., 2015). Poly(amidoamine) (PAMAM) dendrimers, equipped with functional groups such as amine and carboxyl, improve ocular bioavailability and show significant potential for treating eye conditions like glaucoma (Taghavi et al., 2014; Mishra & Jain, 2014).

Niosomes

Niosomes are vesicular nanocarriers composed of non-ionic surfactants and cholesterol, offering improved stability and biocompatibility for glaucoma drug delivery (Razavi et al., 2022). Compared to liposomes, niosomes reduce oxidative degradation and extend shelf life. They can encapsulate both water-soluble and fat-soluble drugs, providing controlled release that helps maintain intraocular pressure over time (Chandel & Kandav, 2024; Tripathi & Yadav, 2024). The vesicular structure of niosomes protects drugs from enzymatic degradation and enhances bioavailability. Surface modifications using mucoadhesive polymers such as chitosan or hyaluronic acid improve corneal retention and drug penetration. Recent advances include stimuli-responsive formulations designed for targeted delivery, promoting better intraocular pressure control and reduced toxicity (Arshad et al., 2024; Abdelmonem et al., 2021).

Micelles

Micelles are nanoscale carriers formed by the self-assembly of amphiphilic block copolymers in aqueous environments, creating a hydrophobic core surrounded by a hydrophilic shell (Imran & Shah, 2018). This architecture enhances the solubility and bioavailability of poorly water-soluble glaucoma drugs such as prostaglandin analogs and beta-blockers. Their small size, typically between 10 and 100 nanometers, facilitates improved corneal penetration and reduces clearance by tears, thereby increasing ocular retention and drug absorption (Cai et al., Jain et al., 2025). Micelles based on polyethylene glycol provide stability and protect drugs from degradation (Zhang et al., 2014). Targeted micelles functionalized with ligands like hyaluronic acid or folic acid enhance delivery to tissues involved in intraocular pressure regulation, including the trabecular meshwork and ciliary body (Wickens et al., 2017). Table 1 summarizes recent ocular nanoformulations.

Table 1. List of nanoformulations explored for managing glaucoma through ocular drug delivery. These formulations aim to overcome the eye's natural barriers and enhance drug bioavailability, retention, and targeted delivery to the anterior and posterior segments

Nanoformulation	API	Composition	Preparation Method	*In-Vitro/In-vivo Model*	Application in Glaucoma Management	References
Nanoparticles	Memantine	PLGA-PEG	Nanoprecipitation method	Rodent model of ocular hypertension	MEM-NP formulation provides prolonged drug release and reducing dosing frequency.	(Sánchez-López et al., 2018)
	siRNA targeting connective tissue growth factor	Lipid, layered hyaluronan, CD44 binding	siRNA encapsulation, layer deposition	-	Targeted delivery, Tissue penetration, gene silencing, intraocular pressure reduction.	(Dillinger et al., 2018)
	Brimonidine	Polymers, emulsifiers and nanoparticles stabilizers	Emulsification solvent diffusion method	*In-vitro* transcorneal permeation drug release	Reduced intraocular pressure, reduced dosing frequency, enhanced penetration, and improved efficacy.	(Prajapati et al., 2024)
	Timolol	Propoxylated glyceryl triacylate, macromer	Thermal polymerization	Albino rats	Controlled release, patient compliance, intraocular pressure reduction	(Jung et al., 2013)
Dendrimer	Acetazolamide	Ethylenediamine, poly(propylene imine)	Equilibrium dialysis technique	Albino Rats	Sustained drug release, ocular safety, extended retention, intraocular pressure reduction	(Mishra & Jain, 2014)
	Brimonidine	PEG,12000 DA, tetrahydrofuran, triethanolamine	Equilibrium dialysis technique	*Ex-vivo* studies: Ocular tissues	Enhanced delivery of antiglaucoma drugs	(Holden et al., 2012)

Nanoformulation	API	Composition	Preparation Method	*In-Vitro/In-vivo Model*	Application in Glaucoma Management	References
Niososmes	Brimonidine	Non-ionic surfactant, cholesterol, silicone hydrogel, ethylene glycol dimethacrylate, phosphate buffer saline	Thin film hydration	Abino rats	Prolonged drug retention, improved drug penetration, enhanced bioavailability, patient compliance	(Tripathi & Yadav, 2024)
	Dipivefrin HCl	Span 20, span 60 and span 80, cholesterol, carbopol 934, locust	Ether injection method	*In-vitro* drug release using Franz diffusion cell	Slow and prolonged drug release from niosomal gel formulations, reduced intraocular pressure	(Jain et al., 2025)
	Acetazolamide and carvediol	Polyoxyethylene 20, sorbitan monolaurate, glycol sorbitan monostearate, sorbitan monolaurate, sorbitan monostearate, cholesterol	Reverse phase evaporation technique	Ocular hypersensitive rabbits	Decreased intraocular pressure, reduced aqueous humor secretion, sustained intraocular pressure, patient compliance	(Abdelmonem et al., 2021)

Key Design Considerations in Developing Nanocarriers

Particle Size, Surface Chemistry and Stability

Particle size plays a crucial role in the performance of nanocarriers for glaucoma therapy by influencing corneal penetration, ocular retention, and clearance rates. Nanoparticles sized between 10 and 100 nanometers demonstrate optimal penetration and reduced elimination through the nasolacrimal duct, whereas larger particles exceeding 200 nanometers are cleared more quickly, which decreases bioavailability (Wu et al., 2011). Particle size also impacts cellular uptake and diffusion through the ocular mucus layer. Surface chemistry is equally important; hydrophilic coatings such as polyethylene glycol enhance stability and extend residence time on the tear film by preventing opsonization (Desai, 2012). Additionally, mucoadhesive polymers like chitosan or hyaluronic acid improve adhesion to the cornea and increase drug retention. Maintaining stability during storage and administration is essential, often achieved through techniques like lyophilization or freeze-drying to preserve formulation integrity.

Drug Loading and Release Kinetics

The effectiveness of nanocarriers in glaucoma therapy depends significantly on drug loading capacity and release kinetics. High drug loading reduces the amount of carrier material needed, thereby lowering toxicity and enhancing patient compliance (Cegielska and Sajkiewicz, 2019). Lipid-based carriers are well suited for hydrophobic drugs such as prostaglandin analogs, while aqueous-core systems like liposomes are more appropriate for hydrophilic drugs (Hnin et al., 2022). Biodegradable polymers, including PLGA and PCL, facilitate sustained drug release. A controlled release profile, ideally following zero-order kinetics, helps maintain therapeutic drug levels over time and reduces the need for frequent dosing. Strategies such as polymer cross-linking, surface modification, and stimuli-responsive designs like pH-sensitive systems enable more precise and targeted drug delivery (Fleige et al., 2012). To address the common issue of burst release, researchers optimize surface characteristics and strengthen drug–carrier interactions.

Biocompatibility and Biodegradability

Biocompatibility and biodegradability are essential for ensuring the safe application of nanocarriers in ocular glaucoma therapy (Sah et al., 2020). Biocompatible carriers must not cause irritation or toxicity in delicate ocular tissues. Biodegradable polymers such as PLGA, PLA, and chitosan are commonly used because they break down into harmless byproducts like lactic and glycolic acid (Bee et al., 2018). The rate of degradation can be adjusted by altering the polymer composition, allowing for controlled and sustained drug release. Although non-biodegradable polymers like PMMA can provide extended drug delivery, they pose a risk of tissue accumulation. PEGylation enhances biocompatibility by minimizing immune reactions and preventing particle aggregation. For regulatory approval, comprehensive ocular toxicity studies are required to confirm long-term safety and therapeutic effectiveness.

Targeted and Stimuli-Responsive Delivery Systems

Targeting Aqueous Humour Dynamics

Targeted delivery systems for glaucoma are designed to modulate aqueous humor dynamics by directing therapeutic agents specifically to the tissues responsible for its production and drainage. Disruption in aqueous humor flow, whether through overproduction or impaired outflow, is a key factor contributing to elevated intraocular pressure in glaucoma (Goel et al., 2010). These targeted nanocarriers aim to improve therapeutic outcomes while minimizing systemic side effects by achieving site-specific delivery. Surface modification of nanoparticles with targeting ligands such as hyaluronic acid, peptides, or antibodies enables selective interaction with receptors found in ocular tissues involved in intraocular pressure regulation. For example, hyaluronic acid-functionalized nanoparticles can bind to CD44 receptors on the trabecular meshwork, allowing localized delivery of drugs like prostaglandin analogs or rho kinase inhibitors (El-Dakdouki & Huang, 2011). Additionally, advances in liposomal and polymeric nanoparticles have led to targeted delivery systems for drugs such as timolol and brimonidine, focusing on the ciliary body to reduce aqueous humor production. Recent innovations include multifunctional carriers that simultaneously target both the trabecular meshwork and the ciliary body, offering a comprehensive strategy to regulate aqueous humor flow and maintain intraocular pressure within normal physiological ranges.

Stimuli-Responsive Nanocarriers

Smart nanocarrier systems that respond to physiological stimuli are emerging as valuable tools in ocular drug delivery, particularly for glaucoma therapy. These systems offer controlled drug release triggered by environmental factors such as changes in pH, enzymatic activity, or temperature (Nguyen & Lai, 2020). This adaptability allows for targeted, on-demand drug release, which enhances therapeutic precision and reduces drug wastage. In glaucoma treatment, pH-responsive carriers exploit the slightly acidic ocular microenvironment or the altered pH present in diseased tissues. Polymers like poly(β-amino esters) and PLGA have been used to encapsulate anti-glaucoma drugs that are released when local pH levels change. Similarly, enzyme-responsive systems target elevated levels of enzymes such as matrix metalloproteinases found in glaucomatous tissues. These carriers are often functionalized with enzyme-specific peptide linkers, enabling selective drug release at the disease site while minimizing off-target exposure. For example, liposomes modified with MMP-cleavable peptides have been developed to deliver brimonidine directly to affected ocular areas, improving localized drug activity and therapeutic outcomes (Tewari et al., 2022).

Temperature-Sensitive Systems

Temperature-responsive nanocarriers offer an innovative strategy for delivering drugs in glaucoma therapy by releasing medication in response to temperature changes associated with ocular inflammation or disease activity (Lin et al., 2021). Polymers like poly(N-isopropylacrylamide) (PNIPAM) have a lower critical solution temperature, enabling drug release when the ocular surface temperature rises above a defined threshold (Wu et al., 2023). Thermosensitive gels containing nanoparticles can shift from a liquid to a gel state upon contact with the eye, enhancing drug retention and reducing dosing frequency. Additionally, dual-responsive systems that react to both temperature and pH allow for more precise control of drug release (Jiang & Xu, 2023; Liu et al., 2024). While issues related to formulation stability, large-scale production, and long-term safety remain, these systems show great potential for enhancing therapeutic outcomes and patient adherence in glaucoma treatment.

Safety, Toxicity, and Regulatory Perspectives

Preclinical and Clinical Safety Profiles of Nanocarriers

Ensuring the safety and minimizing toxicity of nanocarriers in glaucoma management is essential due to the eye's sensitivity to inflammation and oxidative stress. Preclinical safety assessments typically involve in vitro studies using corneal and retinal cell lines, ex vivo testing on porcine or rabbit corneas, and in vivo animal models to evaluate pharmacokinetics and potential inflammatory responses. Biodegradable nanocarriers, such as liposomes and niosomes, generally exhibit good biocompatibility, while surface modifications can improve ocular retention and reduce irritation (Dobrovolskaia, 2015). Clinical evaluations include ophthalmic examinations, intraocular pressure monitoring, and histopathological analysis, with particular attention to limiting systemic absorption and avoiding local or systemic toxicity (Al-Lawati et al., 2018).

Regulatory Challenges in Ophthalmic Nanotechnology

Ocular nanocarriers encounter significant regulatory challenges due to their intricate formulations and the stringent standards applied to ophthalmic products. Regulatory bodies such as the FDA and EMA often classify these systems as combination products or advanced therapy medicinal products (ATMPs), necessitating comprehensive characterization of both the nanocarrier and the encapsulated drug. Critical requirements include detailed physicochemical characterization, such as particle size, morphology, surface charge, and stability testing, to ensure product consistency and shelf-life (Pignatello & Puglisi, 2011; Rahman et al., 2018). A major hurdle lies in the absence of standardized safety evaluation protocols for nanocarriers, given their unique interactions with biological tissues. Therefore, long-term studies and thorough biocompatibility assessments are essential to address concerns related to immunogenicity, inflammation, and potential systemic toxicity.

Future Directions and Opportunities

Integration of Artificial Intelligence (AI) and Machine Learning (ML) in Nanocarrier Design

AI and ML are revolutionizing the design and development of nanocarriers for glaucoma therapy by significantly reducing reliance on traditional trial-and-error methods (Gao et al., 2024). These technologies leverage large datasets to predict the behavior of nanocarriers, enabling the optimization of key parameters such as particle size, surface charge, and material composition to enhance ocular penetration and reduce toxicity. AI-powered in silico models simulate interactions with ocular tissues and forecast drug release profiles under physiological conditions. Additionally, molecular docking techniques assist in identifying targeting ligands specific to glaucoma-affected tissues such as the trabecular meshwork. AI also facilitates the design of multi-drug delivery systems and predicts the long-term stability and safety of formulations. As these technologies continue to advance, they support the development of adaptive, patient-specific nanocarriers that align treatment strategies with individual disease progression and therapeutic needs (Jayasinghe et al., 2022).

Personalized Medicine Approaches

Personalized medicine represents a significant advancement in glaucoma treatment by tailoring therapeutic strategies to an individual's genetic makeup, disease progression, and ocular physiology. Nanocarrier-based drug delivery systems offer a versatile platform for such personalized therapies, as they can be customized to deliver multiple drugs and adjust release profiles based on specific biomarkers. Pharmacogenomic testing identifies genetic variations that influence drug metabolism and response, enabling the formulation of nanocarriers optimized for each patient's unique needs (Baghban et al., 2023). These nanocarriers can also be adapted to account for anatomical differences and coexisting conditions, ensuring more precise drug targeting and improved retention in ocular tissues. Moreover, innovations such as wearable ocular sensors and continuous intraocular pressure (IOP) monitoring provide real-time feedback on disease status, supporting dynamic and responsive drug release from smart nanocarrier systems. The integration of nanotechnology with pharmacogenomics and precision diagnostics paves the way for more effective, individualized glaucoma management.

Challenges and Opportunities in Scaling Up Production

Scaling up nanocarrier production for glaucoma therapy involves challenges in formulation consistency, sterilization, and regulatory compliance, but also holds promise for commercialization (Gorantla et al., 2020). Ensuring batch-to-batch uniformity in size and surface charge is vital, as it affects safety and efficacy. Technologies like microfluidics and high-shear homogenization offer precise, scalable production. Sterilization poses hurdles, as methods like autoclaving may degrade formulations; thus, aseptic processing or filtration is preferred. Regulatory demands require proof of stability and consistency. Innovations like AI-based quality control and collaborations across sectors are advancing standardized manufacturing and regulatory pathways, facilitating broader clinical use of nanocarrier-based glaucoma treatments.

Conclusion

Nanocarrier-based strategies have revolutionized glaucoma therapy by improving drug solubility, stability, and bioavailability, while enabling targeted and sustained drug delivery. Carriers such as liposomes, micelles, dendrimers, and nanoparticles enhance ocular retention, minimize systemic side effects, and allow controlled release, thereby addressing many of the limitations associated with conventional therapies. Advanced systems, including stimuli-responsive and combination nanocarriers, enable site-specific and on-demand drug release in response to environmental cues such as pH changes or enzymatic activity. The simultaneous delivery of both pharmacological agents and genes provides a dual therapeutic benefit by lowering intraocular pressure and offering neuroprotection. Furthermore, the integration of artificial intelligence and machine learning facilitates the optimization of nanocarrier design and supports the development of therapies tailored to individual patient biomarkers. Smart nanocarriers combined with real-time intraocular pressure monitoring may also enable closed-loop drug delivery systems that adjust treatment dynamically. Despite these advances, challenges such as scalable manufacturing, formulation stability, and regulatory approval remain significant. Addressing these barriers will require progress in microfluidic technologies, aseptic processing techniques, and stronger collaboration among academic institutions, industry stakeholders, and regulatory agencies. Ultimately, nanocarrier technologies hold the promise to shift glaucoma management from symptom control to proactive, personalized treatment, improving long-term outcomes and preserving vision.

References

Abdelmonem, R., Elhabal, S. F., Abdelmalak, N. S., El-Nabarawi, M. A., & Teaima, M. H. (2021). Formulation and characterization of acetazolamide/carvedilol niosomal gel for glaucoma treatment: In vitro, and in vivo study. *Pharmaceutics, 13*(2), 221.

Al-Lawati, H., Aliabadi, H. M., Makhmalzadeh, B. S., & Lavasanifar, A. (2018). Nanomedicine for immunosuppressive therapy: Achievements in pre-clinical and clinical research. *Expert Opinion on Drug Delivery, 15*(4), 397–418.

Akhter, M. H., Ahmad, I., Alshahrani, M. Y., Al-Harbi, A. I., Khalilullah, H., Afzal, O., Altamimi, A. S., Najib Ullah, S. N., Ojha, A., & Karim, S. (2022). Drug delivery challenges and current progress in nanocarrier-based ocular therapeutic system. *Gels, 8*(2), 82.

Arshad, N., Shaheen, F., Khan, I. N., Naeem, S., Riaz, M., Siddique, M. I., Ayesha, M., & Waqar, M. A. (2024). A comprehensive review on niosomes: Novel manufacturing techniques, factors influencing formation, applications and recent advances. *International Journal of Polymeric Materials and Polymeric Biomaterials*, 2024, 1–8.

Baghban, R., Talebnejad, M. R., Meshksar, A., Heydari, M., & Khalili, M. R. (2023). Recent advancements in nanomaterial-laden contact lenses for diagnosis and treatment of glaucoma: Review and update. *Journal of Nanobiotechnology, 21*(1), 402.

Baryakova, T. H., Pogostin, B. H., Langer, R., & McHugh, K. J. (2023). Overcoming barriers to patient adherence: The case for developing innovative drug delivery systems. *Nature Reviews Drug Discovery, 22*(5), 387–409.

Bee, S. L., Hamid, Z. A., Mariatti, M., Yahaya, B. H., Lim, K., Bee, S. T., & Sin, L. T. (2018). Approaches to improve therapeutic efficacy of biodegradable PLA/PLGA microspheres: A review. *Polymer Reviews, 58*(3), 495–536.

Cai, R., Zhang, L., & Chi, H. (2023). Recent development of polymer nanomicelles in the treatment of eye diseases. *Frontiers in Bioengineering and Biotechnology, 11*, 1246974.

Chandel, A., & Kandav, G. (2024). Insights into ocular therapeutics: A comprehensive review of anatomy, barriers, diseases and nanoscale formulations for targeted drug delivery. *Journal of Drug Delivery Science and Technology*, 97, 105785.

Cegielska, O., & Sajkiewicz, P. (2019). Targeted drug delivery systems for the treatment of glaucoma: Most advanced systems review. *Polymers, 11*(11), 1742.

Ciulla, L., Harris, A., Geng, J., Ip, C., Serrano, P., Gross, J., & Siesky, B. (2017). The role of hypertension in retinal blood flow alterations in open-angle glaucoma patients. *Acta Ophthalmologica, 95*(8), e794–e795.

Desai, N. (2012). Challenges in development of nanoparticle-based therapeutics. *The AAPS Journal, 14*(2), 282–295.

Dillinger, A. E., Guter, M., Froemel, F., Weber, G. R., Perkumas, K., Stamer, W. D., Ohlmann, A., Fuchshofer, R., & Breunig, M. (2018). Intracameral delivery of layer-by-layer coated siRNA nanoparticles for glaucoma therapy. *Small, 14*(50), 1803239.

Dobrovolskaia, M. A. (2015). Pre-clinical immunotoxicity studies of nanotechnology-formulated drugs: Challenges, considerations and strategy. *Journal of Controlled Release, 220*, 571–583.

El-Dakdouki, M., & Huang, X. (2011). Biological applications of hyaluronic acid functionalized nanomaterials. In *Petite and Sweet: Glyco-Nanotechnology as a Bridge to New Medicines* (pp. 181–213). American Chemical Society.

Fleige, E., Quadir, M. A., & Haag, R. (2012). Stimuli-responsive polymeric nanocarriers for the controlled transport of active compounds: Concepts and applications. *Advanced Drug Delivery Reviews, 64*(9), 866–884.

Gao, X. J., Ciura, K., Ma, Y., Mikolajczyk, A., Jagiello, K., Wan, Y., Gao, Y., Zheng, J., Zhong, S., Puzyn, T., & Gao, X. (2024). Toward the integration of machine learning and molecular modeling for designing drug delivery nanocarriers. *Advanced Materials, 36*(45), 2407793.

Ghate, D., & Edelhauser, H. F. (2008). Barriers to glaucoma drug delivery. *Journal of Glaucoma, 17*(2), 147–156.

Goel, M., Picciani, R. G., Lee, R. K., & Bhattacharya, S. K. (2010). Aqueous humor dynamics: A review. *The Open Ophthalmology Journal, 4*, 52.

Gorantla, S., Rapalli, V. K., Waghule, T., Singh, P. P., Dubey, S. K., Saha, R. N., & Singhvi, G. (2020). Nanocarriers for ocular drug delivery: Current status and translational opportunity. *RSC Advances, 10*(46), 27835–27855.

Gunda, S., Hariharan, S., Mandava, N., & Mitra, A. K. (2008). Barriers in ocular drug delivery. In *Ocular Transporters in Ophthalmic Diseases and Drug Delivery: Ophthalmology Research* (pp. 399–413). Humana Press.

Harasymowycz, P., Birt, C., Gooi, P., Heckler, L., Hutnik, C., Jinapriya, D., Shuba, L., Yan, D., & Day, R. (2016). Medical management of glaucoma in the 21st century from a Canadian perspective. *Journal of Ophthalmology, 2016*, 6509809.

Hnin, H. M., Stefánsson, E., Loftsson, T., Asasutjarit, R., Charnvanich, D., & Jansook, P. (2022). Physicochemical and stability evaluation of topical niosomal encapsulating fosinopril/γ-cyclodextrin complex for ocular delivery. *Pharmaceutics, 14*(6), 1147.

Holden, C. A., Tyagi, P., Thakur, A., Kadam, R., Jadhav, G., Kompella, U. B., & Yang, H. (2012). Polyamidoamine dendrimer hydrogel for enhanced delivery of antiglaucoma drugs. *Nanomedicine: Nanotechnology, Biology, and Medicine*, 8(5), 776–783.

Imran, M., & Shah, M. R. (2018). Amphiphilic block copolymers–based micelles for drug delivery. In *Design and Development of New Nanocarriers* (pp. 365–400). William Andrew Publishing.

Jain, N., Verma, A., Rastogi, V., Jain, N., & Sharma, A. K. (2025). Niosomal gel containing dipivefrin hydrochloride for glaucoma treatment: Development, characterization and in vitro-in vivo assessment. *Food and Pharma International, 2*(1), 17927-17955.

Jayaram, H., Kolko, M., Friedman, D. S., & Gazzard, G. (2023). Glaucoma: Now and beyond. *The Lancet, 402*(10414), 1788–1801.

Jayasinghe, M. K., Lee, C. Y., Tran, T. T., Tan, R., Chew, S. M., Yeo, B. Z., Loh, W. X., Pirisinu, M., & Le, M. T. (2022). The role of in silico research in developing nanoparticle-based therapeutics. *Frontiers in Digital Health, 4*, 838590.

Jiang, H., & Xu, Z. (2023). Hyaluronic acid-based nanoparticles to deliver drugs to the ocular posterior segment. *Drug Delivery, 30*(1), 2204206.

Jung, H. J., Abou-Jaoude, M., Carbia, B. E., Plummer, C., & Chauhan, A. (2013). Glaucoma therapy by extended release of timolol from nanoparticle loaded silicone-hydrogel contact lenses. *Journal of Controlled Release, 165*(1), 82–89.

Khare, A., Grover, K., Pawar, P., & Singh, I. (2015). Mucoadhesive polymers for enhancing retention in ocular drug delivery. In *Progress in Adhesion and Adhesives* (pp. 451–484). John Wiley & Sons.

Leske, M. C. (1983). The epidemiology of open-angle glaucoma: A review. *American Journal of Epidemiology, 118*(2), 166–191.

Lin, X., Wu, X., Chen, X., Wang, B., & Xu, W. (2021). Intellective and stimuli-responsive drug delivery systems in eyes. *International Journal of Pharmaceutics, 602*, 120591.

Liu, L. C., Chen, Y. H., & Lu, D. W. (2024). The application of Rho kinase inhibitors in the management of glaucoma. *International Journal of Molecular Sciences, 25*(11), 5576.

Lipowsky, R. (2019). Understanding giant vesicles: A theoretical perspective. In *The Giant Vesicle Book* (pp. 73–168). CRC Press.

Mishra, G. P., Bagui, M., Tamboli, V., & Mitra, A. K. (2011). Recent applications of liposomes in ophthalmic drug delivery. *Journal of Drug Delivery, 2011*, 863734.

Mishra, V., & Jain, N. K. (2014). Acetazolamide encapsulated dendritic nano-architectures for effective glaucoma management in rabbits. *International Journal of Pharmaceutics, 461*(1–2), 380–390.

Ng, P. C., & Oliver, J. J. (2018). Anatomy of the eye. In *Handbook of Emergency Ophthalmology* (pp. 1–12). Springer International Publishing.

Nguyen, D. D., & Lai, J. Y. (2020). Advancing the stimuli response of polymer-based drug delivery systems for ocular disease treatment. *Polymer Chemistry, 11*(44), 6988–7008.

Niculescu, A. G., & Grumezescu, A. M. (2021). Polymer-based nanosystems—A versatile delivery approach. *Materials, 14*(22), 6812.

Panda, P. K., Verma, A., Saraf, S., Raikwar, S., Da Bidla, P., & Jain, S. K. (2023). Poly (lactic acid) and poly (lactic-co-glycolic acid)-based nanocarrier systems for theranostic applications. In *Polymeric Nanosystems* (pp. 13–36). Academic Press.

Parat, A., & Felder-Flesch, D. (2016). General introduction on dendrimers, classical versus accelerated syntheses and characterizations. In *Dendrimers in Nanomedicine* (pp. 1–22). Jenny Stanford Publishing.

Pignatello, R., & Puglisi, G. (2011). Nanotechnology in ophthalmic drug delivery: A survey of recent developments and patenting activity. *Recent Patents on Nanomedicine, 1*(1), 42–54.

Prajapati, B., Modi, C., Patel, U., & Kendre, P. (2024). Nanoemulsion based in-situ gel for ocular delivery of Brimonidine Tartrate. *Current Drug Therapy, 19*(3), 336–345.

Quigley, H. A. (1999). Neuronal death in glaucoma. *Progress in Retinal and Eye Research, 18*(1), 39–57.

Rahman, Z., Charoo, N. A., Akhter, S., Beg, S., Reddy, I. K., & Khan, M. A. (2018). Nanotechnology-based drug products: Science and regulatory considerations. In *Nanoscale Fabrication, Optimization, Scale-Up and Biological Aspects of Pharmaceutical Nanotechnology* (pp. 619–655). William Andrew Publishing.

Razavi, M. S., Ebrahimnejad, P., Fatahi, Y., D'Emanuele, A., & Dinarvand, R. (2022). Recent developments of nanostructures for the ocular delivery of natural compounds. *Frontiers in Chemistry, 10*, 850757.

Reis, C. P., Neufeld, R. J., Ribeiro, A. J., & Veiga, F. (2006). Nanoencapsulation I. Methods for preparation of drug-loaded polymeric nanoparticles. *Nanomedicine: Nanotechnology, Biology and Medicine, 2*(1), 8–21.

Sah, A. K., Bhuwane, N., Choudhary, I., Ramkar, S., & Suresh, P. K. (2020). Application of biocompatible nanocarriers in glaucoma: Challenges and advances. In *Nanoformulations in Human Health: Challenges and Approaches* (pp. 207–226). Springer.

Sánchez-López, E., Egea, M. A., Davis, B. M., Guo, L., Espina, M., Silva, A. M., Calpena, A. C., Souto, E. M. B., Ravindran, N., Ettcheto, M., Camins, A., García, M. L., & Cordeiro, M. F. (2018). Memantine-loaded PEGylated biodegradable nanoparticles for the treatment of glaucoma. *Small, 14*(2), 1701808.

Sapino, S., Chirio, D., Peira, E., Abellán Rubio, E., Brunella, V., Jadhav, S. A., Chindamo, G., & Gallarate, M. (2019). Ocular drug delivery: A special focus on the thermosensitive approach. *Nanomaterials, 9*(6), 884.

Singh, A., Parikh, S., Sethi, N., Patel, S., Modi, N., & Patel, K. (2024). Nanoparticle formulations: A sustainable approach to biodegradable and non-biodegradable products. In *Nanocarrier Vaccines: Biopharmaceutics-Based Fast Track Development* (pp. 95–151). Wiley.

Soppimath, K. S., Aminabhavi, T. M., Kulkarni, A. R., & Rudzinski, W. E. (2001). Biodegradable polymeric nanoparticles as drug delivery devices. *Journal of Controlled Release, 70*(1–2), 1–20.

Sommer, A., Tielsch, J. M., Katz, J., Quigley, H. A., Gottsch, J. D., Javitt, J., & Singh, K. (1991). Relationship between intraocular pressure and primary open angle glaucoma among White and Black Americans: The Baltimore Eye Survey. *Archives of Ophthalmology, 109*(8), 1090–1095.

Spataro, G., Malecaze, F., Turrin, C. O., Soler, V., Duhayon, C., Elena, P. P., Majoral, J. P., & Caminade, A. M. (2010). Designing dendrimers for ocular drug delivery. *European Journal of Medicinal Chemistry, 45*(1), 326–334.

Suri, R., Beg, S., & Kohli, K. (2020). Target strategies for drug delivery bypassing ocular barriers. *Journal of Drug Delivery Science and Technology, 55*, 101389.

Syama, K., Jakubek, Z. J., Chen, S., Zaifman, J., Tam, Y. Y., & Zou, S. (2022). Development of lipid nanoparticles and liposomes reference materials (II): Cytotoxic profiles. *Scientific Reports, 12*(1), 18071.

Taghavi Pourianazar, N., Mutlu, P., & Gunduz, U. (2014). Bioapplications of poly(amidoamine) (PAMAM) dendrimers in nanomedicine. *Journal of Nanoparticle Research, 16*, 2342.

Tewari, A. K., Upadhyay, S. C., Kumar, M., Pathak, K., Kaushik, D., Verma, R., Bhatt, S., Massoud, E. E., Rahman, M. H., & Cavalu, S. (2022). Insights on development aspects of polymeric nanocarriers: The translation from bench to clinic. *Polymers, 14*(17), 3545.

Tham, Y. C., Li, X., Wong, T. Y., Quigley, H. A., Aung, T., & Cheng, C. Y. (2014). Global prevalence of glaucoma and projections of glaucoma burden through 2040: A systematic review and meta-analysis. *Ophthalmology, 121*(11), 2081–2090.

Tripathi, S., & Yadav, K. S. (2024). Development of brimonidine niosomes laden contact lenses for extended release and promising delivery system in glaucoma treatment. *DARU Journal of Pharmaceutical Sciences, 32*(1), 161–175.

Tsung, T. H., Tsai, Y. C., Lee, H. P., Chen, Y. H., & Lu, D. W. (2023). Biodegradable polymer-based drug-delivery systems for ocular diseases. *International Journal of Molecular Sciences, 24*(16), 12976.

Vijaya, L., George, R., Baskaran, M., Arvind, H., Raju, P., Ramesh, S. V., Kumaramanickavel, G., & McCarty, C. (2008). Prevalence of primary open-angle glaucoma in an urban South Indian population and comparison with a rural population: The Chennai Glaucoma Study. *Ophthalmology, 115*(4), 648–654.

Wickens, J. M., Alsaab, H. O., Kesharwani, P., Bhise, K., Amin, M. C., Tekade, R. K., Gupta, U., & Iyer, A. K. (2017). Recent advances in hyaluronic acid-decorated nanocarriers for targeted cancer therapy. *Drug Discovery Today, 22*(4), 665–680.

Wu, K. Y., Ashkar, S., Jain, S., Marchand, M., & Tran, S. D. (2023). Breaking barriers in eye treatment: Polymeric nano-based drug-delivery system for anterior segment diseases and glaucoma. *Polymers, 15*(6), 1373.

Wu, L., Zhang, J., & Watanabe, W. (2011). Physical and chemical stability of drug nanoparticles. *Advanced Drug Delivery Reviews, 63*(6), 456–469.

Zhai, Z., Cheng, Y., & Hong, J. (2021). Nanomedicines for the treatment of glaucoma: Current status and future perspectives. *Acta Biomaterialia, 125*, 41–56.

Zhang, K., Tang, X., Zhang, J., Lu, W., Lin, X., Zhang, Y., Tian, B., Yang, H., & He, H. (2014). PEG–PLGA copolymers: Their structure and structure-influenced drug delivery applications. *Journal of Controlled Release, 183*, 77–86.

Zeng, S., Chen, Y., Zhou, F., Zhang, T., Fan, X., Chrzanowski, W., Gillies, M. C., & Zhu, L. (2023). Recent advances and prospects for lipid-based nanoparticles as drug carriers in the treatment of human retinal diseases. *Advanced Drug Delivery Reviews, 199*, 114965.

Chapter 16

Nanocarrier-Based Delivery of Biologics for Retinal and Posterior Segment Eye Diseases

Shithin Ann Varghese[1]
M. Sruthi Laakshmi[2]
Srikruthi Kunigal Sridhar[1]
Prakash Goudanavar[1]
Santhosh Fattepur[3]
and Nimbagal Raghavendra Naveen[1,*]
[1]Department of Pharmaceutics, Sri Adichunchanagiri College of Pharmacy, B. G. Nagara, Mandya, India
[2]Department of Pharmaceutics, SRM College of Pharmacy, SRMIST, Kattankulathur, Tamil Nadu, India
[3]School of Pharmacy, Management and Science University, Seksyen 13, Shah Alam, Selangor, Malaysia

Abstract

Biologic therapies have significantly advanced the treatment of various ocular disorders such as age-related macular degeneration, diabetic retinopathy, and uveitis by providing precise and long-lasting therapeutic effects. These biologics, including monoclonal antibodies, gene-based treatments, and recombinant proteins, offer greater specificity and fewer systemic side effects compared to conventional small molecule drugs. However, effectively delivering these large molecules to ocular tissues remains challenging due to physiological barriers such as the blood-retinal barrier, tear dynamics, and rapid clearance. Although intravitreal

[*] Corresponding Author's Email: nrn@accp.co.in; raghavendra.naveen@gmail.com

In: Advances in Nanotechnology for Ocular Drug Delivery
Editors: Anjali Sharma, Ajmer Singh Grewal, Geeta Deswal et al.
ISBN: 979-8-89530-814-1
© 2025 Nova Science Publishers, Inc.

injections remain the most effective delivery method, they are invasive and associated with potential risks. Nanotechnology has emerged as a promising solution by enabling sustained drug release, enhanced tissue penetration, and targeted delivery through platforms like liposomes, dendrimers, and hydrogels. Furthermore, emerging systems that respond to environmental cues, combined with artificial intelligence-driven innovations, are shaping the next generation of ocular therapeutics. Despite ongoing challenges in regulation, manufacturing, and cost-effectiveness, interdisciplinary efforts are rapidly advancing the clinical application of nanotechnology in ophthalmology. This chapter explores current hurdles, innovative delivery strategies, and future directions aimed at improving the therapeutic effectiveness of biologics in ocular treatments.

Keywords: ocular drug delivery, biologics, nanotechnology, sustained release, intravitreal injection, retinal barriers

Abbreviations

AMD: Age-related macular degeneration
BAB: Blood-aqueous barrier
BRB: Blood retinal barrier
CNV: Choroidal neovascularization
NPs: Nanoparticles
PLA: Polylactic acid
PLGA: Poly (lactic-co-glycolic acid)
ROS: Reactive oxygen species
TPGS: D-ά-Tocopheryl polyethylene glycol 1000 succinate
VEGF: Vascular endothelial growth factor

Introduction

Biologic therapies have significantly improved the treatment of eye diseases by offering precise targeting with fewer systemic side effects. Conditions such as age-related macular degeneration (AMD), diabetic retinopathy, uveitis, and retinal vein occlusion now benefit from advanced options including monoclonal antibodies, recombinant proteins, and gene- and cell-based therapies (Neves da Silva et al., 2022). Anti-VEGF agents such as bevacizumab, ranibizumab, and aflibercept are widely used to prevent abnormal blood vessel growth in diseases like AMD and diabetic macular edema. Immunosuppressants, including adalimumab and infliximab, effectively control ocular inflammation. Meanwhile, gene therapy approaches like voretigene neparvovec are demonstrating long-term benefits for inherited retinal conditions (Nicolela Susanna & Pavesio, 2020). Despite these successes, delivering biologics to the eye remains challenging due to natural barriers such as the corneal epithelium, conjunctival clearance, the blood-retinal barrier, and rapid aqueous humor turnover. Although topical drugs are easy to administer, they are generally ineffective for reaching deeper ocular tissues. Systemic administration often requires high doses, which can increase the risk of toxicity. Intravitreal injection is currently the most effective delivery method but is invasive and carries risks such as infection and retinal damage (Joseph et al., 2017). The pharmacokinetics of biologics in the eye are influenced by factors including molecular size, solubility, and sensitivity to enzymatic degradation. Rapid clearance and degradation limit their therapeutic action, and these drugs also require specific storage and handling due to their fragile nature. To address these challenges, researchers are developing sustained-release systems and more stable biologics with improved targeting capabilities (Cuny et al., 2022). This chapter focuses on how biologics are transforming the treatment of ocular diseases, highlighting current delivery challenges, pharmacokinetic limitations, and the potential of nanotechnology to enhance therapeutic outcomes and patient quality of life.

Anatomical and Physiological Barriers in Ocular Drug Delivery

The eye's unique anatomy and physiology present major challenges for effective drug delivery. Its natural defense systems, including tear production, the corneal epithelium, and blood-ocular barriers, are designed to protect against harmful substances but also limit drug absorption. Achieving high local drug concentrations while minimizing systemic exposure is a key objective in ocular therapy. However, maintaining therapeutic levels over time remains difficult due to rapid clearance and limited permeability. A deep understanding of these barriers is essential for developing advanced delivery systems that improve drug bioavailability and enhance outcomes in conditions such as glaucoma, diabetic retinopathy, and age-related macular degeneration. Anatomical barriers significantly hinder ocular drug delivery regardless of whether systemic or local administration routes are used. These barriers include epithelial, vascular, and structural elements within the eye. Drug delivery is reduced by both anatomical and physiological barriers that limit absorption, retention, and overall therapeutic efficacy. The epithelial barriers, which include the corneal and conjunctival layers, selectively restrict drug penetration. The corneal epithelium's lipophilic nature limits the uptake of hydrophilic drugs (Chen et al., 2008; Sweeney et al., 2013), while the conjunctiva allows broader permeability but is prone to systemic absorption due to its higher vascularization. Vascular barriers such as the blood-aqueous barrier and blood retinal barrier tightly regulate exchange between ocular tissues and systemic circulation. The blood-aqueous barrier permits inward flow, whereas the blood retinal barrier primarily supports outward transport to maintain retinal health. Additionally, structural components including the tear film, sclera, aqueous humor, and vitreous humor further influence drug delivery (del Amo & Urtti, 2008). Rapid tear turnover and blinking quickly clear drugs from the ocular surface, while the scleral matrix and the gel-like vitreous humor restrict the movement of macromolecules. These factors make sustained delivery methods, such as intravitreal injections, essential for effective treatment (Yellepeddi and Palakurthi, 2016). On the physiological side, dynamic processes including aqueous humor turnover, efflux pumps, and enzymatic breakdown by cytochrome P450 enzymes and proteases further reduce drug bioavailability (Mannermaa et al., 2006). To overcome these challenges, advanced strategies such as mucoadhesive formulations, prodrug designs, enzyme inhibitors, and nanoparticulate delivery systems are being explored to improve ocular drug delivery.

Types of Nanocarriers Used in Ocular Delivery

Nanocarriers such as liposomes, polymeric nanoparticles, micelles, dendrimers, hydrogels, and in situ gels play a vital role in improving ocular drug delivery by enhancing drug solubility, retention time, and bioavailability (Nsairat et al., 2022). Liposomes encapsulate both hydrophilic and hydrophobic drugs within phospholipid bilayers, with cholesterol helping to stabilize the structure. Biodegradable polymeric nanoparticles provide sustained drug release, while non-biodegradable types may require removal after therapy (Mandal et al., 2017). Micelles enhance the solubility of poorly water-soluble drugs through their core-shell structure. Dendrimers offer precise targeting by utilizing surface functional groups. Hydrogels and in situ gels form viscous networks upon application, which prolong drug contact with ocular tissues (Wu et al., 2019).

Applications and Case Studies

Nanocarrier-Based Delivery of Anti-VEGF Biologics for AMD

Nanocarrier-based drug delivery systems offer a promising alternative for treating age-related macular degeneration, particularly for anti-VEGF biologics targeting wet AMD. Conventional intravitreal injections of anti-VEGF agents such as ranibizumab, aflibercept, and bevacizumab have significantly improved patient outcomes but are associated with a high treatment burden due to frequent injections, potential adverse effects, and poor patient compliance (Uludag et al., 2022). Drug delivery systems designed to reduce injection frequency through extended in vitro release have shown considerable promise. Nanocarriers including liposomes, nanomicelles, nanoemulsions, and nanoparticles improve drug stability, enable controlled release, and facilitate targeted ocular delivery, effectively addressing these challenges (Figure 1). Among these nanocarriers, nanoparticles, liposomes, and nanomicelles have demonstrated notable advantages (Table 2).

Liposomes, owing to their biocompatibility and ability to encapsulate both hydrophilic and hydrophobic drugs, help extend ocular drug retention and reduce injection frequency. Nanomicelles enhance drug solubility and ocular penetration, while nanoemulsions incorporated into in-situ gels provide sustained release and improved bioavailability. Xu et al. formulated ranibizumab-loaded copolymeric nanomicelles (Rabz-CP-NMs), which demonstrated strong cellular uptake, enhanced permeability, and maintained cell viability in D407 and ARPE19 cell lines. These nanomicelles also reduced reactive oxygen species and inflammation, indicating good biocompatibility for AMD treatment (Al Yabhouni et al., 2024). Polymeric nanoparticles such as PLGA and PLA offer controlled drug release and retinal targeting, reducing systemic side effects. Zhang et al. reported bevacizumab-loaded PLGA nanoparticles exhibiting an initial burst release of 40 percent within two hours followed by sustained delivery over 28 days. Peak drug levels were observed on day six, with a six-fold increase in half-life and effective anti-angiogenic activity. Qin et al. developed hyaluronic acid-modified cationic niosomes achieving high retinal pigment epithelium transfection via CD44 targeting (Zhang et al., 2018). Stimuli-responsive polymeric nanoparticles enable site-specific drug release triggered by changes in pH, temperature, or ionic conditions. Ghezzi et al. (2022) utilized TPGS and Solutol HS15 micelles to improve cyclosporine delivery for dry eye disease, successfully overcoming its poor permeability. Ongoing advancements in nanocarrier design and delivery strategies continue to support safer and more effective therapies for age-related macular degeneration (Ghezzi et al., 2022).

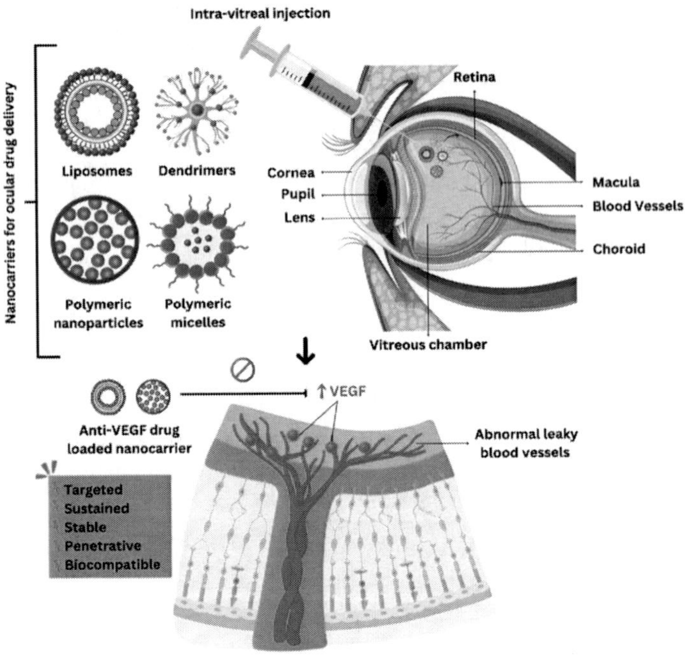

Figure 1. Nanocarrier-based anti-VEGF therapy for targeted AMD treatment.

Table 2. Summary of studies exploring the use of NPs for AMD treatment

Type of Nanoparticle	Drug	Type of AMD	Route of Administration	Ref.
Lipid NPs	Dexamethasone	Dry & wet AMD	Intra-vitreal injection	(Canioni et al., 2021)
Lipid NPs	mRNA-150 & Quercetin	Wet AMD, CNV	Intra-vitreal injection	(Nikolaidou et al., 2025)
Liposomes	N-acetyl cysteine	Dry AMD	Intra-vitreal injection	(Behroozi et al., 2018)
PLGA NPs	Axitinib	Wet AMD	Intra-vitreal injection	(Shen et al., 2018)

Type of Nanoparticle	Drug	Type of AMD	Route of Administration	Ref.
Polymeric NPs	Gelatin-epigallocatechin gallate NPs	Not specified	Subconjunctival	(Tsai et al., 2022)
PolySia-NPs	Immune modulation	Wet AMD, CNV	Intra-vitreal injection	(Peterson et al., 2024)
Quantum dots	Minocycline	Wet AMD	Intra-vitreal injection	(Huang et al., 2023)
Nanoceria	CeO_2	CNV	Intra-vitreal injection	(Tisi et al., 2022)
Light-sensitive NPs	Nintedanib	CNV	Intra-vitreal injection	(Huu et al., 2015)
Gene therapy NPs	Flt23k	Wet AMD	Intra-vitreal injection	(Luo et al., 2013)
Dendrimers	Triamcinolone	Dry and wet AMD	Intra-vitreal injection	(Kambhampati et al., 2021)
Nanofibers	Betamethasone	Wet AMD, CNV	Injectable hydrogel	(Gao et al., 2023)
Gold NPs	Anti-angiogenic therapy	Not specified	Intra-vitreal injection	(De Jong et al., 2008)
Niosomes	Umbelliprenin	Not specified	Not specified	(Dastaviz et al., 2024)
Silica NPs	Anti-VEGF siRNA	Wet AMD, CNV	Eye drops	(Ultimo et al., 2023)

Gene Therapy Approaches Using Nanocarriers for Retinal Degenerative Diseases

Nanocarrier-based gene therapy enables precise and effective gene delivery to retinal cells and offers a safer alternative to viral vectors such as adeno-associated viruses (AAVs). Although AAVs achieve high transfection rates, their use is limited by immunogenicity, restricted cargo size, and potential side effects (Taghdiri and Mussolino, 2024). Non-viral carriers including lipid-based, polymeric, and metallic nanoparticles are biocompatible, non-toxic, and capable of crossing ocular barriers while providing controlled and sustained gene release. Fritsche et al. conducted a genome-wide study on patients with age-related macular degeneration (AMD), identifying variants across 34 loci and linking them to the overexpression of matrix metalloproteases, particularly in wet AMD (Fritsche et al., 2015). This underscores the need for advanced delivery platforms such as lipid and polymer-based nanoparticles. Polyethyleneimine (PEI)-based systems have demonstrated promise in reversing ocular fibrosis and inflammation (Casper et al., 2023; Jin et al., 2023), while lipid nanoparticles (LNPs) are gaining interest due to their biocompatibility and capacity to carry larger genetic tools such as CRISPR. Vasudevan et al. developed LNPs for intracameral gene therapy in the anterior segment of the eye, achieving 14.3 percent gene editing efficiency in the trabecular meshwork without causing tissue damage. Surface modifications like PEGylation and ligand attachment, including hyaluronic acid and transferrin, enhance targeting and stability, improving retinal penetration and minimizing off-target effects. Dillinger et al. designed PEI-coated PLGA nanoparticles functionalized with hyaluronic acid, which improved mobility within the extracellular matrix and targeted CD44-positive cells to silence connective tissue growth factor in glaucoma therapy (Dillinger et al., 2018). Gold nanoparticles conjugated with polyethylene amine have also shown strong gene transfection and anti-fibrotic effects in preclinical models. LNPs delivering CRISPR-Cas9 have enabled accurate retinal gene editing for inherited disorders such as Leber congenital amaurosis. Despite ongoing challenges related to transfection efficiency and sustained gene expression, nanocarriers continue to evolve as safer and more effective platforms for personalized retinal gene therapy.

Protein and Peptide-Based Nanotherapeutics for Corneal Wound Healing and Disorders

Proteins and peptides are promising therapeutics for ocular diseases due to their high potency, selectivity, low toxicity, and reduced off-target effects. However, their clinical application is limited by short half-lives, enzymatic degradation, immunogenicity, poor permeability, and rapid clearance by the reticuloendothelial system. Therefore, novel delivery systems are essential to effectively target ocular tissues. While nanocarriers for small molecules are well established, their use for macromolecules is still emerging. For instance, spermine-conjugated hyaluronic acid-modified nanoceria loaded with insulin-like growth factor-1 reduced oxidative stress and promoted healing in corneal alkali burns (Ger et al., 2023). Yaylaci et al. developed ranibizumab-loaded peptide nanofiber hydrogels that enable sustained drug release and reduce injection frequency in wet age-related macular degeneration (Yaylaci et al., 2023). Kim et al. demonstrated that anionic human serum albumin nanoparticles exhibit better retinal penetration than cationic counterparts (Y. Zhang et al., 2024). Yang et al. prepared bovine serum albumin-loaded nanoparticles using silk fibroin, achieving sustained release and improved ocular retention in rabbit models. Mei et al. reported targeted retinal pigment epithelium localization of rapamycin-loaded high-density lipoprotein nanoparticles following intravitreal injection (Mei et al., 2022). Thiolated gelatin-coated ceria nanoparticles enhanced ocular absorption, retention, and therapeutic efficacy by two to three times (Yang et al., 2023). Surface modifications such as poly(L-histidine) coatings further improved permeability and biocompatibility. These studies confirm that surface-engineered nanocarriers can overcome the challenges of macromolecular delivery and significantly enhance therapeutic outcomes in ocular diseases.

Emerging Biologic Therapies in Clinical Trials: Current Trends and Future Potential

Ocular diseases such as age-related macular degeneration, diabetic retinopathy, and uveitis require targeted and sustained drug delivery systems for effective treatment. Biologic therapies, including monoclonal antibodies, gene therapy, peptides, and proteins, offer promising therapeutic options. Nanocarriers have emerged as important tools to overcome the physiological and anatomical challenges associated with ocular drug delivery. Current trends in ocular biologic delivery focus on monoclonal antibodies like bevacizumab, ranibizumab, and aflibercept; gene transfer using adeno-associated virus vectors; peptides and proteins such as vascular endothelial growth factor inhibitors and neurotrophic peptides; as well as mRNA therapeutics targeting retinal diseases. Key challenges include barriers like the blood-retinal barrier, corneal epithelium, rapid tear turnover, short drug half-life with swift clearance, limited penetration to posterior eye segments, and immunogenicity concerns. Nanocarriers contribute significantly by enabling sustained and controlled drug release, enhancing bioavailability, reducing systemic exposure, and facilitating precise delivery to target tissues. Several clinical trials are currently investigating nanocarrier-based biologic therapies for ocular conditions. Advances in stimuli-responsive nanocarriers, biodegradable polymers, and precision medicine are shaping the future landscape of ocular drug delivery. Furthermore, the integration of artificial intelligence and machine learning is enhancing personalized treatment regimens and optimizing therapeutic outcomes. Nanocarriers hold considerable promise in overcoming the limitations of biologic delivery to the eye, offering targeted, sustained, and efficient therapies. Ongoing clinical research combined with innovations in nanotechnology, biomaterials, and gene editing is expected to revolutionize ocular therapeutics and improve patient care.

Table 3. Current human clinical trials on nanocarrier-based ocular drug delivery systems

Product	Trial ID	Nanocarrier	Indication	Clinical Trial Phase
Cyclosporine	NCT02845674	Nanomicelles	Dry eye disease	Phase 3
Dexamethasone	NCT04130802	Nanoparticles	Inflammation, corneal pain	Phase 2
Urea	NCT03001466	Nanoparticles	Cataracts	Phase 2
Paclitaxel	NCT00738361	Albumin-stabilized nanoparticles	Metastatic ocular melanoma	Phase 2
Latanoprost	NCT01987323	Liposomes	Ocular Hypertension	Phase 1 Phase 2
SeeQ CdSe 655 (Device)	NCT04008771	NPs	Retinitis pigmentosa	Not Applicable
Clobetasol Propionate	NCT04246801	Nanoemulsion	Inflammation and pain associated with Cataract surgery	Phase 3
Brimonidine tartrate	NCT03785340	Nanoemulsion	Dry Eye Disease	Phase 3
Aflibercept	NCT05387837	Dendrimer	Wet AMD & DME	Phase 2
Vincristine	NCT00506142	Liposomes	Metastatic Malignant Uveal Melanoma	Phase 2

Challenges and Limitations

Ocular nanomedicine faces a range of interconnected challenges that complicate its translation from research to clinical application. These challenges include regulatory, manufacturing, safety, economic, and ethical considerations. Regulatory agencies such as the FDA and EMA require thorough preclinical and clinical evaluations to demonstrate safety, efficacy, and biocompatibility. However, existing characterization tools and standardized protocols for nanocarriers, particularly those assessing critical parameters like particle size, surface charge, and drug release, are still underdeveloped. Manufacturing also presents difficulties, including batch-to-batch inconsistencies, the need for sterile and scalable production processes, and the high costs associated with advanced equipment, raw materials, and compliance with good manufacturing practice standards. Additionally, scalability and economic feasibility remain major hurdles because nanocarrier-based formulations are more complex and expensive compared to traditional eye drops or emulsions. To address these issues, researchers are developing eco-friendly and solvent-free synthesis methods as well as multifunctional nanocarriers capable of delivering multiple therapeutics in a single dose (Moghimi et al., 2025).

Future Directions and Emerging Trends

Recent advancements in intelligent nanocarriers have led to the development of pH-sensitive, temperature-sensitive, and enzyme-sensitive drug delivery platforms that respond to changes in ocular pH, temperature, and enzyme activity. These smart systems enable highly precise delivery of biologics with prolonged duration of action, resulting in increased therapeutic efficacy and reduced side effects. The integration of machine learning and artificial intelligence is revolutionizing nanocarrier design by optimizing nanoparticle characteristics, predicting drug release kinetics, and improving penetration into ocular tissues. Personalized medicine is becoming more prevalent in ophthalmology, utilizing biomarker-guided therapies to tailor treatments based on individual genetic and proteomic profiles. Additionally, the combination of bioinspired nanomaterials with surface-functionalized carriers enhances drug penetration across ocular barriers, paving the way for future generations of biologic drugs with superior efficacy and safety.

Conclusion

Nanotechnology has revolutionized ocular biologic drug delivery by enhancing bioavailability, enabling controlled release, and improving targeted efficacy. Traditional intravitreal injections are limited by the need for repeated dosing and issues with drug stability. These challenges are addressed by nanocarriers such as liposomes, nanoparticles, dendrimers, and hydrogels. Stimuli-sensitive nanocarriers that respond to factors like pH, enzymes, or light provide additional precision, while ligand-based targeting enables receptor-mediated uptake. Despite significant promise in preclinical studies, clinical translation faces hurdles including regulatory approval, scalability, and concerns about long-term safety. Progress is being made toward practical applications with biodegradable nanoparticles, sustained-release systems, and minimally invasive delivery methods such as microneedles and implants. Future advancements will integrate artificial intelligence to optimize nanocarrier design, incorporate biomarker-driven precision therapies, and develop hybrid systems that combine drug delivery with diagnostics. Continued interdisciplinary collaboration, innovative regulatory frameworks, and scalable manufacturing will be essential for clinical adoption, ultimately transforming the treatment of ocular diseases, maximizing therapeutic benefits, and enhancing patient quality of life.

References

Al Yabhouni, S. A., Mozumder, M. S., Hassan, N., Mourad, A. H. I., & Issa MD, T. M. A. (2024). Nanocarrier-Based, ocular drug delivery: Challenges, prospects, and the therapeutic landscape in the United Arab Emirates. *International Journal of Pharmaceutics*, *667*, 124899.

Behroozi, F., Abdkhodaie, M. J., Abandansari, H. S., Satarian, L., Ashtiani, M. K., Jaafari, M. R., & Baharvand, H. (2018). Smart liposomal drug delivery for treatment of oxidative stress model in human embryonic stem cell-derived retinal pigment epithelial cells. *International Journal of Pharmaceutics*, *548*(1), 62–72.

Canioni, R., Reynaud, F., Leite-Nascimento, T., Gueutin, C., Guiblin, N., Ghermani, N. E., Jayat, C., Daull, P., Garrigue, J. S., Fattal, E., & Tsapis, N. (2021). Tiny dexamethasone palmitate nanoparticles for intravitreal injection: Optimization and in vivo evaluation. *International Journal of Pharmaceutics*, *600*, 120509.

Casper, J., Schenk, S. H., Parhizkar, E., Detampel, P., Dehshahri, A., & Huwyler, J. (2023). Polyethylenimine (PEI) in gene therapy: Current status and clinical applications. *Journal of Controlled Release*, *362*, 667–691.

Chen, M. S., Hou, P. K., Tai, T. Y., & Lin, B. J. (2008). Blood-ocular barriers. *Tzu Chi Medical Journal*, *20*(1).

Cuny, A., Guillo, L., Baumann, C., Netter, P., Danese, S., Caron, B., Peyrin-Biroulet, L., & Angioi, K. (2022). Ocular Manifestations in Patients with Inflammatory Bowel Disease in the Biologics Era. *Journal of Clinical Medicine*, *11*(15), 4538.

Dastaviz, F., Vahidi, A., Khosravi, T., Khosravi, A., Arabi, M. S., Bagheri, A., Rashidi, M., & Oladnabi, M. (2024). Impact of umbelliprenin-containing niosome nanoparticles on VEGF-A and CTGF genes expression in retinal pigment epithelium cells. *International Journal of Ophthalmology*, *17*(1), 7.

De Jong, W. H., Hagens, W. I., Krystek, P., Burger, M. C., Sips, A. J. A. M., & Geertsma, R. E. (2008). Particle size-dependent organ distribution of gold nanoparticles after intravenous administration. *Biomaterials*, *29*(12), 1912–1919.

del Amo, E. M., & Urtti, A. (2008). Current and future ophthalmic drug delivery systems. A shift to the posterior segment. *Drug Discovery Today,* 13(3–4), 135-143.

Dillinger, A. E., Guter, M., Froemel, F., Weber, G. R., Perkumas, K., Stamer, W. D., Ohlmann, A., Fuchshofer, R., & Breunig, M. (2018). Intracameral Delivery of Layer-by-Layer Coated siRNA Nanoparticles for Glaucoma Therapy. *Small*, *14*(50), 1803239.

Fritsche, L. G., Igl, W., Bailey, J. N., Grassmann, F., Sengupta, S., Bragg-Gresham, J. L., Burdon, K. P., Hebbring, S. J., Wen, C., Gorski, M., Kim, I. K., Cho, D., Zack, D., Souied, E., Scholl, H. P., Bala, E., Lee, K. E., Hunter, D. J., Sardell, R. J., Mitchell, P., Merriam, J. E., Cipriani, V., Hoffman, J. D., Schick, T., Lechanteur, Y. T., Guymer, R. H., Johnson, M. P., Jiang, Y., Stanton, C. M., Buitendijk, G. H., Zhan, X., Kwong, A. M., Boleda, A., Brooks, M., Gieser, L., Ratnapriya, R., Branham, K. E., Foerster, J. R., Heckenlively, J. R., Othman, M. I., Vote, B. J., Liang, H. H., Souzeau, E., McAllister, I. L., Isaacs, T., Hall, J., Lake, S., Mackey, D. A., Constable, I. J., Craig, J. E., Kitchner, T. E., Yang, Z., Su, Z., Luo, H., Chen, D., Ouyang, H., Flagg, K., Lin, D., Mao, G., Ferreyra, H., Stark, K., von Strachwitz, C. N., Wolf, A., Brandl, C., Rudolph, G., Olden, M., Morrison, M. A., Morgan, D. J., Schu, M., Ahn, J., Silvestri, G., Tsironi, E. E., Park, K. H., Farrer, L. A., Orlin, A., Brucker, A., Li, M., Curcio, C. A., Mohand-Saïd, S., Sahel, J. A., Audo, I., Benchaboune, M., Cree, A. J., Rennie, C. A., Goverdhan, S. V., Grunin, M., Hagbi-Levi, S., Campochiaro, P., Katsanis, N., Holz, F. G., Blond, F., Blanché, H., Deleuze, J. F., Igo, R. P., Jr., Truitt, B., Peachey, N. S., Meuer, S. M., Myers, C. E., Moore, E. L., Klein, R., Hauser, M. A., Postel, E. A., Courtenay, M. D., Schwartz, S. G., Kovach, J. L., Scott, W. K., Liew, G., Tan, A. G., Gopinath, B., Merriam, J. C., Smith, R. T., Khan, J. C., Shahid, H., Moore, A. T., McGrath, J. A., Laux, R., Brantley, M. A., Jr., Agarwal, A., Ersoy, L., Caramoy, A., Langmann, T., Saksens, N. T., de Jong, E. K., Hoyng, C. B., Cain, M. S., Richardson, A. J., Martin, T. M., Blangero, J., Weeks, D. E., Dhillon, B., van Duijn, C. M., Doheny, K. F., Romm, J., Klaver, C. C., Hayward, C., Gorin, M. B., Klein, M. L., Baird, P. N., den Hollander, A. I., Fauser, S., Yates, J. R., Allikmets, R., Wang, J. J., Schaumberg, D. A., Klein, B. E., Hagstrom, S. A., Chowers, I., Lotery, A. J., Léveillard, T., Zhang, K., Brilliant, M. H., Hewitt, A. W., Swaroop, A., Chew, E. Y., Pericak-Vance, M. A., DeAngelis, M., Stambolian, D., Haines, J. L., Iyengar, S. K., Weber, B. H., Abecasis, G. R., & Heid, I. M. (2016). A large genome-wide association study of age-related macular degeneration highlights contributions of rare and common variants. *Nature Genetics*, 48(2), 134–143.

Gao, H., Chen, M., Liu, Y., Zhang, D., Shen, J., Ni, N., Tang, Z., Ju, Y., Dai, X., Zhuang, A., Wang, Z., Chen, Q., Fan, X., Liu, Z., & Gu, P. (2023). Injectable Anti-Inflammatory Supramolecular Nanofiber Hydrogel to Promote Anti-VEGF Therapy in Age-Related Macular Degeneration Treatment. *Advanced Materials*, *35*(2), 2204994.

Ger, T. Y., Yang, C. J., Ghosh, S., & Lai, J. Y. (2023). Biofunctionalization of nanoceria with sperminated hyaluronan enhances drug delivery performance for corneal alkali burn therapy. *Chemical Engineering Journal*, *476*, 146864.

Ghezzi, M., Ferraboschi, I., Delledonne, A., Pescina, S., Padula, C., Santi, P., Sissa, C., Terenziani, F., & Nicoli, S. (2022). Cyclosporine-loaded micelles for ocular delivery: Investigating the penetration mechanisms. *Journal of Controlled Release*, *349*, 744–755.

Huang, K., Liu, X., Lv, Z., Zhang, D., Zhou, Y., Lin, Z., & Guo, J. (2023). MMP9-Responsive Graphene Oxide Quantum Dot-Based Nano-in-Micro Drug Delivery System for Combinatorial Therapy of Choroidal Neovascularization. *Small*, *19*(39), 2207335.

Huu, V. A. N., Luo, J., Zhu, J., Zhu, J., Patel, S., Boone, A., Mahmoud, E., McFearin, C., Olejniczak, J., De Gracia Lux, C., Lux, J., Fomina, N., Huynh, M., Zhang, K., & Almutairi, A. (2015). Light-responsive nanoparticle depot to control release of a small molecule angiogenesis inhibitor in the posterior segment of the eye. *Journal of Controlled Release*, *200*, 71–77.

Jin, Y., Adams, F., Möller, J., Isert, L., Zimmermann, C. M., Keul, D., & Merkel, O. M. (2023). Synthesis and Application of Low Molecular Weight PEI-Based Copolymers for siRNA Delivery with Smart Polymer Blends. *Macromolecular Bioscience*, *23*(2), 2200409.

Joseph, M., Trinh, H. M., Cholkar, K., Pal, D., & Mitra, A. K. (2017). Recent perspectives on the delivery of biologics to back of the eye. *Expert Opinion on Drug Delivery*, 14(5), 631-645.

Kambhampati, S. P., Bhutto, I. A., Wu, T., Ho, K., McLeod, D. S., Lutty, G. A., & Kannan, R. M. (2021). Systemic dendrimer nanotherapies for targeted suppression of choroidal inflammation and neovascularization in age-related macular degeneration. *Journal of Controlled Release*, *335*, 527–540.

Luo, L., Zhang, X., Hirano, Y., Tyagi, P., Barabás, P., Uehara, H., Miya, T. R., Singh, N., Archer, B., Qazi, Y., Jackman, K., Das, S. K., Olsen, T., Chennamaneni, S. R., Stagg, B. C., Ahmed, F., Emerson, L., Zygmunt, K., Whitaker, R., Mamalis, C., Huang, W., Gao, G., Srinivas, S. P., Krizaj, D., Baffi, J., Ambati, J., Kompella, U. B., & Ambati, B. K. (2013). Targeted intraceptor nanoparticle therapy reduces angiogenesis and fibrosis in primate and murine macular degeneration. *ACS Nano*, *7*(4), 3264–3275.

Mandal, A., Bisht, R., Rupenthal, I. D., & Mitra, A. K. (2017). Polymeric micelles for ocular drug delivery: From structural frameworks to recent preclinical studies. *Journal of Controlled Release,* 248, 96-116.

Mannermaa, E., Vellonen, K. S., & Urtti, A. (2006). Drug transport in corneal epithelium and blood-retina barrier: Emerging role of transporters in ocular pharmacokinetics. *Advanced Drug Delivery Reviews,* 58(11), 1136-1163.

Mei, L., Yu, M., Liu, Y., Weh, E., Pawar, M., Li, L., Besirli, C. G., & Schwendeman, A. A. (2022). Synthetic high-density lipoprotein nanoparticles delivering rapamycin for the treatment of age-related macular degeneration. *Nanomedicine: Nanotechnology, Biology and Medicine*, *44*, 102571.

Moghimi, M., Savvidou, G., Spyratou, E., Zachou, M.-E., & Efstathopoulos, E. P. (2025). Nanomedicine: transforming the management of ocular neuroinflammatory and neurodegenerative diseases. *Journal of Nanotheranostics, 6,* 6.

Neves da Silva, H. V., Placide, J., Duong, A., Ronquillo, Y., McCabe, S., & Moshirfar, M. (2022). Ocular adverse effects of therapeutic biologics. *Therapeutic Advances in Ophthalmology*, 2022, 14.

Nicolela Susanna, F., & Pavesio, C. (2020). A review of ocular adverse events of biological anti-TNF drugs. *Journal of Ophthalmic Inflammation and Infection*, 10(1), 11.

Nikolaidou, A., Spyratou, E., Sandali, A., Gianni, T., Platoni, K., Lamprogiannis, L., & Efstathopoulos, E. P. (2025). Utilization of Nanoparticles for Treating Age-Related Macular Degeneration. *Pharmaceuticals, 18*(2), 162.

Nsairat, H., Khater, D., Sayed, U., Odeh, F., Al Bawab, A., & Alshaer, W. (2022). Liposomes: structure, composition, types, and clinical applications. Heliyon, 8(5), e09394.

Peterson, S. L., Krishnan, A., Patel, D., Khanehzar, A., Lad, A., Shaughnessy, J., Ram, S., Callanan, D., Kunimoto, D., Genead, M. A., & Tolentino, M. J. (2024). PolySialic Acid Nanoparticles Actuate Complement-Factor-H-Mediated Inhibition of the Alternative Complement Pathway: A Safer Potential Therapy for Age-Related Macular Degeneration. *Pharmaceuticals*, *17*(4), 517.

Shen, N., Zhang, R., Zhang, H. R., Luo, H. Y., Shen, W., Gao, X., Guo, D. Z., & Shen, J. (2018). Inhibition of retinal angiogenesis by gold nanoparticles via inducing autophagy. *International Journal of Ophthalmology*, *11*(8), 1269.

Sweeney, D. F., Millar, T. J., & Raju, S. R. (2013). Tear film stability: a review. *Experimental Eye Research*, 117, 28–38.

Taghdiri, M., & Mussolino, C. (2024). Viral and non-viral systems to deliver gene therapeutics to clinical targets. International Journal of Molecular Sciences, 25(13), 7333.

Tisi, A., Pulcini, F., Carozza, G., Mattei, V., Flati, V., Passacantando, M., Antognelli, C., Maccarone, R., & Monache, S. D. (2022). Antioxidant properties of cerium oxide nanoparticles prevent retinal neovascular alterations in vitro and in vivo. *Antioxidants*, *11*(6), 1133.

Tsai, C. H., Hoang, L. N., Lin, C. C., Pan, L. C., Wu, C. L., Lin, I. C., Wang, P. Y., & Tseng, C. L. (2022). Evaluation of topical and subconjunctival injection of hyaluronic acid-coated nanoparticles for drug delivery to posterior eye. *Pharmaceutics*, 14(6), 1253.

Ultimo, A., Orzaez, M., Santos-Martinez, M. J., Martínez-Máñez, R., Marcos, M. D., Sancenón, F., & Ruiz-Hernández, E. (2023). High-Capacity Mesoporous Silica Nanocarriers of siRNA for Applications in Retinal Delivery. *International Journal of Molecular Sciences*, *24*(3), 2753.

Uludag, G., Hassan, M., Matsumiya, W., Pham, B. H., Chea, S., Trong Tuong Than, N., Doan, H. L., Akhavanrezayat, A., Halim, M. S., Do, D. V., & Nguyen, Q. D. (2022). Efficacy and safety of intravitreal anti-VEGF therapy in diabetic retinopathy: what we have learned and what should we learn further? *Expert Opinion on Biological Therapy*, *22*(10), 1275–1291.

Wu, Y., Liu, Y., Li, X., Kebebe, D., Zhang, B., Ren, J., Lu, J., Li, J., Du, S., & Liu, Z. (2019). Research progress of in-situ gelling ophthalmic drug delivery system. *Asian Journal of Pharmaceutical Sciences*, 14(1), 1–15.

Yang, C. J., Anand, A., Huang, C. C., & Lai, J. Y. (2023). Unveiling the Power of Gabapentin-Loaded Nanoceria with Multiple Therapeutic Capabilities for the Treatment of Dry Eye Disease. *ACS Nano*, *17*(24), 25118–25135.

Yaylaci, S., Dinç, E., Aydın, B., Tekinay, A. B., & Guler, M. O. (2023). Peptide Nanofiber System for Sustained Delivery of Anti-VEGF Proteins to the Eye Vitreous. *Pharmaceutics*, *15*(4), 1264.

Yellepeddi, V. K., & Palakurthi, S. (2016). Recent advances in topical ocular drug delivery. *Journal of Ocular Pharmacology and Therapeutics*, 32(2), 67–82.

Zhang, X. P., Sun, J. G., Yao, J., Shan, K., Liu, B. H., Yao, M. Di, Ge, H. M., Jiang, Q., Zhao, C., & Yan, B. (2018). Effect of nanoencapsulation using poly (lactide-co-glycolide) (PLGA) on anti-angiogenic activity of bevacizumab for ocular angiogenesis therapy. *Biomedicine & Pharmacotherapy*, *107*, 1056–1063.

Zhang, Y., Watson, S., Ramaswamy, Y., & Singh, G. (2024). Intravitreal therapeutic nanoparticles for age-related macular degeneration: Design principles, progress and opportunities. *Advances in Colloid and Interface Science*, *329*, 103200.

Chapter 17

Innovative Nanocarrier Strategies for Ocular Delivery of Biologics: Formulations, Challenges, and Future Directions

Laxmi Gharti
and Hemlata Kaurav*
School of Pharmaceutical Sciences, Shoolini University of Biotechnology and Management Sciences, Bajhol, Solan, Himachal Pradesh, India

Abstract

Ocular diseases significantly impair vision and diminish overall quality of life. Traditional drug delivery methods often face serious limitations due to the eye's intricate anatomical and physiological barriers, such as the cornea, sclera, and blood-retinal barrier. The application of nanotechnology in ocular therapeutics offers a promising approach to enhance the delivery of biologically active molecules, including antibodies, messenger RNA, and peptides. A wide range of nanocarrier systems, including liposomes, emulsions, nanostructured lipid carriers, and polymeric nanoparticles, are currently being explored to overcome the limitations of conventional ocular drug delivery methods for biologics. These innovative nanocarriers provide several advantages, including enhanced drug stability, improved bioavailability, and the ability to bypass ocular barriers such as nasolacrimal clearance and restricted retinal access. As a result, they show great potential in the treatment of diseases affecting both the anterior and posterior segments of the eye. This chapter presents recent advancements in formulation strategies that aim to improve the ocular delivery of biologics, with

* Corresponding Author's Email: kaurav.hem88@gmail.com

In: Advances in Nanotechnology for Ocular Drug Delivery
Editors: Anjali Sharma, Ajmer Singh Grewal, Geeta Deswal et al.
ISBN: 979-8-89530-814-1
© 2025 Nova Science Publishers, Inc.

particular focus on enhancing therapeutic outcomes and patient compliance. It also highlights the transformative potential of nanotechnology-based drug delivery systems in reshaping current treatment modalities by offering more effective, patient-friendly, and sustainable solutions. By synthesizing current research and future perspectives, this chapter aims to guide further innovation in the development of advanced ocular drug delivery systems.

Keywords: ocular drug delivery, nanotechnology, biologics, mRNA delivery, antibody delivery, corneal barrier, ocular nanomedicine

Abbreviations

Apa-BSA-NPs: Apatinib-loaded bovine serum albumin nanoparticles
CMC: Critical micelle concentration
DES: Dry eye syndrome
EPOβ: Epoetin beta
FA: Ferulic acid
HA: Hyaluronic acid
HuR: Human antigen R
Mab: Monoclonal antibody
miRNA: Micro RNA
MNPs: Magnetic nanoparticles
MPDA: Mesoporous polydopamine
mRNA: Messenger RNA
MSNs: Mesoporous silica nanoparticles
NPs: Nanoparticles
OCT: Octreotide
PEG: Polyethylene glycol
siRNA: Small interfering RNA
VEGF: Vascular endothelial growth factor

Introduction

The human eye is a highly complex organ characterized by numerous specialized anatomical and physiological features. Structurally, it is divided into two primary segments: the anterior and the posterior. The anterior segment includes the cornea, conjunctiva, aqueous humor, iris, ciliary body, and lens, while the posterior segment consists of the sclera, choroid, retinal pigment epithelium, neural retina, optic nerve, and vitreous humor (Qi et al., 2021; Ahmed et al., 2023). Ocular drug delivery presents significant challenges due to several anatomical and physiological barriers. For instance, the presence of tight junctions within the epithelial layers restricts paracellular transport, directly limiting drug absorption into ocular tissues (Chandel and Kandav, 2024). Reflex blinking rapidly removes instilled medications from the ocular surface, thereby reducing contact time and therapeutic efficacy. Additionally, the metabolic activity within ocular tissues may deactivate therapeutic agents before they exert their intended effects (Tang et al., 2022). The continuous turnover of tear and aqueous fluids further dilutes drugs, hindering their effectiveness. Moreover, efflux transporters actively remove drugs from ocular tissues, thereby lowering their bioavailability at target sites. These physiological obstacles are especially pronounced when delivering biologics, which are large and structurally complex molecules. Their poor permeability across the corneal barrier significantly limits their therapeutic potential in ocular applications (Wang et al., 2018). Overcoming these challenges remains a critical focus in the development of effective ocular drug delivery systems. Biologics are therapeutic agents derived from biological sources, including proteins, peptides, and nucleic acids. The delivery of biologics to ocular tissues presents substantial challenges due to their intrinsic properties. Proteins and peptides, in particular, possess high molecular weights, which severely limit their ability to permeate ocular barriers (Wu et al., 2023). Additionally, these molecules are susceptible to both physical and chemical degradation, leading to a significant reduction in their therapeutic efficacy before they can exert their intended effects (Bansal et al., 2021). These challenges underscore the urgent need for advanced drug delivery systems to improve the clinical utility of ocular biologics. In recent years, the advancement of nanotechnology-based delivery platforms has addressed many of these limitations. Nanotechnology involves the use of precisely engineered nanoparticles designed to facilitate targeted and controlled delivery of therapeutic agents. This approach enhances drug permeability, stability, and bioavailability, effectively overcoming many of the obstacles associated with conventional delivery systems (Wang et al., 2020; Afzal et al., 2022). Advancements in nanotechnology for ocular drug delivery, particularly for biologics, are progressing rapidly. The development of sophisticated nanocarrier systems is essential, as these platforms can efficiently encapsulate biologics and overcome the limitations of existing delivery methods (Liu et al., 2023). Significant attention is being directed towards responsive nanocarriers that adapt to the ocular environment, thereby enhancing targeted drug release. Additionally, there is increasing interest in non-viral gene delivery systems capable of transporting genetic material directly to retinal cells, presenting promising potential for treating inherited retinal diseases (Sharma et al., 2022). Innovative delivery approaches such as microneedles, hydrogels, and nanowafers are also under investigation to improve the administration and bioavailability of biologics (Tian et al., 2022). Biologics used in ocular delivery include proteins, peptides, and nucleic acids, all of which have an inherently unstable nature. Their instability arises from susceptibility to enzymatic degradation, rapid clearance mechanisms, poor permeability across ocular barriers, and environmental factors that limit their bioavailability. Nanocarrier-based formulations, such as liposomes, metal nanoparticles, and nanofibers, provide a protective environment that enhances the stability of biologics. Encapsulation within these nanocarriers shields biologics from proteolytic enzymes and oxidative stress. Furthermore, surface modifications using stabilizing agents such as polyethylene glycol (PEG) or albumin prevent aggregation and improve colloidal stability (Gugleva & Andonova, 2023). Overall, these advances hold considerable promise for improving therapeutic outcomes in ocular biologic treatments. Lyophilization of biologic-loaded nanocarriers using cryoprotectants such as trehalose has been shown to effectively maintain stability during storage and transport. Nanocarrier-based formulations enhance the bioavailability of biologics by overcoming unique ocular drug delivery barriers, including tear clearance and limited corneal permeability. Surface functionalization with mucoadhesive polymers like chitosan or hyaluronic acid improves retention on the ocular surface, thereby prolonging contact time and enabling sustained drug release. Nanocarriers with ultra-small particle sizes facilitate deeper penetration into ocular tissues, reaching even the posterior segment (Sarkar et al., 2025). Furthermore, ligand-mediated targeting strategies employing aptamers or antibodies allow active targeting of specific ocular receptors, ensuring efficient delivery to retinal tissues. Innovative in situ gelling systems incorporating nanocarriers provide a promising approach for prolonged drug release, significantly reducing the frequency of administration. For instance, bevacizumab-loaded chitosan nanoparticles coated with hyaluronic acid have demonstrated enhanced bioavailability and therapeutic efficacy in the treatment of corneal neovascularization (Pandit et al., 2021).

Overview of Nanocarrier Systems

Nanocarriers represent a significant advancement in addressing the unique anatomical and physiological barriers associated with ocular drug delivery. These submicron-scale delivery systems are specifically designed to enable efficient transport of therapeutic agents to targeted sites within the eye (Gaur and Kumar, 2025). Table 1 summarizes various nanocarrier systems developed for the ocular delivery of biologics. Central to these systems are nanoparticles, which, due to their diverse particle size distributions, provide multifunctional strategies for biologic delivery. Nanoparticles can be broadly classified into different types, including carbon-based, magnetic core, and metal-based nanoparticles (Eker et al., 2024). Among the most versatile nanocarriers for ocular applications are lipid vesicular systems. These possess a spherical structure made of phospholipid bilayers that can encapsulate both hydrophilic and hydrophobic drugs. Their unique configuration enhances corneal permeability by facilitating close interaction with ocular tissues, thereby improving drug uptake and prolonging retention on the ocular surface (Paramshetti et al., 2024). Nanomicelles are another promising class of nanocarriers. These self-assembling colloidal dispersions consist of a hydrophobic core and a hydrophilic shell, making them well suited to enhance the solubility, absorption, and delivery of hydrophobic drug molecules (Gohar et al., 2024). Thermosensitive hydrogels and electrospun nanofibers are also being investigated for ocular drug delivery. These systems provide advantages over conventional approaches by allowing modulation of drug release and offering structural support, which can be particularly beneficial in treating various ocular disorders (Annala et al., 2022; Romeo et al., 2023). Additionally, carbon-based nanocarriers are gaining increased attention due to their biocompatibility and customizable surface properties, making them promising candidates for delivering biologics in ocular disease treatment (Mahor et al., 2021). Figure 1 provides a visual representation of the diverse nanocarrier systems under investigation for ocular biologic delivery.

Nanoparticles (NPs)

Nanoparticles function as nanocarriers to enhance drug delivery systems by modifying drug properties and bioactivity, a capability attributed to their high surface area to volume ratio (Yusuf et al., 2023). Various types of nanoparticles, including magnetic nanoparticles, chitosan nanoparticles, serum albumin nanoparticles, and mesoporous nanoparticles, have been scientifically validated for delivering ocular biologics and are discussed in this section.

Magnetic NPs

Magnetic nanoparticles (MNPs), due to their reactive surfaces, can be easily functionalized to enable targeted ocular therapeutic delivery. When combined with biocompatible coatings and biologics, these systems enhance cellular targeting while reducing effects on surrounding healthy tissues (Vilkelyte et al., 2024). Several studies have shown the potential of targeted nanocarrier systems for ocular delivery of biologics. For example, Amato et al. (2020) developed octreotide (OCT)-loaded MNPs for intraocular administration to treat diabetic retinopathy. Preclinical evaluations confirmed that the MNP-OCT formulation effectively localized OCT within ocular tissues, improved its bioactivity, and provided a modified, sustained release profile (Amato et al., 2020).

Chitosan NPs

Chitosan nanoparticles, derived from the natural polymer chitosan, have attracted significant interest in pharmaceutical applications because of their biocompatibility, biodegradability, and mucoadhesive properties. Rathore et al. (2020) developed chitosan-based nanoparticles for ocular insulin delivery using the ionotropic gelation method, with tripolyphosphate serving as a crosslinking agent. In vitro drug release and ex vivo permeation studies showed a sustained release profile and significantly improved corneal penetration compared to free insulin. Pharmacokinetic evaluations indicated enhanced systemic insulin absorption through ocular delivery, resulting in better bioavailability than conventional insulin administration. The nanoparticles' mucoadhesive nature, prolonged ocular retention, and increased permeation collectively contributed to superior hypoglycemic effects in diabetic models. These findings highlight the potential of chitosan-based nanoparticles as effective carriers for ocular biologic delivery.

Serum Albumin NPs

Serum albumin has been employed as a nanocarrier system offering several advantages such as biodegradability, biocompatibility, straightforward fabrication, prolonged drug release, and reproducibility (Spada et al., 2021). It is a hydrophilic protein with a high drug-loading capacity and strong binding affinity to biologics (Mardikasari et al., 2023). Radwan et al. (2021) developed Apatinib-loaded bovine serum albumin nanoparticles (Apa-BSA-NPs) coated with hyaluronic acid for topical diabetic retinopathy therapy. These HA-Apa-BSA-NPs exhibited sustained drug release, high retinal accumulation, no cytotoxicity, and effective retinal protection comparable to intravitreal injections, positioning them as a promising non-invasive platform for diabetic retinopathy treatment. Another study demonstrated the topical delivery of epoetin beta (EPOβ) using a chitosan-hyaluronic acid-EPOβ nanoparticle system in rats. The nanoparticles efficiently delivered EPOβ to ocular tissues with biological safety, suggesting a valuable approach for ocular administration while reducing risks associated with invasive delivery methods (Silva et al., 2023).

Mesoporous NPs

Mesoporous silica nanoparticles (MSNs) possess distinctive features such as large pore volume, high effective surface area, substantial loading capacity, customizable morphology, surface modification potential, biocompatibility, and enhanced stability under physiological conditions. These characteristics make MSNs excellent nanocarrier systems for delivering biologics (Zhou et al., 2024). Beyond delivering single biologic molecules, there are studies reporting combinations of biologics with other drugs or molecules to achieve synergistic therapeutic effects. For instance, Ma et al. (2024) developed a multifunctional nanodrug targeting various ocular neovascular disorders by coupling dexamethasone with VEGF-siRNA and bioactive mesoporous polydopamine (MPDA). This system demonstrated enhanced therapeutic outcomes in established animal models, showcasing the synergistic potential of RNA and other medicinal agents delivered through MPDA-based carriers. While these findings are promising, further research is required to optimize dosing, explore alternative delivery routes, and clarify the molecular mechanisms involved to facilitate clinical translation.

Liposomes and Liposome Complex

Liposomes are lipid-based vesicular systems increasingly recognized as effective carriers for the ocular delivery of biologics. Recent research has focused on improving liposomal formulations to enhance retention on the ocular surface, tissue penetration, and overall therapeutic efficacy in eye treatments (Batur et al., 2024). For example, Supe et al. (2023) developed liposome-polyethylenimine complexes, known as lipopolyplexes, to deliver Human antigen R (HuR) siRNA for treating diabetic retinopathy (DR). HuR is an RNA-binding protein that regulates VEGF expression by stabilizing its mRNA, and targeting HuR with siRNA-mediated RNA interference offers a novel approach to suppress VEGF in DR. In their study, the lipopolyplexes demonstrated efficient and enhanced cellular uptake in animal models, confirming the potential of siRNA-based therapies for DR. This approach positions lipopolyplexes as a promising nanocarrier platform for targeted gene silencing in ocular diseases.

Nano-Micelles

Nanomicelles improve the solubility of medications through their unique structure, which consists of an outer hydrophilic layer surrounding a lipophilic core that encapsulates hydrophobic drugs or biologics. Due to their amphiphilic nature, nanomicelles can penetrate ocular cells effectively, thereby enhancing the bioavailability of the therapeutic agents. Their small size also facilitates efficient cellular uptake, increasing treatment effectiveness (Sruthi et al., 2023). For example, Guo et al. (2024) developed a triblock copolymer, Poly-aminoethyl methacrylate (PAMA), specifically PEG-b-PAMA-b-P(C7A-r-DBA) (PACD), designed for siRNA delivery to retinal cells. These nanomicelle-based polyplexes exhibit pH-sensitive behavior, allowing controlled and efficient drug release both in vitro and in vivo. The formulation demonstrated prolonged retention for over a week after a single injection, attributed to the structural properties of the nanomicelles. In vivo studies highlighted these nanomicelles as a promising platform for ocular gene delivery.

Another study by Xu et al. (2024) showed enhanced therapeutic activity of ranibizumab-loaded copolymeric nanomicelles in treating age-related macular degeneration. These findings support the growing evidence that nanomicelles are a promising nanocarrier system for ocular drug delivery.

Thermosensitive Hydrogels

Thermosensitive hydrogels are gaining attention for their ability to transition states in response to temperature changes, making them highly effective for delivering biologics via ocular routes. One study focused on ocu-miRNA 24-3p, a microRNA that promotes corneal epithelial cell migration and regeneration in rabbits. Exosomes derived from adipose-derived mesenchymal stem cells were loaded with miRNA 24-3p (Exos-miRNA 24-3p) to enhance corneal permeability both in vitro and in vivo. To optimize ocular application, a thermosensitive hydrogel was created by modifying hyaluronic acid (HA) with di(ethylene glycol) monomethyl ether methacrylate, enabling controlled and prolonged release of Exos-miRNA 24-3p. This formulation formed a stable layer on the ocular surface, extending drug contact time and accelerating corneal healing after 28 days of treatment. The system improved ocular bioavailability, stability, and biocompatibility, representing a promising miRNA-based, cell-free therapeutic approach for corneal regeneration (Sun et al., 2023). Another study investigated a novel in situ forming thermosensitive hydrogel system for ocular drug delivery. These hydrogels rapidly underwent sol-gel transition within one minute due to physical self-assembly and crosslinking. They demonstrated controlled release of an anti-VEGF antibody fragment over 13 days and dexamethasone over 35 days, with stability maintained for over a year. Cytocompatibility assessments confirmed the safety and potential of this system for ocular applications (Ilochonwu et al., 2023).

Electrospun Nanofibers

Electrospun nanofibers are gaining prominence as ocular drug delivery platforms because of their adaptability to the eye's surface and their ability to provide sustained drug release. One study developed an ophthalmic insert composed of hyaluronan and ferulic acid (FA) nanofibers, where FA acted as an antioxidant and antimicrobial agent. The ferulic acid and ε-polylysine (ε-PL) were fully released within 20 to 30 minutes, demonstrating rapid drug availability. These inserts effectively inhibited bacterial strains such as *Pseudomonas aeruginosa* and *Staphylococcus aureus*, confirming their antimicrobial potential. The crosslinked inserts showed appropriate biocompatibility, thickness, and drug release profiles suitable for ocular application. However, further ex vivo and in vivo studies focusing on retention time and permeability are necessary to assess their clinical potential for ocular drug delivery (Grimaudo et al., 2020).

Another study by Lokhande et al. developed a biodegradable nanofiber insert combining hyaluronic acid and polyvinyl alcohol using electrospinning techniques to enable sustained ocular drug delivery. In vitro studies demonstrated enhanced drug loading efficiency with a controlled release pattern extending over 12 hours. In vivo inflammatory and pharmacodynamic evaluations indicated that this nanofiber insert was more effective than conventional eye drops, suggesting its promise as an alternative for long-term treatment of dry eye disease (Lokhande et al., 2023).

Carbonized Nanogels

Carbonized nanogel (CNG) is a type of nanocarrier synthesized by carbonizing various organic compounds, typically through pyrolysis. These nanogels improve ocular retention, thereby enhancing bioavailability and therapeutic effectiveness compared to conventional eye drops. Dry eye syndrome (DES) commonly requires frequent dosing and high medication volumes due to the short precorneal residence time of treatments.

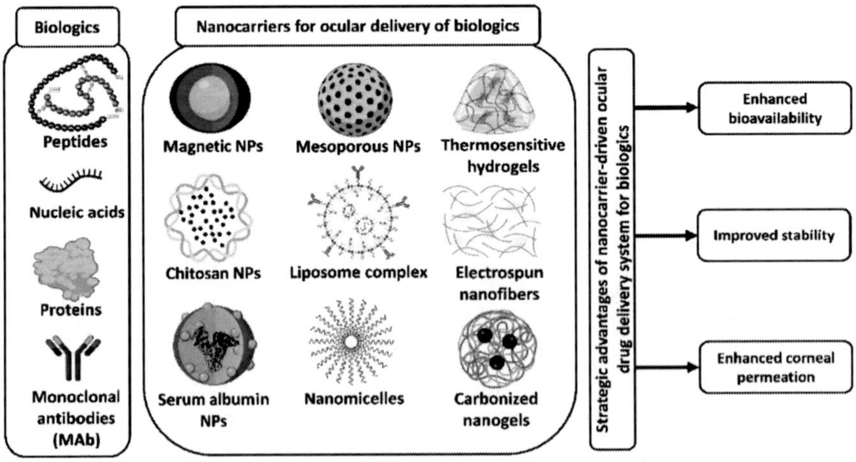

Figure 1. Representation of nanocarriers for ocular delivery of biologics.

To overcome these challenges, CNGs have been developed using highly biocompatible carbon-based nanoparticle formulations. These formulations exhibit notable antioxidant and anti-inflammatory properties while maintaining relatively low side effects (Lin et al., 2021). Specifically, carbonized nanogels derived from lysine hydrochloride through a simple pyrolysis process form nitrogen-doped, crosslinked polymers embedded with nanographitic structures. Both in vitro and in vivo studies demonstrated excellent biocompatibility with ocular epithelial cells and showed enhanced therapeutic efficacy following a single application compared to conventional doses (Lin et al., 2022).

Table 1. Summary of nanocarrier based biologics for ocular delivery

NDDS	Composition	Structural characteristics	Therapeutic Outcome	Ocular Biologics	Reference
Magnetic nanoparticles	Magnetic core of iron oxide and an organic shell linked to carboxylic groups	Variable sizes with round-to-oval morphology	Enhanced bioactivity for upto 5 days post injection	Octreotide	Amato et al., 2020
Chitosan nanoparticles	Chitosan + Tripolyphosphate	Nanosized particles	Enhanced stability and ocular interaction	Insulin	Rathore et al., 2020
Serum albumin nanoparticles	Bovine serum albumin + glutaraldehyde+ and coated with HA	Smooth solid spherical nanoparticles with narrow distribution	Increased topical mucoadhesive property, an alternative to invasive delivery system	Apatinib	Radwan et al., 2021
Mesoporous nanoparticles	PAMA as siRNA binding segment and 2-(dibutyl amino) ethyl methacrylate as hydrophobic segments, and PEG as hydrophilic segments	Organic shell with octreotide around the magnetite core formed	Increased efficacy and stability	Dexameth asone with VEFG-siRNA	Xiaochen et al., 2024

Table 1. (Continued)

NDDS	Composition	Structural characteristics	Therapeutic Outcome	Ocular Biologics	Reference
Liposomes and liposome complex	Phospholipids + Cholesterol + Additives	Bilayer formation, size and type variability	Biocompatibility, enhanced efficacy against ocular diseases	TAT-peptide	Wu et al., 2021
Nanomicelles	Amphiphilic Molecules + Hydrophobic core + Hydrophilic shell	Shape variability, Formation occurs above specific concentration known as CMC	Improved ocular bioavailability along with the sustained release	Ranibizumab	Xu et al., 2024
Thermosensitive hydrogels	Modified HA with methacrylic anhydride grafted with temperature-sensitive di (ethylene glycol) monomethyl ether methacrylate	Temperature scanning showed the thermosensitive nature of the gel	Increase bioavailability, stability and controlled release	Ocu-miRNA 24-3p	Sun et al., 2023
Electrospun nanofibres	Polyvinylpyrrolidone facilitated electrospinning of FA and then crosslinked with ε-PL	Fibers with diameters of approx. 100nm	Increased ocular residence and controlled release	FA, hyaluronan and ε-PL	Grimaudo et al., 2020
Carbonized nanogel	Pyrolysis of Quercetin and Lysine	Enhanced permeation and antibacterial activity	Prolonged retention, sustained release, dose reduction	Quercetin and Lysine	Lin et al., 2022

Clinical Implications and Applications

In recent years, the development of nanocarriers for biologics has transformed drug delivery by enhancing the stability, bioavailability, and targeted delivery of therapeutic molecules such as proteins, nucleic acids, and other biologics. These advancements hold great potential to improve clinical outcomes and patient compliance significantly. However, comprehensive long-term safety and efficacy monitoring remain essential, especially for biologic-based therapies. Affordability and accessibility continue to present major challenges, particularly in resource-limited settings. Broader access to these innovative therapies may be facilitated by cost-effective manufacturing strategies combined with strong public and private partnerships. The design and development of nanocarriers are inherently complex processes that require multidisciplinary collaboration across scientific and engineering fields. The integration of artificial intelligence and machine learning offers promising opportunities for optimizing formulations, predicting clinical outcomes, and enabling personalized therapeutic approaches. Emerging technologies such as stimuli-responsive systems, hybrid nanoparticles, and gene-editing delivery platforms are further enhancing therapeutic efficacy and reducing dosing frequency, especially in the management of chronic diseases. Nevertheless, the deployment of these advanced drug delivery systems must carefully consider ethical, economic, and regulatory implications. Robust regulatory frameworks are critical to ensure the long-term safety and effectiveness of these innovations, particularly those involving gene-editing technologies.

Future Perspectives

Encouraging preclinical findings highlight the potential of nanoparticles and other nanocarrier-based delivery systems in advancing retinal gene editing and anti-VEGF therapies. However, additional research is necessary to optimize formulations and validate long-term safety and efficacy. Future work should focus on developing multifunctional nanocarriers that integrate gene therapy with anti-inflammatory and neuroprotective approaches to provide comprehensive treatment for ocular disorders. Furthermore, scaling up manufacturing processes, improving in vivo stability, and enhancing patient compliance through non-invasive delivery methods such as topical and minimally invasive injectable formulations will be essential for successful clinical translation. By bridging the gap between cutting-edge nanocarrier technologies and clinical applicability, ongoing research can accelerate the creation of more effective, safe, and accessible biologic therapies for ocular diseases.

Conclusion

The field of ocular drug delivery for biologics is increasingly focused on developing innovative strategies to overcome challenges related to delivering monoclonal antibodies, peptides, and gene therapies. Nanocarrier-based systems have emerged as transformative platforms, providing solutions to major limitations such as poor stability, low bioavailability, and inefficient targeting of retinal tissues. By employing nanoparticles, liposomes, and hybrid delivery systems, researchers have demonstrated enhanced drug retention, deeper ocular penetration, and improved therapeutic efficacy across various ocular disease models. However, despite these promising advancements, the clinical translation of nanocarrier-based therapies remains limited due to concerns about long-term stability, potential immunogenicity, and the complexity of achieving effective penetration into ocular tissues.

References

Afzal, O., Altamimi, A. S. A., Nadeem, M. S., Alzarea, S. I., Almalki, W. H., Tariq, A., Mubeen, B., Murtaza, B. N., Iftikhar, S., Riaz, N., & Kazmi, I. (2022). Nanoparticles in Drug Delivery: From History to Therapeutic Applications. *Nanomaterials*, 12(24), 4494.

Ahmed, S., Amin, M. M., & Sayed, S. (2023). Ocular drug delivery: a comprehensive review. *AAPS PharmSciTech*, 24(2), 66.

Amato, R., Giannaccini, M., Dal Monte, M., Cammalleri, M., Pini, A., Raffa, V., Lulli, M., & Casini, G. (2020). Association of the somatostatin analog octreotide with magnetic nanoparticles for intraocular delivery: A possible approach for the treatment of diabetic retinopathy. *Frontiers in Bioengineering and Biotechnology,* 8, 144.

Annala, A., Ilochonwu, B. C., Wilbie, D., Sadeghi, A., Hennink, W. E., & Vermonden, T. (2022). Self-Healing Thermosensitive Hydrogel for Sustained Release of Dexamethasone for Ocular Therapy. *ACS Polymers Au*, *3*(1), 118-131.

Bansal, R., Jha, S. K., & Jha, N. K. (2021). Size-based Degradation of Therapeutic Proteins-Mechanisms, Modelling and Control. *Biomolecular Concepts*, 12(1), 68-84.

Batur, E., Özdemir, S., Durgun, M. E., & Özsoy, Y. (2024). Vesicular Drug Delivery Systems: Promising Approaches in Ocular Drug Delivery. Pharmaceuticals, 17(4), 511.

Chandel, A., & Kandav, G. (2024). Insights into ocular therapeutics: a comprehensive review of anatomy, barriers, diseases and nanoscale formulations for targeted drug delivery. Journal of Drug Delivery Science and Technology, 97, 105785.

Eker, F., Duman, H., Akdaşçi, E., Bolat, E., Sarıtaş, S., Karav, S., & Witkowska, A. M. (2024). A comprehensive review of nanoparticles: from classification to application and toxicity. *Molecules*, 29(15), 3482.

Gohar, S., Iqbal, Z., Nasir, F., Khattak, M. A., E Maryam, G., Pervez, S., Alasmari, F., Neau, S. H., Zainab, S. R., Ali, A. T., & Ur Rahman, A. (2024). Self-assembled latanoprost loaded soluplus nanomicelles as an ophthalmic drug delivery system for the management of glaucoma. *Scientific Reports*, 14(1), 27051.

Grimaudo, M. A., Concheiro, A., & Alvarez-Lorenzo, C. (2020). Crosslinked hyaluronan electrospun nanofibers for ferulic acid ocular delivery. *Pharmaceutics*, 12(3), 274.

Gugleva, V., & Andonova, V. (2023). Recent progress of solid lipid nanoparticles and nanostructured lipid carriers as ocular drug delivery platforms. *Pharmaceuticals*, 16(3), 474.

Guo, S., Li, C., Wang, C., Cao, X., Liu, X., Liang, X. J., & Weng, Y. (2024). pH-responsive polymer boosts cytosolic siRNA release for retinal neovascularization therapy. *Acta Pharmaceutica Sinica B*, 14(2), 781–794.

Ilochonwu, B. C., Van Der Lugt, S. A., Annala, A., Di Marco, G., Sampon, T., Siepmann, J., Siepmann, F., Hennink, W. E., & Vermonden, T. (2023). Thermo-responsive Diels-Alder stabilized hydrogels for ocular drug delivery of a corticosteroid and an anti-VEGF Fab fragment. *Journal of Controlled Release*, 361, 334-349.

Lin, H.-Y., Wang, S.-W., Mao, J.-Y., Chang, H.-T., Harroun, S. G., Lin, H.-J., Huang, C.-C., & Lai, J.-Y. (2021). Carbonized nanogels for simultaneous antibacterial and antioxidant treatment of bacterial keratitis. *Chemical Engineering Journal*, 411, 128469.

Lin, P. H., Jian, H. J., Li, Y. J., Huang, Y. F., Anand, A., Huang, C. C., Lin, H. J., & Lai, J. Y. (2022). Alleviation of dry eye syndrome with one dose of antioxidant, anti-inflammatory, and mucoadhesive lysine-carbonized nanogels. *Acta Biomaterialia*, 141, 140-150.

Lokhande, S., Bhandari, Y., Ansari, M., Khairnar, S., & Singh, K. (2023). Development and evaluation of biodegradable nanofiber inserts containing hyaluronic acid for treatment of dry eye syndrome. *Journal of Drug Delivery Science and Technology*, 87, 104847.

Ma, X., Cui, Y., Zhang, M., Lyu, Q., & Zhao, J. (2024). A Multifunctional Nanodrug Co-Delivering VEGF-siRNA and Dexamethasone for Synergistic Therapy in Ocular Neovascular Diseases. *International Journal of Nanomedicine*, 19, 12369-12387.

Mahor, A., Singh, P. P., Bharadwaj, P., Sharma, N., Yadav, S., Rosenholm, J. M., & Bansal, K. K. (2021). Carbon-based nanomaterials for delivery of biologicals and therapeutics: A cutting-edge technology. *Journal of Carbon Research*, 7(1), 19.

Mardikasari, S. A., Katona, G., Budai-Szűcs, M., Sipos, B., Orosz, L., Burián, K., Rovó, L., & Csóka, I. (2023). Quality by design-based optimization of in situ ionic-sensitive gels of amoxicillin-loaded bovine serum albumin nanoparticles for enhanced local nasal delivery. *International Journal of Pharmaceutics*, 645, 123435.

Pandit, J., Sultana, Y., & Aqil, M. (2021). Chitosan coated nanoparticles for efficient delivery of bevacizumab in the posterior ocular tissues via subconjunctival administration. *Carbohydrate Polymers*, 267, 118217.

Paramshetti, S., Angolkar, M., Talath, S., Osmani, R. A. M., Spandana, A., Al Fatease, A., Hani, U., Ramesh, K. V. R. N. S., & Singh, E. (2024). Unravelling the in vivo dynamics of liposomes: Insights into biodistribution and cellular membrane interactions. *Life Sciences*, 346, 122616.

Gaur, P. K., & Kumar, S. (2025). Nanocarrier-Mediated Dermal Drug Delivery System of Antimicrobial Agents for Targeting Skin and Soft Tissue Infections. *ASSAY and Drug Development Technologies*, 23(1), 2-28.

Qi, J., He, W., Meng, J., Wei, L., Qian, D., Lu, Y., & Zhu, X. (2021). Distribution of Ocular Anterior and Posterior Segment Lengths Among a Cataract Surgical Population in Shanghai. *Frontiers in Medicine*, 8, 688805.

Radwan, S. E. S., El-Kamel, A., Zaki, E. I., Burgalassi, S., Zucchetti, E., & El-Moslemany, R. M. (2021). Hyaluronic-coated albumin nanoparticles for the non-invasive delivery of apatinib in diabetic retinopathy. *International Journal of Nanomedicine*, 16, 4481-4494.

Rathore, P., Mahor, A., Jain, S., Haque, A., & Kesharwani, P. (2020). Formulation development, in vitro and in vivo evaluation of chitosan engineered nanoparticles for ocular delivery of insulin. *RSC Advances*, 10(71), 43629-43639.

Romeo, A., Kazsoki, A., Omer, S., Pinke, B., Mészáros, L., Musumeci, T., & Zelkó, R. (2023). Formulation and characterization of electrospun nanofibers for melatonin ocular delivery. *Pharmaceutics*, 15(4), 1296.

Sarkar, T., Gogoi, N. R., Jana, B. K., & Mazumder, B. (2025). Formulation advances in posterior segment ocular drug delivery. Journal of Ocular Pharmacology and Therapeutics, 41(3), 101-130.

Sharma, J., & Paschalis, E. I. (2022). The future of non-viral gene delivery for the treatment of inherited retinal diseases. *Molecular Therapy. Nucleic Acids*, 30, 354.

Silva, B., Gonçalves, L. M., São Braz, B., & Delgado, E. (2023). Topical ocular delivery of nanoparticles with epoetin beta in Wistar Hannover rats. *Scientific Reports*, 13(1), 1559.

Spada, A., Emami, J., Tuszynski, J. A., & Lavasanifar, A. (2021). The uniqueness of albumin as a carrier in nanodrug delivery. *Molecular Pharmaceutics*, 18(5), 1862-1894.

Sruthi, K. S., Rajan, M., Saravanan, K. M., & Shakila, H. (2023). Eye Posterior segment drug delivery and Nanomicelles as a drug delivery system for Age Related Macular Degeneration. *Eye*, 12, 59.

Sun, X., Song, W., Teng, L., Huang, Y., Liu, J., Peng, Y., Lu, X., Yuan, J., Zhao, X., Zhao, Q., & Xu, Y. (2023). MiRNA 24-3p-rich exosomes functionalized DEGMA-modified hyaluronic acid hydrogels for corneal epithelial healing. *Bioactive Materials*, 25, 640-656.

Supe, S., Upadhya, A., & Tripathi, S. (2023). Liposome-polyethylenimine complexes for the effective delivery of HuR siRNA in the treatment of diabetic retinopathy. *Drug Delivery and Translational Research*, 13, 1675-1698.

Tang, Z., Fan, X., Chen, Y., & Gu, P. (2022). Ocular nanomedicine. *Advanced Science*, 9(15), 2003699.

Tian, B., Bilsbury, E., Doherty, S., Teebagy, S., Wood, E., Su, W., Gao, G., & Lin, H. (2022). Ocular Drug Delivery: Advancements and Innovations. *Pharmaceutics*, 14(9), 1931.

Vilkelyte, V., Thompson, P., Coelho, M., Woronkowicz, M., Skopinski, P., & Roberts, H. (2024). Challenges and advances in magnetic nanoparticle-guided delivery of cultured human corneal endothelial cells—A review. *Applied Sciences*, 14(13), 5877.

Wang, Y., Pi, C., Feng, X., Hou, Y., Zhao, L., & Wei, Y. (2020). The influence of nanoparticle properties on oral bioavailability of drugs. *International Journal of Nanomedicine*, 15, 6295-6310.

Liu, L. C., Chen, Y. H., & Lu, D. W. (2023). Overview of Recent Advances in Nano-Based Ocular Drug Delivery. *International Journal of Molecular Sciences*, 24(20), 15352.

Wu, B., Li, M., Li, K., Hong, W., Lv, Q., Li, Y., Xie, S., Han, J., & Tian, B. (2021). Cell penetrating peptide TAT-functionalized liposomes for efficient ophthalmic delivery of flurbiprofen: Penetration and its underlying mechanism, retention, anti-inflammation and biocompatibility. *International Journal of Pharmaceutics*, 598, 120405.

Wu, K. Y., Tan, K., Akbar, D., Choulakian, M. Y., & Tran, S. D. (2023). A new era in ocular therapeutics: advanced drug delivery systems for uveitis and neuro-ophthalmologic conditions. *Pharmaceutics*, 15(7), 1952.

Xu, J. F., Wang, Y. P., & Liu, X. H. (2024). Novel fabrication of anti-VEGF drug ranibizumab loaded PLGA/PLA co-polymeric nanomicelles for long-acting intraocular delivery in the treatment of age-related macular degeneration therapy. *Regenerative Therapy*, 26, 620-634.

Yusuf, A., Almotairy, A. R. Z., Henidi, H., Alshehri, O. Y., & Aldughaim, M. S. (2023). Nanoparticles as drug delivery systems: a review of the implication of nanoparticles' physicochemical properties on responses in biological systems. *Polymers*, 15(7), 1596.

Zhou, Y., Xu, M., Shen, W., Xu, Y., Shao, A., Xu, P., Yao, K., Han, H., & Ye, J. (2024). Recent advances in nanomedicine for ocular fundus neovascularization disease management. *Advanced Healthcare Materials*, 13(17), e2304626.

Chapter 18

Smart Nanocarriers for Targeted Ocular Drug Delivery

Athrv Arora[1]
Sapna Kumari[2]
and Madhukar Garg[1,*]
[1]Chitkara College of Pharmacy, Chitkara University, Punjab, India
[2]Shiva Institute of Pharmacy, Bilaspur, Himachal Pradesh, India

Abstract

Eye diseases significantly impact vision and overall quality of life for people around the world. However, delivering drugs effectively to the eye is challenging due to various anatomical and physiological barriers that limit drug absorption and reduce therapeutic effectiveness. The most commonly used method for ocular drug delivery is the topical route, but achieving adequate drug concentrations in the targeted ocular tissues remains difficult. To address these challenges, advancements in nanotechnology-based drug delivery systems have been explored. These systems offer targeted and sustained drug release, improving treatment outcomes. Formulation strategies such as the use of viscosity enhancers, mucoadhesive agents, and penetration enhancers have been applied to increase drug availability and retention in ocular tissues. This review highlights several advanced drug delivery systems that improve the delivery of medications to both the anterior and posterior segments of the eye. Although these techniques are promising, challenges still exist, including maintaining long-term stability and ensuring safety. Further

* Corresponding Author's Email: madhukar.garg@chitkara.edu.in

In: Advances in Nanotechnology for Ocular Drug Delivery
Editors: Anjali Sharma, Ajmer Singh Grewal, Geeta Deswal et al.
ISBN: 979-8-89530-814-1
© 2025 Nova Science Publishers, Inc.

research is necessary to advance this field and develop more effective, safe, and user-friendly treatment options.

Keywords: ocular drug delivery, anatomical and physiological barriers, nanocarriers, targeted and sustained drug release, formulation strategies

Abbreviations

APBs: Anatomical and Physiological Barrier
BAB: Blood-Aqueous Barrier
BRB: Blood-Retinal Barrier
NDDSs: Nano-Based Drug Delivery Systems
Nes: Nanoemulsions
TR: Topical Route
WHO: World Health Organization

Introduction

According to the World Health Organization (WHO, 2019), billions of people around the world are affected by vision problems. The human eye has complex anatomical structures and protective barriers that make it difficult for medications to reach targeted areas effectively (Taylor, 2009). Anatomically, the eye is divided into two main segments: the anterior segment and the posterior segment (Dave et al., 2021). Natural defense mechanisms such as tear production, blinking, and nasolacrimal drainage further limit the amount of drug that successfully penetrates ocular tissues (Gaudana et al., 2010; Gote et al., 2019; Tsai et al., 2018). Although topical eye drops and ointments are widely used due to their ease of administration, they often suffer from poor tissue penetration and limited therapeutic effectiveness (Adrianto et al., 2022). To overcome these challenges, researchers are developing advanced drug delivery strategies to ensure that medications reach the desired intraocular sites. One promising approach involves the use of nanocarrier-based drug delivery systems, which have gained considerable attention for their potential to enhance drug absorption, prolong release, and provide targeted delivery to specific ocular tissues (Kaushal et al., 2023). Various types of nanocarriers, including nanoparticles and nanomicelles, have been investigated to overcome the limitations of conventional drug delivery methods (Qamar et al., 2019).

These advanced systems offer improved drug penetration, reduced systemic side effects, and sustained therapeutic effects when compared to traditional approaches. This article discusses the role of nanocarriers in addressing ocular drug delivery challenges and enhancing treatment outcomes (Zhang et al., 2021).

Understanding Ocular Drug Delivery: Anatomy and Challenges

The human eye measures approximately 22 to 27 millimeters in length and 69 to 85 millimeters in circumference (Kels et al., 2015). As mentioned in the introduction, the eye is primarily divided into two main regions: the anterior segment, which includes the front structures, and the posterior segment, which encompasses the back portion. These segments are separated by the ciliary body and the lens. Due to its intricate anatomy and multiple protective barriers, effective drug delivery to the eye presents a considerable challenge. Much like the brain, the eye features specialized barriers, most notably the blood-retinal barrier (BRB), which limits the passage of therapeutic agents from the bloodstream into ocular tissues. This selective permeability complicates the treatment of ocular diseases, especially those affecting the posterior segment, such as age-related macular degeneration and diabetic retinopathy.

Tear Film Barrier

On the surface of the eye, the tear film acts as a major barrier to drug delivery by limiting drug penetration and washing away medications through the nasolacrimal duct. It is composed of three distinct layers: an outer lipid layer, a middle aqueous layer, and an inner mucous layer (Dartt et al., 2013). The lipid layer serves a dual function as it prevents drug molecules from entering the eye and also reduces evaporation of the underlying aqueous layer (Mandal et al., 2017). Although essential for maintaining ocular health, the tear film is very thin, measuring approximately 3 micrometers in thickness and containing a small volume of around 3 microliters. This limited volume, along with the continuous turnover of tears, leads to rapid elimination of drugs from the ocular surface, significantly reducing their retention time. Additionally, drugs that cause irritation or have a non-physiological pH can stimulate reflex tearing, which may expel more than 85 percent of the medication before it reaches the corneal surface (Gorantla et al., 2020). As a result, only a small fraction, estimated between 0.1 percent and 5 percent, of the administered drug actually penetrates into the intraocular tissues (Subrizi et al., 2019).

Cornea and Conjunctival Barrier

The cornea is a transparent outer layer of the eye that plays a crucial role in focusing light and providing protection. It is primarily composed of three layers: an outer lipid layer, a middle aqueous layer, and an inner lipid layer (Mofidfar et al., 2021). The corneal epithelium contains tightly connected cells, which create a barrier that limits the passage of drugs. Additionally, the presence of blood vessels and lymphatic channels in the conjunctiva facilitates the rapid removal of drugs into the systemic circulation, thereby reducing their local effectiveness. The sclera, which is the white portion of the eye, permits the passage of small molecules but restricts larger ones from entering (Gaudana et al., 2010).

Blood-Aqueous Barrier (BAB)

The movement of substances between the eye's front chambers and the bloodstream is controlled by this barrier. It is composed of tight cell junctions in the ciliary body, iris blood vessels, and Schlemm's canal (Coca-Prados, 2014).

Vitreal Barrier

The vitreous is a gel-like substance inside the eye that helps maintain its shape. It is composed of water, collagen, and hyaluronic acid, which can hinder the movement of positively charged drug particles (Peynshaert et al., 2018). Negatively charged drugs are often used in specialized ocular treatments because they can more effectively move through the vitreous humor (Tavakoli et al., 2020).

Blood-Retinal Barrier (BRB)

The flow of nutrients and waste between the retina and the blood is controlled by the blood-retinal barrier (BRB). It consists of two parts: the outer layer and the inner layer formed by retinal blood vessels (Cunha-Vaz et al., 2011). The outer BRB permits only small molecules to pass through, while the inner BRB regulates the passage of molecules based on their size, charge, and lipid solubility. Researchers are continuously working on designing drugs that can penetrate these barriers more effectively (Peynshaert et al., 2018).

Sclera and Choroid Barrier

The choroid is a layer rich in blood vessels that supplies nutrients to the retina and acts as a barrier to hydrophilic drugs. The sclera, composed of collagen, allows small, hydrophilic drugs to pass through (Varela-Fernández et al., 2020). The movement of drugs through the sclera depends on factors such as molecular size, charge, and solubility. An illustration of the eye structure and its barriers can be seen in Figure 1.

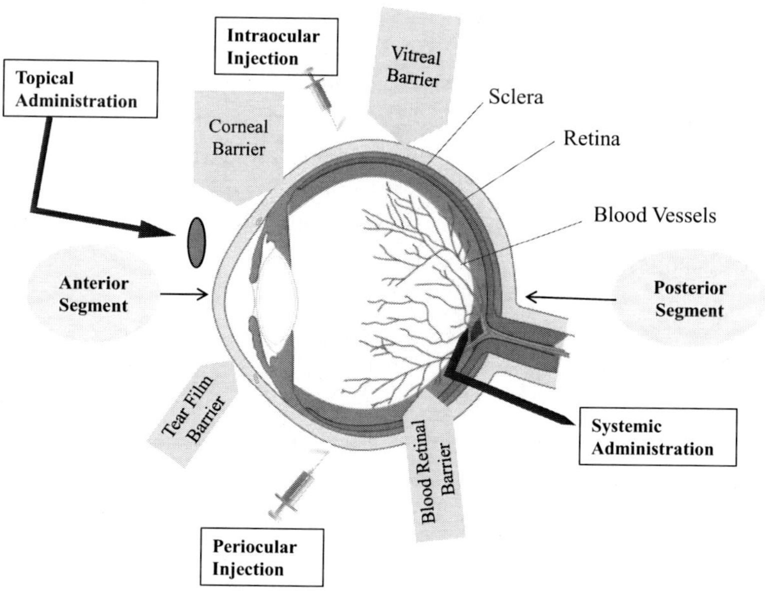

Figure 1. Eye structure and its barriers.

Nanotechnology

Nanotechnology employs tiny carriers called nanoscale systems to deliver medicines to the eye with precision. These carriers assist in transporting drugs to the target site, enhance solubility and stability, and enable controlled release. Over the past decades, this approach has become increasingly popular for improving drug effectiveness and overcoming ocular barriers (Li et al., 2023).

Nanomicelles

Micelles are tiny self-assembling particles, ranging from 5 to 100 nm, composed of a hydrophobic core and a hydrophilic shell. They enhance drug delivery to specific tissues by dissolving and transporting water-insoluble drugs. Micelles are mainly classified into three types: polymer micelles, surfactant micelles, and multi-ion composite micelles (Bose et al., 2021). These structures can easily form in both aqueous and non-polar solvents, adapting their configuration accordingly (Trivedi et al., 2010). Natural polymers such as hyaluronic acid and chitosan are biocompatible and safe, while synthetic polymers like PEG and PVP contribute to stability and controlled drug release (Tawfik et al., 2020). Micelle-based formulations have demonstrated effectiveness in ocular treatments, with some already approved for clinical use. Notable examples include rapamycin nanomicelle eye drops designed to prevent immune rejection (Shah et al., 2017; Zhou et al., 2017; Nagai et al., 2022). However, comprehensive safety evaluations are necessary before clinical application to assess potential toxicity, genetic effects, and immune responses.

Nanoparticles (NPs)

Nanoparticles are tiny particles ranging from 1 to 1000 nanometers, composed of materials such as metals, polymers, lipids, or ceramics. They can be engineered in various shapes, including spheres and rods, and may have solid or hollow structures. Nanoparticles are broadly classified into nanospheres, which have a solid polymeric structure, and nanocapsules, which consist of a thin polymeric shell surrounding an oily core. These particles protect drugs from degradation, enhance absorption, and provide controlled release, thereby increasing their effectiveness in ocular drug delivery (Zielińska et al., 2020). In ocular applications, nanoparticles are frequently made from natural polymers like cellulose and hyaluronic acid due to their biocompatibility and safety. Chitosan offers excellent adhesion and antimicrobial properties but is water-insoluble, whereas hyaluronic acid provides superior hydration but tends to be expensive (Tsung et al., 2023; Sánchez-López et al., 2017). Smaller nanoparticles, typically between 50 and 400 nanometers, exhibit better absorption and penetration, with sizes around 200 nanometers considered ideal for uptake by the cornea and conjunctiva (Han et al., 2023; Yetisgin et al., 2020; Akhter et al., 2016). Nanoparticles are widely utilized in both the diagnosis and treatment of eye diseases. For instance, gold nanoparticles have been employed to detect and treat retinoblastoma, where their combination with brachytherapy and hyperthermia contributed to tumor size reduction (Moradi et al., 2020). Additionally, PLGA-based formulations represent polymeric nanoparticles that improve ocular drug delivery by providing prolonged release and reducing dosing frequency (Abrego et al., 2015; Varshochian et al., 2015).

Nanosuspension (NSs)

Nanosuspensions are tiny particles ranging from 200 to 500 nanometers, stabilized within a liquid medium (Khanuja et al., 2021; Manca et al., 2020). They can be administered in liquid form or processed into powders, tablets, capsules, or gels (Ma et al., 2023; Pınar et al., 2023). In ocular treatments, nanosuspensions improve drug absorption and retention, although they require thorough stability and safety evaluations. These systems have been explored for drug delivery through various routes, including oral, ocular, brain-targeted, and transdermal applications (Jacob et al., 2020). Research has demonstrated their effectiveness in managing conditions such as glaucoma and macular degeneration, enhancing both drug efficacy and patient comfort.

Nanoemulsions (NEs) and Microemulsions (MEs)

Nanoemulsions are small droplets ranging from 2 to 500 nanometers, composed of two immiscible liquids stabilized by surfactants (Singh et al., 2017). Due to their ability to encapsulate lipophilic drugs and easily mix with tear fluid, oil-in-water nanoemulsions are widely used for ocular drug delivery

(Hegde et al., 2013). Similar to nanoemulsions, microemulsions require a higher concentration of surfactants and can have spherical or non-spherical particles. Nanoemulsions offer longer retention times, improved drug absorption, and higher bioavailability compared to gels or ointments in eye treatments. Their small droplet size and stability make them suitable for topical, ocular, and intravenous drug delivery (Gupta et al., 2020). They have demonstrated effectiveness in treating anterior segment eye disorders. For delivering hydrophobic drugs to the posterior segment of the eye, gel-in-water nanoemulsions have been developed, showing biocompatibility and no irritation in preclinical tests (Fardous et al., 2022). However, challenges remain, including the need for high surfactant concentrations, potential toxicity, stability concerns, and solubilization limits, all of which require careful formulation and safety evaluation.

Nanofibers

Nanostructures are classified into four types, among which one-dimensional nanofibers are particularly notable due to their high surface area, porosity, mechanical strength, and flexibility. These characteristics allow them to mimic the structure of the extracellular matrix (Uzel et al., 2023). Nanofibers, typically ranging from tens to hundreds of nanometers in diameter, are composed of natural or synthetic polymers, carbon, or ceramics. Nanofiber composites are used in medical applications such as implants and wound dressings. They can be loaded with a variety of drugs, including antibiotics, anticancer agents, and proteins (Hu et al., 2014). Due to their unique properties, nanofibers have shown promise in the treatment of both anterior and posterior segment eye disorders.

Dendrimers

Dendrimers are nanoscale, highly branched polymer structures with a three-dimensional framework. Their surfaces contain multiple functional groups, making them highly suitable for drug delivery. They are generally classified into two main types: poly(amidoamines) and polypropyleneimines. For specific applications, properties such as size, charge, and solubility can be tailored to enhance their effectiveness (Sherje et al., 2018; Wang et al., 2022). Dendrimers have shown potential in targeted drug delivery by minimizing side effects and extending drug activity. Some dendrimers naturally possess antimicrobial, antifungal, or anticancer properties (Janaszewska et al., 2019). However, larger dendrimers or those with a strong positive charge may exhibit toxicity. To mitigate these effects, surface modifications such as PEGylation and acetylation are often employed (Qin et al., 2020). In terms of medical applications, DenTimo, a dendrimer-based timolol formulation, has been shown to effectively reduce intraocular pressure (Lancina et al., 2018). Additionally, dendrimers are used in corneal regeneration, acting as scaffolds that support corneal cell adhesion and growth (Duan et al., 2006).

Liposomes and Niosomes

Liposomes and niosomes are types of drug delivery systems designed to carry and release medicines in a controlled manner. Although they have similar structural characteristics, their compositions differ. Liposomes are spherical vesicles composed of phospholipid bilayers that form small, water-filled compartments. Their size typically ranges from 0.025 to 10 micrometers (Ge et al., 2019), and they are capable of carrying both water-soluble (hydrophilic) and fat-soluble (lipophilic) drugs. By modifying their surface with specific molecules or attaching targeting ligands, liposomes can be engineered to deliver drugs to targeted cells or tissues with greater precision. This makes them especially valuable in the delivery of chemotherapy drugs, vaccines, and even some cosmetic formulations. However, liposomes face certain limitations, such as instability and the need for special storage conditions (De Leo et al., 2022). Niosomes, which resemble liposomes in structure, are made from non-ionic amphiphilic molecules rather than phospholipids. These molecules can self-assemble in aqueous solutions to form stable vesicles. Compared to liposomes, niosomes are generally more stable and less susceptible to degradation or oxidation. Like liposomes, they can encapsulate both hydrophilic and lipophilic drugs within the same vesicle. Moreover, niosomes are more cost-effective and easier to manufacture, making them a promising alternative for drug delivery (López-Cano et al., 2021). The choice between these two systems in clinical applications depends on factors such as the nature of the drug and the required delivery characteristics. Both liposomes and niosomes are currently used in various FDA-approved products, including Visudyne® for the treatment of ocular diseases (Moghtaderi et al., 2022; Ghosh et al., 2019).

Table 1. Characteristics of different nanotechnologies

Category	Size Range	Composition	Characteristics	Applications	References
Nanomicelles	5-100 nm	Hydrophobic core, hydrophilic shell; Natural (hyaluronic acid, chitosan), Synthetic (PEG, PVP)	Improves solubility, dissolves and transports water-insoluble drugs, self-assembling in water/non-polar solvents	Terbinafine hydrochloride nanomicelles	(Bose et al., 2021)
NPs	1-1000 nm	Metals, polymers (natural: cellulose, chitosan, hyaluronic acid; synthetic: PLA, PLGA, PCL), lipids, ceramics	Protects drugs from degradation, improves absorption, enables controlled release	Gold NPs (retinoblastoma treatment), PLGA NPs (prolonged ocular drug delivery)	(Zielińska et al., 2020)
NSs	200-500 nm	Drug particles stabilized in a liquid	Improves solubility, suitable for various dosage forms (liquid, powder, tablet, capsule, gel)	Used for glaucoma, macular degeneration, diabetic retinopathy, uveitis	(Khanuja et al., 2021)
NEs	2-500 nm	Oil-water systems stabilized by surfactants	Oil-in-water NEs encapsulate lipophilic drugs; better bioavailability than gels/ointments	Treat anterior segment disorders, gel-in-water NEs for posterior segment	(Sing et al., 2017)
MEs	5-200 nm	Water, oil, surfactants, cosurfactants	Thermodynamically stable, requires more surfactants than NEs	Similar benefits as NEs but higher surfactant use	(Hegde et al., 2013)
Nanofibers	Tens to hundreds of nm	Natural (hyaluronic acid, chitosan, collagen), Synthetic (PLA, PLGA, PCL), Carbon, Ceramics	High surface area, porous, mechanically strong, resembles extracellular matrix	Used for medical implants, wound dressings, tissue scaffolds, ocular drug delivery	(Bavatharani et al., 2021)

Category	Size Range	Composition	Characteristics	Applications	References
Dendrimers	Nanoscale (varied)	Poly(amidoamines), polypropyleneimines	Highly branched, 3D polymer structures with functional groups on the surface, customizable size, charge, solubility	Targeted drug delivery, reduces side effects, prolongs drug activity; example: DenTimo (lowers intraocular pressure), corneal regeneration	(Sherje et al., 2018)
Liposomes	0.025-10 μm	Phospholipid layers forming water-filled compartments	Spherical, can carry hydrophilic and lipophilic drugs; surface can be modified for targeting	Used in chemotherapy drugs, vaccines; example: Visudyne® for AMD and retinal diseases	(Moghtaderi et al., 2021)
Niosomes	Similar to liposomes	Non-ionic amphiphilic molecules	More stable than liposomes, cost-effective	Used for controlled drug release; potential for drug leakage may affect drug release rate and effectiveness	(López-Cano et al., 2021)
Nanowafers	Small, transparent discs	PVA	Slow drug release, dissolves naturally, reduces dosing frequency, biodegradable	Treats corneal neovascular	(Marcano et al., 2016)

Nanowafers

Nanowafers are small, transparent discs that contain nanoreservoirs designed to store and gradually release drugs into the eye, thereby enhancing drug absorption. After the drug is completely released, the nanowafer naturally dissolves. Polyvinyl alcohol (PVA) is commonly used in their production. Nanowafers offer multiple benefits, including sustained drug release, which reduces the need for frequent administration; lower systemic exposure, which minimizes side effects; and biodegradability, which prevents the accumulation of materials in the eye. These properties make nanowafers highly effective in the treatment of various ocular conditions. For instance, a nanowafer loaded with axitinib has demonstrated therapeutic effectiveness in managing corneal neovascularization. Once-daily application was found to be more beneficial than twice-daily eye drops and did not interfere with the wound healing process (Marcano et al., 2016; Coursey et al., 2015).

The characteristics of different nanotechnologies can be seen in Table 1.

Future Perspective

The complex anatomy of the posterior segment of the eye poses a significant challenge in developing effective drug delivery systems. Conditions affecting the retina and choroid are often chronic and require prolonged treatment, making sustained and efficient drug delivery essential for better disease management and improved patient outcomes. However, current delivery methods are limited in their ability to specifically target these regions (Wang et al., 2003). Typically, drugs are released into surrounding tissues such as the vitreous chamber, sclera, suprachoroidal space, or ocular surface, rather than being directed precisely to the retina or choroid. Direct implantation of drug carriers is not feasible due to the sensitive neurogenic nature of the retina. While topical nanoformulations and periocular routes are available, they offer limited drug diffusion through the sclera and non-corneal pathways, rendering them less effective and rarely used. There is a growing need for drug delivery systems that are not only more effective but also minimally invasive, patient-friendly, and suitable for long-term use. Such systems should be capable of delivering drugs directly to the retina or choroid in a controlled and sustained manner to reduce the frequency of dosing. Currently, the intravitreal (IVT) route is considered the most effective option, as it places the drug in close proximity to the retina and avoids rapid drug clearance (Jimenez et al., 2016).

In addition, the development of new biocompatible materials with advanced delivery capabilities is vital. Researchers are investigating smart drug delivery systems that respond to external stimuli such as light. Other promising strategies include hydrogel systems, nanoparticles, and ligand-modified liposomes, all of which can extend the retention time of drugs in the tear film and increase their bioavailability. Preclinical studies have shown that enhancing the retention of drug delivery systems at the target site significantly improves drug concentration and therapeutic efficacy. Ongoing research is focused on creating more reliable treatments that not only enhance delivery efficiency but also improve patient outcomes and reduce adverse effects (Gaudana et al., 2010).

Conclusion

Diseases that affect the anterior segment of the eye are commonly treated using eye drops, ointments, or ocular inserts. In contrast, disorders involving the posterior segment require specialized treatment approaches, as topical therapies are generally ineffective in reaching the deeper ocular tissues. With ongoing advancements in medical science, nanotechnology-based drug delivery systems have emerged as promising tools in the management of ocular diseases. These systems enhance drug absorption, improve stability, and facilitate targeted delivery to specific ocular tissues. By enabling precise and controlled drug release, they significantly increase therapeutic efficacy while reducing the frequency of administration. Such innovative approaches help overcome the inherent limitations of conventional drug formulations, particularly in delivering medications to the posterior segment. This targeted strategy not only improves treatment outcomes but also minimizes systemic exposure and associated side effects. As a result, patient compliance is improved, overall treatment costs are lowered, and quality of life is enhanced for individuals suffering from ocular disorders. This review highlights the various nanotechnology-based drug delivery platforms currently being explored and emphasizes ongoing research efforts aimed at optimizing these technologies for more effective and patient-friendly eye care solutions.

References

Abrego, G., Alvarado, H., Souto, E. B., Guevara, B., Bellowa, L. H., Parra, A., Calpena, A., & Garcia, M. L. (2015). Biopharmaceutical profile of pranoprofen-loaded PLGA nanoparticles containing hydrogels for ocular administration. *European Journal of Pharmaceutics and Biopharmaceutics, 95*, 261–270.

Adrianto, M. F., Annuryanti, F., Wilson, C. G., Sheshala, R., & Thakur, R. R. S. (2022). In vitro dissolution testing models of ocular implants for posterior segment drug delivery. *Drug Delivery and Translational Research, 12*(5), 1355–1375.

Akhter, S., Anwar, M., Siddiqui, M. A., Ahmad, I., Ahmad, J., Ahmad, M. Z., Bhatnagar, A., & Ahmad, F. J. (2016). Improving the topical ocular pharmacokinetics of an immunosuppressant agent with mucoadhesive nanoemulsions: Formulation development, in-vitro and in-vivo studies. *Colloids and Surfaces B: Biointerfaces, 148*, 19–29.

Bavatharani, C., Muthusankar, E., Wabaidur, S. M., Alothman, Z. A., Alsheetan, K. M., Al-Anazy, M., & Ragupathy, D. (2021). Electrospinning technique for production of polyaniline nanocomposites/nanofibres for multi-functional applications: A review. *Synthetic Metals, 271*, 116609.

Bose, A., Roy Burman, D., Sikdar, B., & Patra, P. (2021). Nanomicelles: Types, properties, and applications in drug delivery. *IET Nanobiotechnology, 15*(1), 19-27.

Coca-Prados, M. (2014). The blood-aqueous barrier in health and disease. *Journal of Glaucoma, 23*(Suppl. 1), S36-S38.

Coursey, T. G., Henriksson, J. T., Marcano, D. C., Shin, C. S., Isenhart, L. C., Ahmed, F., De Paiva, C. S., Pflugfelder, S. C., & Acharya, G. (2015). Dexamethasone nanowafer as an effective therapy for dry eye disease. *Journal of Controlled Release, 213*, 168–174.

Cunha-Vaz, J., Bernardes, R., & Lobo, C. (2011). Blood-retinal barrier. *European Journal of Ophthalmology, 21*(Suppl. 6), S3–S9.

Dartt, D. A., & Willcox, M. D. (2013). Complexity of the tear film: Importance in homeostasis and dysfunction during disease. *Experimental Eye Research, 117*, 1–3.

Dave, R. S., Goostrey, T. C., Ziolkowska, M., Czerny-Holownia, S., Hoare, T., &Sheardown, H. (2021). Ocular drug delivery to the anterior segment using nanocarriers: A mucoadhesive/mucopenetrative perspective. *Journal of Controlled Release, 336*, 71–88.

De Leo, V., Maurelli, A. M., Giotta, L., & Catucci, L. (2022). Liposomes containing nanoparticles: Preparation and applications. *Colloids and Surfaces B: Biointerfaces, 218*, 112737.

Duan, X., &Sheardown, H. (2006). Dendrimer crosslinked collagen as a corneal tissue engineering scaffold: Mechanical properties and corneal epithelial cell interactions. *Biomaterials, 27*(28), 4608–4617.

Fardous, J., Inoue, Y., Yoshida, K., Ono, F., Higuchi, A., & Ijima, H. (2022). Delivery of hydrophobic drugs to the posterior ocular region by gel-in-water nanoemulsion. *Translational Vision Science & Technology, 11*(6), 16.

Gaudana, R., Ananthula, H. K., Parenky, A., & Mitra, A. K. (2010). Ocular drug delivery. *AAPS Journal, 12*(3), 348-360.

Gaudana, R., Ananthula, H. K., Parenky, A., & Mitra, A. K. (2010). Ocular drug delivery. *AAPS Journal, 12*(3), 348–360.

Ge, X., Wei, M., He, S., & Yuan, W. E. (2019). Advances of non-ionic surfactant vesicles (niosomes) and their application in drug delivery. *Pharmaceutics, 11*(2), 55.

Ghosh, S., Carter, K. A., & Lovell, J. F. (2019). Liposomal formulations of photosensitizers. *Biomaterials, 218*, 119341.

Gorantla, S., Rapalli, V. K., Waghule, T., Singh, P. P., Dubey, S. K., Saha, R. N., & Singhvi, G. (2020). Nanocarriers for ocular drug delivery: Current status and translational opportunity. *RSC Advances, 10*, 27835-27855.

Gote, V., Sikder, S., Sicotte, J., & Pal, D. (2019). Ocular drug delivery: Present innovations and future challenges. *Journal of Pharmacology and Experimental Therapeutics, 370*(3), 602–624.

Gupta, A. (2020). Chapter 21—Nanoemulsions. In E. J. Chung, L. Leon, & C. Rinaldi (Eds.), *Nanoparticles for Biomedical Applications* (pp. 371–384). Elsevier.

Han, H., Li, S., Xu, M., Zhong, Y., Fan, W., Xu, J., Zhou, T., Ji, J., Ye, J., & Yao, K. (2023). Polymer- and lipid-based nanocarriers for ocular drug delivery: Current status and future perspectives. *Advanced Drug Delivery Reviews, 196*, 114770.

Hegde, R. R., Verma, A., & Ghosh, A. (2013). Microemulsion: New insights into the ocular drug delivery. *ISRN Pharmaceutics, 2013*, 826798.

Hu, X., Liu, S., Zhou, G., Huang, Y., Xie, Z., & Jing, X. (2014). Electrospinning of polymeric nanofibers for drug delivery applications. *Journal of Controlled Release, 185*, 12-21.

Jacob, S., Nair, A. B., & Shah, J. (2020). Emerging role of nanosuspensions in drug delivery systems. *Biomaterials Research, 24*, 3.

Janaszewska, A., Lazniewska, J., Trzepiński, P., Marcinkowska, M., & Klajnert-Maculewicz, B. (2019). Cytotoxicity of dendrimers. *Biomolecules, 9*(9), 330.

Jimenez, A. I., Paneda, C., & Martinez, T. (2016). siRNA and their use in methods and compositions for inhibiting the expression of the ORAI1 gene. *U.S. Patent No. 20160304880 A1.*

Kaushal, U., Kaur, M., Nagpal, M., Bhuyan, M., & Gounder, K. P. (2023). Nanocarriers-based ocular therapeutics: Updates, challenges, and future perspectives. *Current Drug Research and Reviews, 15*, 15-28.

Kels, B. D., Grzybowski, A., & Grant-Kels, J. M. (2015). Human ocular anatomy. *Clinical Dermatology, 33*(2), 140-146.

Khanuja, H. K., Awasthi, R., Mehta, M., Satija, S., Aljabali, A. A. A., Tambuwala, M. M., Chellappan, D. K., Dua, K., & Dureja, H. (2021). Nanosuspensions—An update on recent patents, methods of preparation, and evaluation parameters. *Recent Patents on Nanotechnology, 15*, 351-366.

Lancina, M. G., III, Wang, J., Williamson, G. S., & Yang, H. (2018). DenTimol as a dendrimeric timolol analogue for glaucoma therapy: Synthesis and preliminary efficacy and safety assessment. *Molecular Pharmaceutics, 15*(8), 2883-2889.

Li, S., Chen, L., & Fu, Y. (2023). Nanotechnology-based ocular drug delivery systems: Recent advances and future prospects. *Journal of Nanobiotechnology, 21*(1), 232.

López-Cano, J. J., González-Cela-Casamayor, M. A., Andrés-Guerrero, V., Herrero-Vanrell, R., & Molina-Martínez, I. T. (2021). Liposomes as vehicles for topical ophthalmic drug delivery and ocular surface protection. *Expert Opinion on Drug Delivery, 18*(6), 819–847.

Ma, Y., Cong, Z., Gao, P., & Wang, Y. (2023). Nanosuspensions technology as a master key for nature products drug delivery and in vivo fate. *European Journal of Pharmaceutical Sciences, 185*, 106425.

Manca, M. L., Lai, F., Pireddu, R., Valenti, D., Schlich, M., Pini, E., Ailuno, G., Fadda, A. M., & Sinico, C. (2020). Impact of nanosizing on dermal delivery and antioxidant activity of quercetin nanocrystals. *Journal of Drug Delivery Science and Technology, 55*, 101482.

Mandal, A., Bisht, R., Rupenthal, I. D., & Mitra, A. K. (2017). Polymeric micelles for ocular drug delivery: From structural frameworks to recent preclinical studies. *Journal of Controlled Release, 248*, 96–116.

Marcano, D. C., Shin, C. S., Lee, B., Isenhart, L. C., Liu, X., Li, F., Jester, J. V., Pflugfelder, S. C., Simpson, J., & Acharya, G. (2016). Synergistic cysteamine delivery nanowafer as an efficacious treatment modality for corneal cystinosis. *Molecular Pharmaceutics, 13*(11), 3468–3477.

Mofidfar, M., Abdi, B., Ahadian, S., Mostafavi, E., Desai, T. A., Abbasi, F., Sun, Y., Manche, E. E., Ta, C. N., & Flowers, C. W. (2021). Drug delivery to the anterior segment of the eye: A review of current and future treatment strategies. *International Journal of Pharmaceutics, 607*, 120924.

Moghtaderi, M., Sedaghatnia, K., Bourbour, M., Fatemizadeh, M., Salehi Moghaddam, Z., Hejabi, F., Heidari, F., Quazi, S., &Farasati Far, B. (2022). Niosomes: A novel targeted drug delivery system for cancer. *Medical Oncology, 39*(10), 240.

Moradi, S., Mokhtari-Dizaji, M., Ghassemi, F., Sheibani, S., & Asadi Amoli, F. (2020). Increasing the efficiency of the retinoblastoma brachytherapy protocol with ultrasonic hyperthermia and gold nanoparticles: A rabbit model. *International Journal of Radiation Biology, 96*(12), 1614-1627.

Nagai, N., & Otake, H. (2022). Novel drug delivery systems for the management of dry eye. *Advanced Drug Delivery Reviews, 191*, 114582.

Peynshaert, K., Devoldere, J., De Smedt, S. C., & Remaut, K. (2018). In vitro and ex vivo models to study drug delivery barriers in the posterior segment of the eye. *Advanced Drug Delivery Reviews, 126*, 44–57.

Pınar, S. G., Oktay, A. N., Karaküçük, A. E., & Çelebi, N. (2023). Formulation strategies of nanosuspensions for various administration routes. *Pharmaceutics, 15*(6), 1520.

Qamar, Z., Qizilbash, F. F., Iqubal, M. K., Ali, A., Narang, J. K., Ali, J., & Baboota, S. (2019). Nano-based drug delivery system: Recent strategies for the treatment of ocular disease and future perspectives. *Recent Patents on Drug Delivery & Formulation, 13*(4), 246-254.

Qin, C., Wen, S., Zhu, S., Liu, D., Chen, S., Qie, J., Chen, H., & Lin, Q. (2020). Are poly(amidoamine) dendrimers safe for ocular applications? Toxicological evaluation in ocular cells and tissues. *Journal of Ocular Pharmacology and Therapeutics, 36*(9), 715-724.

Sánchez-López, E., Espina, M., Doktorovova, S., Souto, E. B., & García, M. L. (2017). Lipid nanoparticles (SLN, NLC): Overcoming the anatomical and physiological barriers of the eye—Part I—Barriers and determining factors in ocular delivery. *European Journal of Pharmaceutics and Biopharmaceutics, 110*, 70-75.

Shah, M., Edman, M. C., Reddy Janga, S., Yarber, F., Meng, Z., Klinngam, W., Bushman, J., Ma, T., Liu, S., Louie, S., Mehta, A., Ding, C., MacKay, J. A., & Hamm-Alvarez, S. F. (2017). Rapamycin eye drops suppress lacrimal gland inflammation in a murine model of Sjögren's syndrome. *Investigative Ophthalmology & Visual Science, 58*(1), 372-385.

Sherje, A. P., Jadhav, M., Dravyakar, B. R., & Kadam, D. (2018). Dendrimers: A versatile nanocarrier for drug delivery and targeting. *International Journal of Pharmaceutics, 548*(1), 707-720.

Singh, Y., Meher, J. G., Raval, K., Khan, F. A., Chaurasia, M., Jain, N. K., & Chourasia, M. K. (2017). Nanoemulsion: Concepts, development and applications in drug delivery. *Journal of Controlled Release, 252*, 28-49.

Subrizi, A., Del Amo, E. M., Korzhikov-Vlakh, V., Tennikova, T., Ruponen, M., &Urtti, A. (2019). Design principles of ocular drug delivery systems: Importance of drug payload, release rate, and material properties. *Drug Discovery Today, 24*, 1446-1457.

Tavakoli, S., Peynshaert, K., Lajunen, T., Devoldere, J., Del Amo, E. M., Ruponen, M., De Smedt, S. C., Remaut, K., &Urtti, A. (2020). Ocular barriers to retinal delivery of intravitreal liposomes: Impact of vitreoretinal interface. *Journal of Controlled Release, 328*, 952-961.

Tawfik, S. M., Azizov, S., Elmasry, M. R., Sharipov, M., & Lee, Y. I. (2020). Recent advances in nanomicelles delivery systems. *Nanomaterials, 11*(1), 70.

Taylor, A. W. (2009). Ocular immune privilege. *Eye, 23*(10), 1885-1899.

Trivedi, R., & Kompella, U. B. (2010). Nanomicellar formulations for sustained drug delivery: Strategies and underlying principles. *Nanomedicine, 5*(4), 485–505.

Tsai, C. H., Wang, P. Y., Lin, I. C., Huang, H., Liu, G. S., & Tseng, C. L. (2018). Ocular drug delivery: Role of degradable polymeric nanocarriers for ophthalmic application. *International Journal of Molecular Sciences, 19*(9), 2830.

Tsung, T. H., Chen, Y. H., & Lu, D. W. (2023). Updates on biodegradable formulations for ocular drug delivery. *Pharmaceutics, 15*(5), 734.

Uzel, E., Durgun, M. E., Esentürk-Güzel, İ., Güngör, S., & Özsoy, Y. (2023). Nanofibers in ocular drug targeting and tissue engineering: Their importance, advantages, advances, and future perspectives. *Pharmaceutics, 15*(4), 1062.

Varela-Fernández, R., Díaz-Tomé, V., Luaces-Rodríguez, A., Conde-Penedo, A., García-Otero, X., Luzardo-Álvarez, A., Fernández-Ferreiro, A., & Otero-Espinar, F. J. (2020). Drug delivery to the posterior segment of the eye: Biopharmaceutic and pharmacokinetic considerations. *Pharmaceutics, 12*(4), 269.

Varshochian, R., Riazi-Esfahani, M., Jeddi-Tehrani, M., Mahmoudi, A. R., Aghazadeh, S., Mahbod, M., Movassat, M., Atyabi, F., Sabzevari, A., &Dinarvand, R. (2015). Albuminated PLGA nanoparticles containing bevacizumab intended for ocular neovascularization treatment. *Journal of Biomedical Materials Research Part A, 103*(9), 3148-3156.

Wang, J., Fonn, D., Simpson, T. L., & Jones, L. (2003). Precorneal and pre- and postlens tear film thickness measured indirectly with optical coherence tomography. *Investigative Ophthalmology & Visual Science, 44*(6), 2524–2528.

Wang, J., Li, B., Qiu, L., Qiao, X., & Yang, H. (2022). Dendrimer-based drug delivery systems: History, challenges, and latest developments. *Journal of Biological Engineering, 16*, 18.

World Health Organization. (2019). *World report on vision.* https://www.who.int/publications/i/item/world-report-on-vision.

Yetisgin, A. A., Cetinel, S., Zuvin, M., Kosar, A., & Kutlu, O. (2020). Therapeutic nanoparticles and their targeted delivery applications. *Molecules, 25*(10), 2193.

Zhang, J., Jiao, J., Niu, M., Gao, X., Zhang, G., Yu, H., Yang, X., & Liu, L. (2021). Ten years of knowledge of nano-carrier based drug delivery systems in ophthalmology: Current evidence, challenges, and future prospects. *International Journal of Nanomedicine, 16*, 6497–6530.

Zhou, T., Zhu, L., Xia, H., He, J., Liu, S., He, S., Wang, L., & Zhang, J. (2017). Micelle carriers based on macrogol 15 hydroxystearate for ocular delivery of terbinafine hydrochloride: In vitro characterization and in vivo permeation. *European Journal of Pharmaceutical Sciences, 109*, 288–296.

Zielińska, A., Carreiró, F., Oliveira, A. M., Neves, A., Pires, B., Venkatesh, D. N., Durazzo, A., Lucarini, M., Eder, P., Silva, A. M., Santini, A., & Souto, E. B. (2020). Polymeric nanoparticles: Production, characterization, toxicology and ecotoxicology. *Molecules, 25*(16), 3731.

Chapter 19

Sustained- and Controlled-Release Nanocarriers for Ocular Drug Delivery

Khurshid Iqbal Molvi[1],*
Taufik Mulla[2]
Drashti Dave[3]
and Ambika Nand Jha[4]
[1]Ismail Mehta College of Pharmacy, Ambad, Maharashtra, India
[2]Institute of Pharmaceutical Sciences, Faculty of Pharmacy, Parul University, Gujarat, India
[3]Department of Pharmaceutics, Sigma Institute of Pharmacy, Sigma University, Gujarat, India
[4]School of Pharmacy, Sharda University, Greater Noida, Uttar Pradesh, India

Abstract

Ocular drug delivery poses significant challenges due to the complex anatomical and physiological barriers that hinder effective drug penetration and retention. To address these limitations, sustained and controlled-release nanocarriers have emerged as innovative solutions. Advanced delivery systems such as hydrogels, dendrimers, liposomes, and polymeric nanoparticles facilitate targeted delivery to ocular tissues, improve drug bioavailability, and extend the duration of therapeutic action. These technologies provide important clinical benefits, including a reduced frequency of drug administration and the maintenance of therapeutic drug levels over prolonged periods. This approach enhances patient adherence and reduces systemic exposure and related side effects. Considering the intricate nature of ocular pharmacotherapy, the

* Corresponding Author's Email: khurshidmolvi@gmail.com

In: Advances in Nanotechnology for Ocular Drug Delivery
Editors: Anjali Sharma, Ajmer Singh Grewal, Geeta Deswal et al.
ISBN: 979-8-89530-814-1
© 2025 Nova Science Publishers, Inc.

development of nanocarriers represents a transformative advancement. These systems offer more convenient and effective treatment modalities for a broad range of eye disorders. This chapter aims to examine the design, benefits, and therapeutic potential of sustained and controlled-release nanocarriers in optimizing ocular drug delivery and promoting patient compliance.

Keywords: sustained-release, controlled-release, nanocarriers, ocular drug delivery, polymeric nanoparticles, stimuli-responsive systems

Abbreviations

AMD: Age-Related Macular Degeneration
BAB: Blood-Aqueous Barrier
BRB: Blood-Retinal Barrier
CO_2: Carbon Dioxide
DR: Diabetic Retinopathy
FDA: Food and Drug Administration
PEG: Polyethylene Glycol
RNA: Ribonucleic Acid
SLNs: Solid Lipid Nanoparticles
VEGF: Vascular Endothelial Growth Factor

Introduction

For the effective management of chronic eye conditions, the steady and controlled delivery of medication is essential. Traditional treatments, such as eye drops, often prove inadequate due to rapid clearance by tears and poor absorption into ocular tissues. As a result, patients are required to apply these medications frequently, which is inconvenient and increases the risk of improper usage (Gupta et al., 2015). To address these challenges, researchers have developed advanced drug delivery systems known as nanocarriers. These microscopic carriers gradually release medications, enhancing therapeutic efficacy while reducing the need for frequent administration. The term "sustained release" refers to the continuous and prolonged delivery of drugs over time, helping to maintain consistent therapeutic levels. In comparison, "controlled release" ensures that the drug is delivered precisely where and when it is needed, thereby optimizing treatment outcomes. Nanocarriers can be engineered to achieve sustained release, controlled release, or a combination of both. This versatility makes them highly effective in targeting specific ocular regions, improving drug stability, and maximizing therapeutic concentrations at the site of action. Consequently, side effects in non-target tissues are minimized, and patient adherence to treatment regimens is significantly improved due to the reduced dosing frequency. These innovative drug delivery systems hold great promise in the treatment of chronic eye diseases such as glaucoma, diabetic retinopathy (DR), and age-related macular degeneration (AMD). Compared to conventional therapies, nanocarrier-based approaches offer significant advantages by enabling precise and long-lasting drug delivery. With ongoing research and technological progress, nanocarrier platforms are poised to transform ocular therapeutics, leading to better clinical outcomes and an improved quality of life for patients (Gorantla et al., 2020).

Barriers to Eye Drug Delivery

The anatomical and physiological structure of the eye presents substantial challenges for effective drug delivery. While the eye's natural defense mechanisms serve to protect it from harmful external agents, they also limit the ability of therapeutic compounds to reach their intended intraocular targets. As a result, the development of specialized drug delivery systems is essential to overcome these barriers and improve treatment outcomes (Hosoya et al., 2005). The following section provides a comprehensive overview of the major barriers present in both the anterior and posterior segments of the eye, along with their implications for ocular drug delivery.

Cornea

The cornea, the transparent outermost layer of the eye, serves as a primary barrier to drug penetration. Its structure includes a lipophilic epithelium and a hydrophilic stroma, creating a biphasic composition that restricts the permeability of both hydrophilic and lipophilic drugs. This dual nature significantly hinders the diffusion of therapeutic agents, making topical ocular drug administration particularly challenging.

Conjunctiva

The conjunctiva is a thin, vascularized membrane covering the sclera and inner eyelids. Although it provides an alternative route for drug administration, its high vascularity and extensive surface area lead to rapid systemic absorption of drugs, thereby reducing their bioavailability in intraocular tissues.

Sclera

The sclera, commonly known as the white part of the eye, is composed of densely packed collagen fibers. This compact structure restricts the permeability of drug molecules, especially those that are large or hydrophobic. Although transscleral delivery has been investigated as a potential approach for targeting diseases in the posterior segment of the eye, the limited permeability of the sclera continues to pose a significant challenge to effective drug delivery.

Blood-Aqueous Barrier (BAB)

BAB is formed by tight junctions between the endothelial cells of the iris vasculature and the non-pigmented epithelial cells of the ciliary body. This barrier plays a crucial role in regulating the movement of solutes between the systemic circulation and the aqueous humor. As a result, it significantly restricts the intraocular bioavailability of systemically administered drugs (Jimenez et al., 2016).

Blood-Retinal Barrier (BRB)

BRB is one of the most restrictive barriers in the human body. It consists of tight junctions between the retinal pigment epithelium, which forms the outer BRB, and the retinal capillary endothelium, which constitutes the inner BRB. This highly selective barrier limits the penetration of therapeutic agents into the retina, thereby complicating the treatment of posterior segment disorders such as diabetic retinopathy and age-related macular degeneration (Nie et al., 2009).

Vitreous Humour

The vitreous humor is a gel-like matrix that occupies the space between the lens and the retina. Its viscoelastic properties hinder the diffusion of drug molecules, particularly those with high molecular weight. As a result, the efficiency of intravitreal drug delivery is significantly reduced.

Tear Film Dynamics

The tear film is continuously renewed with every blink, at an approximate rate of one microliter per minute. This dynamic turnover rapidly clears topically applied medications from the ocular surface, thereby limiting drug residence time and reducing therapeutic effectiveness.

Nasolacrimal Drainage

Drugs administered topically can be quickly cleared from the ocular surface through the nasolacrimal duct into the nasal cavity. This drainage reduces ocular drug concentrations and may also result in systemic absorption, potentially causing unwanted side effects.

Ocular Drug Metabolism

The eye contains various metabolic enzymes, such as esterases and cytochrome P450 isoforms, which can degrade drug molecules before they reach effective intraocular concentrations. This enzymatic activity further reduces the bioavailability of administered drugs.

Protein Binding

Drugs may bind to proteins present in the tear film or ocular tissues. This protein binding sequesters the drugs in inactive complexes, thereby decreasing the concentration of free, pharmacologically active molecules.

Limited Corneal Permeability

The structural complexity of the corneal epithelium and stroma results in inherently low corneal permeability. The epithelium acts as a lipophilic barrier, while the stroma, being hydrophilic, resists the passage of lipophilic molecules. This dual characteristic significantly restricts the penetration of both water-soluble and fat-soluble drugs.

Effective drug delivery to the eye remains challenging due to these complex barriers, which influence the absorption and utilization of medications by ocular tissues. Table 1 presents a detailed overview of these barriers along with strategies to overcome them.

Table 1. Ocular barriers and their impact on drug delivery (Gupta et al., 2015)

Barrier	Location	Impact on Drug Delivery	Possible Solution
Tear Film	Ocular Surface	Rapid drug clearance due to tear turnover	Mucoadhesive systems, viscosity enhancers
Cornea	Anterior Segment	Limited permeability for hydrophilic drugs	Liposomes, nanoparticles
Conjunctiva	Outer Eye Layer	Systemic drug absorption reduces ocular bioavailability	Nano emulsions, sustained-release formulations
Blood-Aqueous Barrier	Anterior Segment	Restricts drug entry to aqueous humor	Prodrug approach, carrier-mediated transport
Blood-Retinal Barrier	Posterior Segment	Limits drug penetration into the retina	Intravitreal injections, nanoparticle-based drug carriers

Nanocarrier-Based Systems for Sustained Ocular Drug Delivery

Chronic ocular conditions often require prolonged and consistent therapeutic intervention. However, conventional formulations such as eye drops have limited effectiveness due to rapid precorneal elimination, poor corneal penetration, and short residence time on the ocular surface (Gupta et al., 2015). Consequently, frequent administration is usually necessary to maintain therapeutic drug levels, which often leads to suboptimal patient compliance and treatment outcomes. To overcome these challenges, nanotechnology-based drug delivery systems have emerged as a promising strategy.

Nanocarriers are engineered to provide sustained and controlled drug release, enhancing ocular bioavailability while reducing the frequency of dosing. These nanosystems offer precise temporal and spatial control over drug release, improving therapeutic accuracy and minimizing off-target effects. By prolonging drug retention within ocular tissues and enabling gradual diffusion, nanocarriers not only stabilize drug molecules but also improve their ability to reach target sites effectively. This approach supports better patient adherence to treatment regimens and reduces systemic side effects commonly associated with conventional ophthalmic therapies (Gorantla et al., 2020). The development of nanocarrier-based ocular drug delivery platforms is transforming the treatment landscape for eye diseases by offering enhanced penetration through ocular barriers, extended drug release, improved therapeutic efficacy, and a reduction in adverse systemic effects.

Polymeric Nanoparticles

Polymeric nanoparticles, typically ranging in size from 10 to 1000 nanometers, are highly valued in ocular drug delivery due to their biocompatibility, biodegradability, and capacity to encapsulate or conjugate therapeutic agents (Kim, 2009). These nanoparticles enable drug release through various mechanisms, including diffusion through the polymer matrix, gradual polymer degradation, and surface erosion. Each of these mechanisms contributes to a controlled and sustained release profile. The use of polymeric nanoparticles offers multiple advantages: they enhance drug stability by protecting sensitive compounds from degradation, reduce ocular toxicity, improve drug penetration into ocular tissues, and provide prolonged therapeutic effects. Together, these benefits increase treatment efficacy and promote better patient compliance.

Liposomes and Niosomes

Liposomes are spherical vesicles composed of lipid bilayers that can encapsulate both hydrophilic and lipophilic drugs. Their structural similarity to biological membranes enables them to fuse with cell membranes or be internalized by cells, facilitating targeted and controlled drug release at specific sites. Niosomes, which share a similar structure to liposomes but are made from non-ionic surfactants instead of phospholipids, offer additional advantages such as greater chemical stability, cost-effectiveness, and easier production (Raghava et al., 2004). Both liposomes and niosomes release their drug payloads through mechanisms such as leakage, fusion with biological membranes, or in response to environmental triggers like pH, temperature, or enzymatic activity (Alcon Inc., 2020). These vesicular carriers improve ocular drug bioavailability, protect therapeutic agents from degradation, and enable more efficient and sustained delivery to ocular tissues.

Dendrimers

Dendrimers are highly branched, tree-like nanostructures that resemble spherical molecules with numerous surface functional groups, allowing for precise drug attachment and high drug-loading capacity. Drug release from dendrimers can be tailored according to factors such as the number of branching layers, the types of chemical linkages employed, and environmental conditions like pH or ionic strength. Their unique architecture facilitates controlled and sustained drug release, targeted delivery to specific tissues, and improved therapeutic efficacy. Moreover, dendrimers can be chemically modified to enhance biocompatibility and to selectively target diseased tissues, making them versatile carriers for ocular drug delivery and other biomedical applications.

Hydrogels

Hydrogels are hydrophilic polymer networks that can absorb substantial amounts of water or physiological fluids, making them highly suitable for ocular drug delivery. Their ability to swell and release drugs gradually enables sustained and controlled delivery, which is especially compatible with the sensitive tissues of the eye, such as the cornea and conjunctiva. Hydrogels are biocompatible, exhibit excellent adhesion to ocular surfaces, and can be engineered to respond to specific stimuli for tailored drug release. These properties make hydrogels ideal candidates for applications such as ocular inserts, contact lenses, and implantable devices (Nirmal et al., 2016).

Systems That Form in Place

In situ-forming drug delivery systems are innovative formulations administered as liquids that transform into gels or solid depots upon exposure to physiological triggers such as temperature changes, pH variations, or ionic strength within the eye (Rivers et al., 2015). This transformation enables the formulation to remain in the ocular environment for extended periods, allowing sustained drug release and enhanced therapeutic efficacy. These systems are minimally invasive, easy to administer, and significantly increase drug retention time, thereby reducing the frequency of dosing (Mann et al., 2018). By prolonging drug residence time and improving bioavailability, in situ-forming systems offer a promising approach to optimizing ocular drug therapy. They improve patient adherence, minimize systemic side effects, and contribute to better clinical outcomes. To tackle the many challenges of ocular drug delivery, researchers have developed various nanoscale carrier systems, each with unique advantages and limitations. A comparative overview of these nanocarriers is provided in Table 2.

Table 2. Comparison of different nanocarriers for ocular drug delivery

Nanocarrier Type	Advantages	Disadvantages	Examples of Drugs Delivered
Liposomes	Biocompatible, good drug entrapment	Short shelf-life, potential leakage	Anti-glaucoma drugs, anti-VEGF
Niosomes	Increased stability over liposomes	Limited scalability	Anti-inflammatory agents
Polymeric Nanoparticles	Controlled drug release, biodegradable	Possible toxicity, complex synthesis	Antibiotics, steroids
Solid Lipid Nanoparticles (SLNs)	High drug loading, stability	Poor drug release profile	Anti-viral, corticosteroids
Dendrimers	High drug-loading capacity	Expensive, potential cytotoxicity	Gene therapy, anti-cancer agents
Hydrogels	Prolonged drug release	Limited drug loading	Anti-inflammatory drugs

Mechanisms of Drug Release from Nanocarriers for Ocular Delivery

The effectiveness of nanocarrier-based ocular drug delivery largely depends on the carrier's ability to release therapeutic agents at the target site in a controlled and optimal manner. Controlled drug release not only improves drug bioavailability but also reduces dosing frequency and minimizes side effects. The release kinetics can be precisely adjusted through different mechanisms, each designed to meet specific therapeutic needs. The three primary release mechanisms include diffusion-controlled release, where the drug gradually diffuses through the carrier matrix; degradation-controlled release, in which the nanocarrier slowly breaks down to release the encapsulated drug; and stimuli-responsive release, where drug delivery is triggered by external or internal factors such as pH, temperature, or enzymatic activity (Nayak & Misra, 2018). Each of these mechanisms provides unique advantages for achieving sustained, targeted, and patient-friendly ocular therapy.

Diffusion-Controlled Release

One of the most commonly used mechanisms for drug release from nanocarriers is diffusion-controlled release. In this process, drug molecules gradually move from the interior of the nanocarrier into the surrounding ocular tissues. The rate of diffusion depends on several factors, including the drug's molecular mobility, the size and shape of the nanocarrier, and the physicochemical properties of the carrier matrix. This mechanism follows Fick's laws of diffusion, which explain that the movement of drug molecules is driven by concentration gradients and slows down as equilibrium is reached. Diffusion-based systems provide a predictable and sustained drug release profile, making them well suited for managing chronic eye conditions that require long-term treatment.

Degradation-Controlled Release

In a degradation-controlled release system, the drug is released as the nanocarrier gradually breaks down, typically through processes such as hydrolysis, enzymatic degradation, or polymer erosion (Gautam & Kesavan, 2017). The rate and mechanism of degradation depend largely on the chemical structure of the carrier material, especially the nature of its chemical bonds. For example, polymers containing ester linkages are particularly susceptible to hydrolysis in aqueous environments. Moreover, external factors such as pH, temperature, and the presence of specific enzymes can significantly influence the degradation rate. This release mechanism allows precise control over the timing of drug delivery and is especially valuable for designing systems that respond to physiological conditions or target tissues with specific enzymatic profiles.

Stimuli-Responsive Release

Stimuli-responsive release is an advanced drug delivery approach where therapeutic agents are released in response to specific physiological or external triggers such as changes in pH, temperature, light, magnetic fields, or enzymatic activity. This strategy provides precise spatial and temporal control over drug release, thereby enhancing the effectiveness and specificity of ocular therapies.

pH-Sensitive Systems

These nanocarriers utilize polymers that undergo structural changes in response to variations in pH, allowing for targeted drug release in specific ocular regions where the pH deviates from the norm.

Temperature-Sensitive Systems

Some materials transition between liquid and gel states with temperature changes, enabling in situ gelation and sustained drug release upon contact with ocular tissues.

Light-Sensitive Systems

Certain polymers are engineered to release their drug payload upon exposure to specific wavelengths of light, allowing externally controlled activation (Jaiswal et al., 2016).

Magnetic-Sensitive Systems

Magnetic nanoparticles can be guided to target areas using an external magnetic field, and release drugs upon magnetic stimulation.

Enzyme-Sensitive Systems

These nanocarriers respond to specific enzymes that are elevated in diseased tissues, enabling site-specific drug release. While stimuli-responsive systems provide highly controlled and targeted drug delivery, they also pose challenges related to design complexity, scalability, and the risk of premature or unintended drug release. Despite these challenges, they represent a promising frontier in ocular nanomedicine, offering potential for personalized and efficient therapeutic outcomes.

Formulation Strategies for Enhanced Ocular Drug Delivery with Nanocarriers

Delivering drugs to the eye requires overcoming its natural barriers, including the cornea, sclera, and blood-ocular barriers. To achieve effective treatment while minimizing side effects, formulation strategies play a crucial role. These strategies focus on optimizing drug loading, surface modification, and the physicochemical properties of nanocarriers to enhance drug delivery to ocular tissues (Hwang et al., 2016).

Drug Loading Techniques

Efficient drug loading is critical for maximizing the therapeutic potential of nanocarriers. Several methods are employed to incorporate drugs into nanocarriers, each offering unique advantages.

Encapsulation Methods

Solvent Evaporation
The drug and polymer are dissolved in a volatile solvent, which is subsequently evaporated to form solid nanocarriers, encapsulating the drug (Lancina & Yang, 2017).

Nanoprecipitation
A drug-polymer solution is introduced into a non-solvent, causing the polymer to precipitate and form nanoparticles (Taskar et al., 2017).

Emulsification-Diffusion
In this method, a drug-polymer solution is mixed with water, followed by solvent evaporation to form nanoparticles (Phase II Study of DA-6034 Eye Drops, 2020).

Supercritical Fluid Technology
Drugs and polymers are dissolved in supercritical CO_2, which evaporates, leaving behind drug-loaded nanocarriers (Study of Brimonidine Tartrate Nanoemulsion Eye Drop Solution, 2020).

Surface Adsorption Methods

Physical Adsorption
The drug is weakly attached to the surface of the nanocarrier through non-covalent interactions, which allows for controlled release (OCS-01 in Treating Inflammation and Pain, 2020).

Layer-by-Layer Deposition
This technique involves depositing multiple polymer layers onto the nanocarrier, incorporating drugs within these layers for gradual release (HU007 Eye Drops in Patients with Dry Eye Syndrome, 2020).

Covalent Linking
Drugs are chemically bonded to the nanocarrier, providing higher drug loading and precise control over release kinetics (Subramanian et al., 2018).

Surface Modification for Targeted Delivery

Surface modification of nanocarriers enhances their ability to target specific regions of the eye, thereby improving drug delivery and minimizing systemic side effects. By functionalizing nanocarriers with targeting ligands such as antibodies, peptides, or vitamins, these carriers can be directed precisely to specific ocular tissues or cells. Additionally, surface modifications with polymers like polyethylene glycol (PEG) or mucoadhesive polymers help extend the residence time of nanocarriers on the ocular surface, ensuring prolonged drug retention and more effective therapy.

Optimization of Physicochemical Properties

The physicochemical properties of nanocarriers, including particle size, surface charge, and morphology, play a critical role in determining their efficiency in ocular drug delivery. Smaller nanoparticles generally penetrate ocular tissues more effectively, thereby enhancing bioavailability. Furthermore, hydrophilic and amphiphilic carriers improve interactions with the aqueous ocular surfaces, resulting in better adhesion and prolonged therapeutic effects.

Challenges in Development and Translation

While ocular nanocarriers hold great promise for improving eye disease treatments, several challenges remain in translating these innovations from the laboratory to commercial markets.

Scale-up and Manufacturing Challenges

One of the primary obstacles is the difficulty in scaling up the production process. Transitioning from lab-scale synthesis to large-scale manufacturing often results in inconsistencies in particle size, shape, and drug loading, which can impact the efficacy of the treatment. Developing robust and scalable manufacturing processes that adhere to stringent regulatory standards is essential. Additionally, the high production costs associated with nanocarriers make it challenging to produce these treatments at a price point that would make them widely accessible.

Regulatory and Safety Considerations

Regulatory agencies, such as the FDA, require comprehensive testing to confirm the safety and effectiveness of new drug delivery systems. This involves a thorough evaluation of potential ocular toxicity, pharmacokinetics, and the system's interaction with the eye tissues. Ensuring that these treatments comply with regulatory requirements and have no harmful side effects is critical to obtaining approval.

Maintaining Germ-Free Conditions

For ocular treatments to be safe, they must be free of microbial contamination. However, sterilization methods, such as high-temperature treatments, can damage the delicate nanocarriers used in these systems. Finding non-invasive sterilization methods that preserve the integrity of the nanocarriers while maintaining sterility is a key challenge.

Ensuring Stability

Nanocarriers are prone to instability, with risks of aggregation, degradation, or premature drug release, which can reduce the therapeutic efficacy of the treatment. Ensuring the long-term stability of these systems during storage and use is essential for maintaining consistent drug release and ensuring patient safety.

Targeting Specific Areas

Precise targeting remains a major challenge in ocular drug delivery. Effectively delivering drugs to specific regions of the eye, such as the retina or cornea, requires overcoming complex physiological barriers and achieving spatial accuracy. Enhancing the targeting capability of nanocarriers while minimizing off-target effects is essential for improving therapeutic outcomes and reducing potential side effects.

Reducing Costs

The cost of developing and manufacturing ocular nanocarrier-based therapies is high due to complex fabrication processes, specialized materials, and regulatory requirements. Identifying strategies to reduce production costs without compromising the quality of the nanocarriers is crucial for making these treatments accessible to a larger population.

Enhancing Patient Compliance

Finally, improving patient compliance is essential for the success of ocular treatments. Developing user-friendly drug delivery systems and providing clear usage instructions can significantly enhance adherence to prescribed regimens. Ensuring that these systems are easy to administer, non-invasive, and effective over extended periods is crucial for optimizing therapeutic outcomes and enhancing the overall patient experience. Addressing these challenges requires a multidisciplinary approach, involving advances in nanotechnology, manufacturing processes, regulatory frameworks, and patient education. By overcoming these obstacles, ocular nanocarriers have the potential to revolutionize the treatment of various eye diseases, offering more effective, targeted, and patient-friendly therapies.

Conclusion

The potential of nanoscale carriers in ocular therapy is highly promising. Smart and multifunctional delivery systems have the capacity to transform the way medications are administered to the eye. When integrated with gene and cell therapies, these carriers offer a powerful platform for treating previously intractable ocular diseases. Ongoing research is expected to yield innovative strategies for preventing vision loss, thereby enhancing the quality of life for many individuals. Achieving these advances will require close collaboration among researchers, clinicians, biomedical engineers, and regulatory authorities. The future of ocular therapy lies in personalized and targeted treatment approaches that can effectively address a broad spectrum of eye disorders.

References

Alcon Inc. (2020). TobraDex® ST. Retrieved December 20, 2020, from https://www.drugs.com/pro/tobradex.html.

Alonso, F. M. J., Calvo, S. P., Remunan, L. C., & Vila, J. J. L. (2003). Application of nanoparticles based on hydrophilic polymers as pharmaceutical forms (U.S. Patent No. 6,649,192).

Bachu, R. D., Chowdhury, P., Al-Saedi, Z. H. F., Karla, P. K., & Boddu, S. H. S. (2018). Ocular drug delivery barriers—Role of nanocarriers in the treatment of anterior segment ocular diseases. *Pharmaceutics*, 10(1), 28.

Banerjee, R., & Carvalho, E. (2014). Nanoparticulate in-situ gels as vitreous humor substitutes for ocular disease (U.S. Patent No. 8,889,175 B2).

Boddu, S. H., Gupta, H., & Patel, S. (2014). Drug delivery to the back of the eye following topical administration: An update on research and patenting activity. *Recent Patents on Drug Delivery & Formulation*, 8(1), 27–36.

Chauhan, A., & Gulsen, D. (2009). Ophthalmic drug delivery system (U.S. Patent No. 7,638,137 B2).

Cholkar, K., Patel, A., Vadlapudi, A. D., & Mitra, A. K. (2012). Novel nanomicellar formulation approaches for anterior and posterior segment ocular drug delivery. *Recent Patents on Nanomedicine*, 2(2), 82–95.

ClinicalTrials.gov. (n.d.). *HU007 Eye Drops in Patients with Dry Eye Syndrome*. https://clinicaltrials.gov/ct2/show/NCT04384991.

ClinicalTrials.gov. (n.d.). *OCS-01 in Treating Inflammation and Pain in Post-cataract Patients (SKYGGN)*. https://clinicaltrials.gov/ct2/show/NCT04130802.

ClinicalTrials.gov. (n.d.). *Phase II Study of DA-6034 Eye Drops in Dry Eye Syndrome (DES)*. https://clinicaltrials.gov/ct2/show/NCT01670357.

ClinicalTrials.gov. (n.d.). *Study of Brimonidine Tartrate Nanoemulsion Eye Drop Solution in the Treatment of Dry Eye Disease (DED)*. https://clinicaltrials.gov/ct2/show/NCT03598699.

ClinicalTrials.gov. (n.d.). *Study of Brimonidine Tartrate Nanoemulsion Eye Drop Solution in the Treatment of Dry Eye Disease (DED)*. https://clinicaltrials.gov/ct2/show/NCT03551808.

Diebold, Y., & Calonge, M. (2010). Applications of nanoparticles in ophthalmology. *Progress in Retinal and Eye Research*, 29(6), 596–609.

Edelhauser, H. F., Rowe-Rendleman, C. L., Robinson, M. R., Dawson, D. G., Chader, G. J., Grossniklaus, H. E., Rittenhouse, K. D., Wilson, C. G., Weber, D. A., Kuppermann, B. D., Csaky, K. G., Olsen, T. W., Kompella, U. B., Holers, V. M., Hageman, G. S., Gilger, B. C., Campochiaro, P. A., Whitcup, S. M., & Wong, W. T. (2010). Ophthalmic drug delivery systems for the treatment of retinal diseases: Basic research to clinical applications. *Investigative Ophthalmology & Visual Science*, 51(11), 5403–5420.

Enzmann, V., Hengerer, A., & Langmann, T. (2019). Exosomes: A promising nanocarrier for ocular drug delivery. *Journal of Controlled Release*, 318, 1–12.

Gao, M., & Robinson, J. R. (2005). Controlled ocular drug delivery. *Pharmaceutical Research*, 22, 1551–1561.

Gaudana, R., Ananthula, H. K., Parenky, A., & Mitra, A. K. (2010). Ocular drug delivery. *The AAPS Journal*, 12(3), 348–360.

Gautam, N., & Kesavan, K. (2017). Development of microemulsions for ocular delivery. *Therapeutic Delivery*, 8(5), 313–330.

Geroski, D. H., & Edelhauser, H. F. (2000). Drug delivery for posterior segment eye disease. *Investigative Ophthalmology & Visual Science*, 41(5), 961–964.

Ghate, D., & Edelhauser, H. F. (2008). Ocular drug delivery. *Expert Opinion on Drug Delivery*, 5(5), 443–455.

Gorantla, S., Rapalli, V. K., Waghule, T., Singh, P. P., Dubey, S. K., Saha, R. N., & Singhvi, G. (2020). Nanocarriers for ocular drug delivery: Current status and translational opportunity. *RSC Advances*, 10(46), 27835–27855.

Gregoriadis, G. (2016). Liposomes in drug delivery: Recent advances and challenges. *Pharmaceuticals*, 9(1), 4.

Gupta, M. P., Herzlich, A. A., Sauer, R., & Chan, C. C. (2015). Retinal anatomy and pathology. *Developments in Ophthalmology*, 55, 7–17.

Hara, H., & Takeuchi, H. (2009). Liposome for delivery to posterior segment of eye and pharmaceutical composition for disease in posterior segment of eye (WO Patent No. 2009107753).

Hosoya, K., Lee, V. H. L., & Kim, K. J. (2005). Roles of the conjunctiva in ocular drug delivery: A review of conjunctival transport mechanisms and their regulation. *European Journal of Pharmaceutics and Biopharmaceutics*, 60(2), 227–240.

Hwang, J. Y., Li, Z., & Loh, X. J. (2016). Small molecule therapeutic-loaded liposomes as therapeutic carriers: From development to clinical applications. *RSC Advances*, 6, 70592–70615.

Jacob, S., Nair, A. B., & Shah, J. (2020). Emerging role of nanosuspensions in drug delivery systems. *Journal of Biomaterials Research*, 24, 3.

Jaiswal, P., Gidwani, B., & Vyas, A. (2016). Nanostructured lipid carriers and their current application in targeted drug delivery. *Artificial Cells, Nanomedicine, and Biotechnology*, 44(1), 27–40.

Jimenez, A. I., Paneda, C., & Martinez, T. (2016). siRNA and their use in methods and compositions for inhibiting the expression of the ORAI1 gene. U.S. Patent No. 20160304880 A1.

Kim, H., Robinson, S. B., & Csaky, K. G. (2009). Investigating the movement of intravitreal human serum albumin nanoparticles in the vitreous and retina. *Pharmaceutical Research*, 26(2), 329–337.

Kuno, N., & Fujii, S. (2011). Recent advances in ocular drug delivery systems. *Polymers*, 3(1), 193–221.

Lancina, M. G., & Yang, H. (2017). Dendrimers for ocular drug delivery. *Canadian Journal of Chemistry*, 95(9), 897–902.

Lee, S. S., Hughes, P. M., & Robinson, M. R. (2009). Recent advances in drug delivery systems for treating ocular complications of systemic diseases. *Current Opinion in Ophthalmology*, 20(6), 511–519.

Mann, B. K., Stirland, D. L., Lee, H. K., & Wirostko, B. M. (2018). Ocular translational science: A review of development steps and paths. *Advanced Drug Delivery Reviews*, 126, 195–203.

Maurice, D. M., & Mishima, S. (1984). Ocular pharmacokinetics. In M. L. Sears (Ed.), *Handbook of Experimental Pharmacology* (pp. 16–119). Springer.

Millar, T. J., & Schuett, B. S. (2015). The real reason for having a meibomian lipid layer covering the outer surface of the tear film—A review. *Experimental Eye Research*, 137, 125–138.

Nayak, K., & Misra, M. (2018). A review on recent drug delivery systems for posterior segment of eye. *Biomedicine & Pharmacotherapy*, 107, 1564–1582.

Nie, Y., Pan, K., & Wu, K. (2009). The potential therapeutic of siRNA eye drops in ocular diseases. *Bioscience Hypotheses,* 2(4), 223–225.

Nirmal, J., Radhakrishnan, K., Moreno, M., Natarajan, J. V., Laude, A., Lim, T. H., Venkatraman, S., & Agrawal, R. (2016). Drug, delivery, and devices for diabetic retinopathy (3Ds in DR). *Expert Opinion on Drug Delivery*, 13(11), 1625–1637.

Nordstrom, B. L., Friedman, D. S., Mozaffari, E., Quigley, H. A., & Walker, A. M. (2005). Persistence and adherence with topical glaucoma therapy. *American Journal of Ophthalmology*, 140(4), 598.e1–598.e11.

Prausnitz, M. R., & Noonan, J. S. (1998). Permeability of cornea, sclera, and conjunctiva: A literature analysis for drug delivery to the eye. *Journal of Pharmaceutical Sciences*, 87(12), 1479–1488.

Raghava, S., Hammond, M., & Kompella, U. B. (2004). Periocular routes for retinal drug delivery. *Expert Opinion on Drug Delivery*, 1(1), 99–114.

Rivers, H. M., Chaudhuri, S. R., Shah, J. C., & Mittal, S. (2015). *A new vision for the eye: Unmet ocular drug delivery needs*. Springer.

Sidney, L., Weiss, W. E., Mitra, A. K., & McNally, E. J. (2019). Topical cyclosporine-containing formulations and uses thereof (U.S. Patent No. 20190060397 A1).

Subramanian, V., Natarajan, J. V., Houden, T., & Boey, F. (2018). Stable liposomal formulation for ophthalmic drug delivery. U.S. Patent No. 9,956,195B2.

Taskar, P., Tatke, A., & Majumdar, S. (2017). Advances in the use of prodrugs for drug delivery to the eye. *Expert Opinion on Drug Delivery*, 14(1), 49–63.

Yellepeddi, V. K., & Palakurthi, S. (2016). Recent advances in topical ocular drug delivery. *Journal of Ocular Pharmacology and Therapeutics,* 32(2), 67–82.

Chapter 20

Clinical Translation and Commercialization of Ocular Nanotechnology

Akash Jain[1]
Vaishali Kamboj[1]
Nikhil[1]
Bhawna Chopra[2]
Ashwani Dhingra[3]
Minky Mukhija[4]
and Jasmine Chaudhary[1,*]
[1]M. M. College of Pharmacy, M. M. (Deemed to be University), Mullana, Ambala, Haryana, India
[2]Guru Gobind Singh College of Pharmacy, Yamuna Nagar, Haryana, India
[3]Global Research Institute of Pharmacy, Radaur, Yamuna Nagar, Haryana, India
[4]Ch. Devi Lal College of Pharmacy, Jagadhri, Yamuna Nagar, Haryana, India

Abstract

The management of ocular diseases has always been challenging due to the physiological and anatomical barriers inherent to the eye's high sensitivity, which, if damaged, can lead to visual impairment or blindness. With advancements in nanotechnology, various nano-based ocular formulations have been developed that efficiently deliver drugs, particularly to the posterior segment of the eye. These formulations are target-specific and enhance drug penetration across ocular barriers while increasing precorneal retention time and enabling controlled drug

* Corresponding Author's Email: jasmine.jain@mmumullana.org

In: Advances in Nanotechnology for Ocular Drug Delivery
Editors: Anjali Sharma, Ajmer Singh Grewal, Geeta Deswal et al.
ISBN: 979-8-89530-814-1
© 2025 Nova Science Publishers, Inc.

release. Many nano-based ocular systems have shown promising results in preclinical studies, with several still under investigation. Consequently, persistent efforts are underway to facilitate their clinical translation, aiming to convert preclinical research findings into effective treatment options for practical healthcare use, thereby improving patient outcomes. Some of these formulations have already received clinical approval and entered the commercial market. This chapter therefore aims to compile and review nanotechnology-based ocular drug delivery systems that have undergone clinical translation and are either commercialized or in the process of commercialization.

Keywords: controlled drug release, FDA-approved ocular nanotechnology, nanomedicine in ophthalmology, nanoparticles, ocular drug delivery, ocular nanotechnology, ophthalmic nanocarriers, posterior eye segment, targeted ocular delivery

Abbreviations

AMD: Age-Related Macular Degeneration
BRB: Blood-Retinal Barrier
CsA: Cyclosporine A
DED: Dry Eye Disease
DRZ: Dorzolamide
EMA: European Medicines Agency
FDA: Food and Drug Administration
IRDs: Inherited Retinal Diseases
MHC: Major Histocompatibility Complex
NLCs: Nanostructured Lipid Carriers
NPs: Nanoparticles
PAA: Polyacrylic Acid
PEG: Polyethylene Glycol
PLGA: Poly(lactic-co-glycolic acid)
PVCL-PVA-PEG: Poly(N-vinylcaprolactam-co-vinyl acetate)-graft-polyethylene glycol
SLNs: Solid Lipid Nanoparticles
STZ: Streptozotocin
VEGF: Vascular Endothelial Growth Factor

Introduction

The eye is an organ of remarkable complexity in terms of both structure and function. The anterior portion of the eye, which includes the cornea, conjunctiva, aqueous humor, iris, ciliary body, and lens, constitutes approximately one-third of the entire eye structure (Le Bourlais et al., 1998). The remaining posterior portion comprises the vitreous humor, neural retina, sclera, choroid, and retinal pigment epithelium (Gulsen and Chauhan, 2004). According to reports from the World Health Organization, more than two billion people globally suffer from some form of vision impairment, with one billion of these cases being preventable or treatable. Population growth and aging are key contributors to the rising prevalence of visual disorders. In 2020, approximately 43 million individuals were blind, with over 90 percent of these cases resulting from preventable or treatable conditions. By 2050, it is projected that there will be 61 million cases of blindness and 474 million individuals with moderate to severe vision impairment. Major causes include cataracts, uncorrected refractive errors, glaucoma, and conjunctivitis, which primarily affect the anterior segment of the eye. In contrast, age-related macular degeneration and diabetic retinopathy predominantly affect the posterior segment, together impacting millions of people worldwide, particularly women and those living in economically disadvantaged regions (Patel et al., 2013; Bourne et al., 2021). Most ocular treatments are administered in the form of eye drops, making it essential for a sufficient amount of medication to reach the anterior segment for effective disease management. However, despite the advantages of topical administration, drug absorption through this route is severely limited. The bioavailability of ocular drugs is influenced by factors such as the physicochemical properties of the drug, ocular metabolism, and various pharmacokinetic variables. The eye's natural protective mechanism, known as ocular clearance, removes approximately ninety percent of the administered medication, significantly limiting therapeutic efficacy.

Physiological and Anatomical Barriers in Ocular Drug Delivery

Corneal and Scleral Barriers

The cornea and sclera are two essential structures that limit drug absorption and significantly impact the effectiveness of therapeutic interventions (Varela-Fernandez et al., 2020). In addition to these physical barriers, the eye also presents several static and dynamic barriers that further complicate drug delivery. Static barriers include the blood-aqueous and blood-retinal barriers, while dynamic barriers consist of choroidal and conjunctival blood flow, lymphatic clearance, tear dilution, and the action of efflux pumps. Together, these obstacles present a major challenge for delivering drugs effectively, whether administered directly or as part of a formulated dosage, particularly when targeting the posterior segment of the eye (Gaudana et al., 2010).

Blood-Retinal Barrier (BRB)

The establishment and maintenance of the blood-retinal barrier are crucial for preserving optimal vision. This barrier also restricts the passage of large or hydrophilic molecules, and its breakdown contributes to the development of various retinal diseases (Runkle and Antonetti, 2011). Traditional pharmaceutical agents are generally unable to cross the blood-retinal barrier, which necessitates the use of advanced drug delivery strategies such as nanoparticles or intravitreal injections (Naylor et al., 2019).

Tear Film and Pre-Corneal Clearance

The tear film, consisting of lipid, aqueous, and mucin layers, plays a vital role in ocular drug delivery by influencing drug retention and absorption. Rapid tear turnover and frequent blinking eliminate most topically administered medications, leaving only about 15 percent of the drug on the ocular surface within minutes (Sebbag et al., 2020). Additionally, the nasolacrimal drainage system reduces drug bioavailability by directing the medication into systemic circulation. Proteins present in the tear film may also bind to drugs, further limiting their absorption. Enhancing drug retention through higher-viscosity formulations or advanced delivery technologies is essential for improving therapeutic outcomes in ophthalmology (Chang and Purt, 2023; Sebbag et al., 2020).

Immune Privilege of the Eye

The eye's immunological privilege, maintained by mechanisms such as the blood-retinal barrier, reduced expression of major histocompatibility complex (MHC) molecules, and a locally immunosuppressive environment, protects against inflammation while allowing selective immune responses (Benhar et al., 2012). This unique condition influences therapeutic delivery, as many drugs are at risk of immune clearance or reduced efficacy (Keino et al., 2018). Therefore, formulations such as nanoparticles or sustained-release systems are required to minimize immunogenicity and enhance local drug retention (Hori, 2008).

Nanotechnology-Based Ocular Drug Delivery Systems

Nanotechnology has emerged as a transformative technology in ocular drug administration, offering innovative options to overcome the constraints of conventional methods and improve treatment effectiveness and patient compliance (Gorantla et al., 2020).

Nanoparticles

Nanoparticles are tiny structures ranging in size from 1 to 1000 nanometers, composed of various materials such as metals, polymers, lipids, ceramics, or other substances. They offer innovative solutions to the challenges of ocular drug delivery by enhancing drug stability, bioavailability, and targeted distribution (Zhou et al., 2013). Nanoparticles sized between 50 and 400 nanometers are commonly used for delivering medications to the anterior segment of the eye, as they can overcome physiological barriers and direct drugs to specific ocular tissues with precision (Almeida et al., 2014). These nanoparticles improve the solubility and retention of drugs in anterior eye disorders and enable sustained and targeted delivery to the retina for posterior segment diseases. Both lipophilic and hydrophilic drugs can be delivered over extended periods using solid lipid nanoparticles (SLNs), nanostructured lipid carriers (NLCs), and biodegradable polymeric nanoparticles. Metallic nanoparticles, such as gold and silver, enhance drug transport due to their unique optical properties and photothermal effects, while carriers like liposomes and dendrimers support targeted and efficient drug encapsulation (Silva et al., 2023). Functionalization with ligands allows for precise targeting of ocular tissues, reducing systemic side effects. Moreover, advancements such as mucoadhesive nanoparticles made from chitosan or hyaluronic acid significantly increase corneal residence time and permeability across ocular barriers, thereby improving therapeutic outcomes.

Liposomes and Niosomes

Liposomes, composed of phospholipid bilayers, and niosomes, formed from non-ionic surfactants such as Span or Tween, are advanced drug delivery systems capable of encapsulating both hydrophilic and hydrophobic drugs (Momekova et al., 2021). These systems enhance the effectiveness of therapeutic agents for ocular applications (Wen et al., 2018). Their unique structure contributes to prolonged pre-corneal retention, improved drug absorption, and sustained drug release. In addition, they enhance drug stability, bioavailability, and solubility, especially for hydrophobic compounds, while minimizing systemic absorption and associated side effects. These attributes make liposomes and niosomes versatile platforms for treating ocular disorders, with ongoing research aimed at further improving their clinical performance (Bartelds et al., 2018).

Dendrimers

Dendrimers are nanoscale, customizable structures (1–10 nm) that facilitate the encapsulation or attachment of therapeutic agents, enabling controlled drug release, enhanced bioavailability, and targeted localization with minimal systemic side effects (Madaan et al., 2014). They allow the precise delivery of nucleic acids to retinal cells, making them effective carriers for ocular gene therapy. Dendrimers are particularly useful in treating inflammation (such as uveitis) and retinal disorders (including age-related macular degeneration and diabetic retinopathy) by delivering anti-inflammatory or anti-VEGF agents directly to the affected tissues. As a result, dendrimers represent a valuable tool in the development of more effective treatments for ocular diseases (Abbasi et al., 2014).

Nanomicelles and Nanoemulsions

Nanomicelles are self-assembling colloidal carriers ranging from 10 to 100 nm, characterized by a hydrophobic core and a hydrophilic shell. They offer innovative strategies for drug delivery by enhancing efficacy, safety, and ease of administration (Vadlapudi & Mitra, 2013). Their high drug-loading capacity and small particle size improve ocular bioavailability by minimizing degradation and facilitating drug absorption through ocular epithelial barriers (Wang et al., 2016). For instance, voclosporin nanomicelles, composed of non-ionic surfactants, can achieve high concentrations (up to 0.2%) in a clear aqueous solution, showing promise for treating dry eye syndrome by improving ocular tissue absorption with minimal side effects, including a lower risk of cataract formation. Similarly, cyclosporine A nanomicelles formulated with PVCL-PVA-PEG enhance cellular uptake and deliver higher drug concentrations to the cornea compared to traditional oil-based formulations, resulting in improved bioavailability and tolerance. Nanoemulsions, also known as nanometric emulsions, are stabilized droplets ranging from 20 to 200 nm, formed using suitable amphiphilic emulsifiers (Zhang et al., 2014). Their small droplet size and high-energy formulation process provide prolonged kinetic stability (Preeti et al., 2023). Dorzolamide nanoemulsions have been shown to enhance ocular bioavailability and significantly reduce intraocular pressure in glaucoma (Kassem et al., 2022). Brinzolamide nanoemulsions improve tissue penetration and aqueous humor concentrations, allowing for effective pressure control with reduced dosing frequency (Mahboobian et al., 2017). Moxifloxacin nanoemulsions increase ocular retention time and offer rapid antibacterial action against conjunctivitis with minimal irritation (Youssef et al., 2022). Notably, Cequa (0.09% cyclosporine A), a commercially approved nanomicellar formulation, effectively treats dry eye disease by significantly improving symptoms and tear production compared to conventional therapies (Burade et al., 2020).

Hydrogels and in Situ Gelling Systems

Hydrogels offer significant advantages for effective ocular drug delivery by initially being administered in liquid form and subsequently transforming into a gel upon contact with the eye (Hamcerencu et al., 2020). Their unique properties, particularly mucoadhesiveness, provide numerous benefits in the treatment of ocular surface disorders. These include prolonged drug retention, enhanced therapeutic efficacy, reduced dosing frequency, improved patient compliance, minimized side effects, and consistent treatment outcomes (Truong et al., 2024). For instance, levocetirizine formulated in a thermosensitive chitosan-based hydrogel demonstrated superior drug retention, sustained release, and enhanced ocular penetration. Similarly, hydrogels composed of polyacrylic acid (PAA) and hydroxyethyl cellulose have been successfully used to deliver ofloxacin, reducing ocular irritation and enabling effective gel formation at physiological pH. Another example involves alginate and methylcellulose-based hydrogels for gatifloxacin delivery, which significantly improve bioavailability and prolong drug release. Ultrasound-responsive hydrogels, which release drugs upon exposure to ultrasonic waves, offer advanced precision and control in drug delivery, potentially enhancing therapeutic outcomes (Cooper & Yang, 2019). Additionally, hydrogels allow for the co-delivery of multiple therapeutic agents, promoting synergistic effects in the treatment of complex ocular diseases.

Clinical Translation of Ocular Nanotechnology

From Preclinical to Clinical Studies

Animal models, particularly rodents, are widely used in ocular research due to their numerous advantages in evaluating potential therapeutics for various eye diseases. Their key benefits include genetic manipulability, the availability of knockout and knock-in models, short lifespans, cost-effectiveness, and the ability to replicate critical features of human ocular pathologies (Jeong et al., 2016). Mice are commonly employed to assess nanomedicines for ocular conditions; however, their smaller eye size and anatomical differences in retinal and corneal structures pose challenges in translating dosage and delivery methods to humans. Despite these limitations, their affordability, ease of handling, and well-established disease models make them invaluable in preclinical studies. For example, streptozotocin (STZ)-induced diabetic rats develop hyperglycemia and retinal changes that closely resemble human diabetic retinopathy (Mahaling et al., 2018). In animal models, nanoparticle-based formulations have demonstrated prolonged therapeutic effects, indicating the potential for reduced injection frequency without sacrificing efficacy in controlling neovascularization and retinal edema. Recently, ocular nanomedicine research has shifted from in vitro studies toward early in vivo evaluations, with a growing emphasis on selecting appropriate animal models. This shift highlights the rising demand for novel nanomedicine formulations aimed at treating ocular diseases more effectively. Numerous critical challenges related to nanomedicines including their pharmacokinetics and pharmacodynamics within the ocular system, regulatory hurdles, and manufacturing complexities must be addressed to enable their successful use in managing eye diseases. Nanoparticles offer a promising approach to overcoming these obstacles and enhancing drug delivery to ocular tissues. Their ability to penetrate various eye structures depends on factors such as size, surface charge, and hydrophilicity. For example, nanoparticles smaller than 200 nanometers and coated with hydrophilic molecules have a greater likelihood of penetrating the conjunctival layers, whereas targeting the vitreous body requires a careful balance between size and charge to navigate its gel-like matrix (Bachu et al., 2018). Although in vivo animal models can approximate human outcomes in ocular disorders, anatomical differences pose translational challenges. The vitreous body varies significantly in size and proximity to the retina between rodents and humans which complicates the extrapolation of preclinical findings. Additionally, mice and rats possess thicker retinas and smaller vitreous volumes compared to humans which makes precise assessment of ocular nanomedicine delivery difficult. Rabbits, another widely used model, have eye sizes closer to humans but differ physiologically by exhibiting increased mucus production, heightened surface sensitivity, and reduced tear secretion. Importantly, administering an equivalent nanoparticle dose across species does not guarantee safety or efficacy in humans (Tang et al., 2022).

Case Studies of Successful Clinical Translation

Nanotechnology in ocular drug delivery has led to the development of several effective formulations that have advanced from preclinical testing to clinical application. The following are notable examples of nanotechnology-based ocular formulations currently in clinical use, which have shown success in treating various eye diseases.

Ikervis®

This ophthalmic emulsion containing cyclosporine A (CsA) was the first eye drop approved by the European Medicines Agency (EMA) for the treatment of severe keratitis and dry eye disease, which involve corneal surface damage and ocular surface inflammation, in adult patients. In global phase III clinical trials involving this patient population, administration of ciclosporin 0.1% once daily over a 12-month period was associated with statistically significant and clinically meaningful improvements in dry eye symptoms within the first six months. These beneficial effects were maintained or further enhanced during the subsequent six months. Additionally, a phase III extension study demonstrated sustained long-term efficacy even after discontinuation of the medication (Hoy, 2017).

Cequa®

Another formulation is an aqueous nanomicellar ophthalmic preparation containing cyclosporine A, designed to enhance tear production in individuals with dry eye disease (DED) (Schechter et al., 2022). Cyclosporine is a neutrally charged, hydrophobic compound with limited water solubility, which poses challenges for developing a safe and effective ocular drug delivery system (de Oliveira & Wilson, 2019). Nevertheless, its immunomodulatory properties help reduce ocular inflammation caused by DED, leading to increased conjunctival goblet cell density and improved tear production.

Dorzolamide Nanoemulsion

This ocular drug contains dorzolamide (DRZ), which is used to reduce intraocular pressure. It demonstrates high efficacy with reduced dosing frequency, enhanced bioavailability, and minimized side effects. Liposomal-encapsulated DRZ offers prolonged retention in ocular tissues, while nanoparticle formulations provide improved penetration and sustained drug release (Pardeshi et al., 2024).

Brinzolamide Nanoemulsion

This commercial formulation, Azopt (Alcon Laboratories, USA), is an aqueous multi-dose suspension containing 1% (w/v) brinzolamide. Brinzolamide is a selective, non-competitive, reversible inhibitor of carbonic anhydrase II that reduces aqueous humor production and lowers intraocular pressure. Additionally, it improves ocular blood circulation, making it a primary treatment for glaucoma patients with vascular dysregulation. However, the medical use of brinzolamide was limited due to its low water solubility (Bhalerao et al., 2020).

Moxifloxacin Nanoemulsion

Moxifloxacin, marketed as a 0.5%w/v eye solution, has been authorized by the US FDA as a singular treatment for ocular bacterial infections. Encapsulating moxifloxacin in droplets to form a nanoemulsion presents a superior option for its ophthalmic application (Youssef et al., 2022).

Travoprost Nanoemulsion

Travatan® contains travoprost, a synthetic analog of prostaglandin F2α, formulated as a nanoemulsion that offers a promising sustained-release approach for glaucoma treatment. This formulation allows for twice-daily dosing and improves patient compliance. The medicated liquid crystalline nanostructures demonstrate effective drug penetration through the ocular layers, enhanced bioavailability, strong stability, and high entrapment efficiency of travoprost (El-Gendy et al., 2023).

Commercialization of Nano-Based Ocular Formulations

Market Potential and Growth

Nanotechnology involves the design and fabrication of materials at the nanoscale in at least one dimension (Lin & He, 2008). Various nanotechnology-based approaches have been developed to treat eye disorders by enhancing bioadhesion, enabling prolonged drug release, providing stealth properties, allowing selective targeted delivery, and offering stimuli-responsive release (Zarbin et al., 2010). The commercial potential and investment landscape for ocular nanotechnology are rapidly evolving, driven by increasing demand for advanced drug delivery systems that overcome the limitations of conventional eye treatments. Traditional methods provide a thorough analysis of the economic factors, investment opportunities, and expected returns associated with the commercialization of nanotechnology in ophthalmology. Advances in nanotechnology are projected to expand the global ocular drug delivery market from $15.3 billion in 2023 to approximately $25.5 billion by 2030 (Strategic Market Research, 2025).

Challenges in Commercialization

Ocular drug delivery faces significant challenges due to the complex and diverse physiological barriers of the eye. To overcome these obstacles, advanced nanotechnology-based delivery systems such as liposomes, nanomicelles, nanocrystals, nanoemulsions, and hydrogel-based nanoparticles have been developed. These formulations are now commercially available as nano-formulation delivery systems that meet required standards for patient acceptability, safety, efficacy, and stability (Gorantla et al., 2020). However, research indicates that scaling up nano-based delivery systems from laboratory to industrial production can hinder the clinical translation of many drugs. For example, increasing nanoparticle size using less energy-intensive methods such as phase inversion temperature/composition and emulsion inversion point often alters their physicochemical properties. To overcome the limitations of these low-energy processes, high-energy methods like heat-assisted homogenization and ultrasonic treatment are necessary; yet, these aggressive techniques can cause nanoparticle coalescence, leading to thermodynamic instability (Li et al., 2023).

Patents and Intellectual Property Studies

A U.S. patent (No. 11,000,509) detailed the composition and formulation methods of phentolamine aimed at enhancing visual function and reducing ocular redness with regular use (Martell & Chelates, 2017). Babizhayev et al. (2004) developed an aqueous ophthalmic solution of N-acetylcarnosine using cellulose to improve intraocular absorption for treating ocular diseases and corneal conditions, including glaucoma and blurred vision. This medication was incorporated into polymeric hydrogel contact lenses and ocular inserts for topical delivery. Patent CN110664757A describes a nanocrystalline eye drop formulated with chitosan or hyaluronic acid designed for congenital macular degeneration treatment. This formulation enhances penetration through ocular barriers via passive targeting or intercellular pinocytosis, ensuring effective drug concentration in the cul-de-sac cavity (Akhter et al., 2022). The U.S. patent US9295693B2 outlines a composition used for dry eye disease and contact lens sensitivity treatment. It contains a water-soluble graft copolymer, specifically poly(l-lysine)-graft-poly(ethylene glycol), with sterically stabilized PEG coatings that demonstrate efficient adhesion to biological surfaces (Cooper, 2018). Another patent describes topical formulations of olopatadine (0.17 to 0.62% w/v) combined with polyvinylpyrrolidone or polystyrene sulfonic acid as stabilizers for managing ocular and nasal allergy or inflammatory conditions (Tai et al., 2018).

Successful Commercial Products

Various nanoformulations for ocular treatments have been developed and commercialized; however, certain formulations, such as drug-loaded emulsions, are advancing slowly in therapeutic applications despite the approval of drug-free nanoemulsions for dry eye treatment. Restasis® is a nanoemulsion formulation of cyclosporine A used for managing chronic dry eye (Reimondez-Troitiño et al., 2015). Similarly, Ozurdex® (Allergan Inc., Irvine, CA, USA) is a notable biodegradable intravitreal implant composed of poly(lactic-co-glycolic acid) conjugate. These biodegradable implants provide therapeutic efficacy for three to six months, due to the exponential decline in drug concentration caused by three-dimensional drug release and gradual scaffold degradation within the vitreous cavity (Poornachandra et al., 2017). Retisert® is a non-biodegradable implant created through polymer conjugation, offering sustained medication release over two to three years by a steady linear decrease in drug concentration within the limited diffusion zone (Mohan & Ratra, 2023).

Future Perspectives

Owing to mechanisms such as lacrimation, tear dilution, and tear turnover, typically less than five percent of the administered dose penetrates the aqueous humor (Gaudana et al., 2009). Nanoformulations delivered via intraocular or periocular routes enable targeted drug delivery while minimizing systemic side effects (Choi et al., 2023; Sen et al., 2014). Intravitreal injections have become a common treatment for retinal disorders such as retinal vein occlusion, endophthalmitis, and age-related macular degeneration. Other routes for intraocular delivery include intrastromal, subconjunctival, retrobulbar, peribulbar, subretinal, and subchoroidal injections (Liu et al., 2023). The eye is considered a favorable target for gene transfer compared to other organs due to its compartmentalized and small structure. Recently, gene therapy has been employed to treat several ocular diseases and inherited retinal disorders. Inherited retinal diseases (IRDs) lead to progressive degeneration of photoreceptor cells and vision loss, with considerable genetic and phenotypic heterogeneity, particularly in conditions such as corneal dystrophies and retinoblastoma (Choi et al., 2023). Patients with ocular hypertension or open-angle glaucoma, including secondary types, may show variable responses to treatments, which raises concerns about their inclusion affecting therapeutic outcomes (Baiza-Durán et al., 2023). Therefore, targeted patient selection is critical for accurately assessing therapy effectiveness, as individuals at different disease stages or with distinct genetic markers may respond differently to nanoparticle-based therapies (Varner et al., 2008). Study endpoints must be clearly defined, emphasizing both efficacy and safety (Foulks, 2003; del Amo et al., 2017). Despite extensive cellular studies and animal research aimed at understanding the therapeutic effects of tailored ocular nanomedicine, fully elucidating the underlying mechanisms on ocular cells and tissues remains challenging. Continuous evaluation of therapeutic safety and efficacy is critical before clinical application, especially since most ocular nanomedicines are novel and still under development. Biosafety considerations, particularly for inorganic nanoparticles, are essential and must be prioritized. These concerns include potential genotoxicity, neurotoxicity, and reproductive toxicity. A primary and urgent challenge for ocular nanomedicine lies in successful commercialization and future clinical translation (Anjum et al., 2021).

Conclusion

Ocular disorders represent a major global health challenge, impacting millions with vision impairment and blindness. The eye's distinctive anatomy and physiology create significant obstacles for effective drug delivery. Conventional treatments, such as eye drops, suffer from low bioavailability due to multiple physiological barriers that limit drug penetration to target tissues. This limitation has driven the development of novel approaches to overcome these challenges. Advances in nanotechnology have led to the creation of sophisticated ocular drug delivery systems including nanoparticles, liposomes, dendrimers, nanomicelles, nanoemulsions, and hydrogels. These nanocarriers enhance drug stability, bioavailability, and enable targeted delivery, effectively addressing the limitations of traditional therapies. Nanoparticles can penetrate ocular barriers and offer potential for sustained drug release, while liposomes and niosomes improve pre-corneal drug retention and permeability, thereby optimizing therapeutic outcomes. Dendrimers enable targeted and controlled drug release, while hydrogels offer mucoadhesive properties that prolong drug retention on the ocular surface. Although nanotechnology-based ocular formulations show promising potential for clinical translation, the number of commercially available products remains limited. Approved nanoemulsion-based ocular products include Ikervis® (sikaproct) and Cequa® (cyclosporine ophthalmic emulsion) for dry eye disease, dorzolamide and brinzolamide nanoemulsions for glaucoma, and moxifloxacin nanoemulsion for bacterial conjunctivitis. However, several challenges hinder the commercialization and clinical translation of ocular nanomedicines. Issues such as scalability, formulation stability, evolving regulatory frameworks, and the demand for personalized medicine need to be addressed concurrently. Additionally, anatomical and physiological differences between animal models and humans complicate the extrapolation of preclinical results to clinical settings. Despite these obstacles, experts remain optimistic that ongoing advances in nanotechnology will usher in a new era of ocular therapeutics. The integration of these innovations, alongside gene therapy and targeted delivery systems, promises a more personalized and effective approach to treating ocular diseases, ultimately benefiting millions of patients worldwide.

References

Abbasi, E., Aval, S. F., Akbarzadeh, A., Milani, M., Nasrabadi, H. T., Joo, S. W., Hanifehpour, Y., Nejati-Koshki, K., & Pashaei-Asl, R. (2014). Dendrimers: synthesis, applications, and properties. *Nanoscale Research Letters*, *9*(1), 247.

Akhter, M. H., Ahmad, I., Alshahrani, M. Y., Al-Harbi, A. I., Khalilullah, H., Afzal, O., Altamimi, A. S. A., Najib Ullah, S. N. M., Ojha, A., & Karim, S. (2022). Drug Delivery Challenges and Current Progress in Nanocarrier-Based Ocular Therapeutic System. *Gels*, *2022, 8*(2), 82.

Almeida, H., Amaral, M. H., Lobao, P., Silva, A. C., & Lobo, J. M. S. (2014). Applications of polymeric and lipid nanoparticles in ophthalmic pharmaceutical formulations: Present and future considerations. *Journal of Pharmacy & Pharmaceutical Sciences*, *17*(3), 278–293.

Anjum, S., Ishaque, S., Fatima, H., Farooq, W., Hano, C., Abbasi, B. H., & Anjum, I. (2021). Emerging applications of nanotechnology in healthcare systems: Grand challenges and perspectives. *Pharmaceuticals*, *14*(8), 707.

Babizhayev, M. A. (2004). Method for topical treatment of eye disease and composition and device for said treatment (WO2004028536A1). World Intellectual Property Organization.

Bachu, R. D., Chowdhury, P., Al-Saedi, Z. H. F., Karla, P. K., & Boddu, S. H. S. (2018). Ocular drug delivery barriers—Role of nanocarriers in the treatment of anterior segment ocular diseases. *Pharmaceutics*, *10*(1), 28.

Baiza-Duran, L. M., Munoz-Villegas, P., Sanchez-Rios, A., & Olvera-Montano, O. (2023). Efficacy and safety of an ophthalmic DMPC-based nanoemulsion in patients with dry eye disease: A Phase I/II randomized clinical trial. *Journal of Ophthalmology*, *2023*(1), 1431473.

Bartelds, R., Nematollahi, M. H., Pols, T., Stuart, M. C. A., Pardakhty, A., Asadikaram, G., & Poolman, B. (2018). Niosomes, an alternative for liposomal delivery. *PLoS ONE*, *13*(4), e0194179.

Benhar, I., London, A., & Schwartz, M. (2012). The privileged immunity of immune privileged organs: The case of the eye. *Frontiers in Immunology*, *3*(SEP), 32898.

Bhalerao, H., Koteshwara, K., & Chandran, S. (2020). Design, optimisation and evaluation of in situ gelling nanoemulsion formulations of brinzolamide. *Drug Delivery and Translational Research*, *10*(2), 529–547.

Bourne, R. R. A., Steinmetz, J. D., Saylan, M., Mersha, A. M., Weldemariam, A. H., Wondmeneh, T. G., Sreeramareddy, C. T., Pinheiro, M., Yaseri, M., Yu, C., Zastrozhin, M. S., Zastrozhina, A., Zhang, Z. J., Zimsen, S. R. M., Yonemoto, N., Tsegaye, G. W., Vu, G. T., Vongpradith, A., Renzaho, A. M. N., Vos, T. (2021). Causes of blindness and vision impairment in 2020 and trends over 30 years, and prevalence of avoidable blindness in relation to VISION 2020: The Right to Sight: An analysis for the Global Burden of Disease Study. *The Lancet Global Health*, *9*(2), e144–e160.

Burade, V., Zalawadia, R., Patel, A., & Ogundele, A. (2020). Preclinical efficacy comparison of cyclosporine ophthalmic solution 0.09% Vs cyclosporine ophthalmic emulsion 0.05% vs Ciclosporin ophthalmic emulsion 0.1% in a NOD mouse model of dry eye disease. *Clinical Ophthalmology (Auckland, N.Z.)*, *14*, 2747.

Chang, A. Y., & Purt, B. (2023). Biochemistry, Tear Film. StatPearls Publishing.

Choi, E. H., Suh, S., Sears, A. E., Hołubowicz, R., Kedhar, S. R., Browne, A. W., & Palczewski, K. (2023). Genome editing in the treatment of ocular diseases. *Experimental & Molecular Medicine 55*(8), 1678–1690.

Cooper. (2018). Bi-functional co-polymer use for ophthalmic and other topical and local applications. *Chemical Reviews*, *3147*(60), 287–328.

Cooper, R. C., & Yang, H. (2019). Hydrogel-based ocular drug delivery systems: Emerging fabrication strategies, applications, and bench-to-bedside manufacturing considerations. *Journal of Controlled Release*, *306*, 29.

de Oliveira, R. C., & Wilson, S. E. (2019). Practical guidance for the use of cyclosporine ophthalmic solutions in the management of dry eye disease. *Clinical Ophthalmology*, *13*, 1115.

del Amo, E. M., Rimpela, A. K., Heikkinen, E., Kari, O. K., Ramsay, E., Lajunen, T., Schmitt, M., Pelkonen, L., Bhattacharya, M., Richardson, D., Subrizi, A., Turunen, T., Reinisalo, M., Itkonen, J., Toropainen, E., Casteleijn, M., Kidron, H., Antopolsky, M., Vellonen, K. S., Urtti, A. (2017). Pharmacokinetic aspects of retinal drug delivery. *Progress in Retinal and Eye Research*, *57*, 134–185.

El-Gendy, M. A., Mansour, M., El-Assal, M. I. A., Ishak, R. A. H., & Mortada, N. D. (2023). Travoprost liquid nanocrystals: An innovative armamentarium for effective glaucoma therapy. *Pharmaceutics*, *15*(3), 954.

Foulks, G. N. (2003). Challenges and pitfalls in clinical trials of treatments for dry eye. *The Ocular Surface*, *1*(1), 20–30.

Gaudana, R., Ananthula, H. K., Parenky, A., & Mitra, A. K. (2010). Ocular Drug Delivery. *The AAPS Journal*, *12*(3), 348. https://doi.org/10.1208/S12248-010-9183-3.

Gaudana, R., Jwala, J., Boddu, S. H. S., & Mitra, A. K. (2009). Recent perspectives in ocular drug delivery. *Pharmaceutical Research*, *26*(5), 1197–1216. https://doi.org/10.1007/S11095-008-9694-0/FIGURES/2.

Gorantla, S., Rapalli, V. K., Waghule, T., Singh, P. P., Dubey, S. K., Saha, R. N., & Singhvi, G. (2020). Nanocarriers for ocular drug delivery: current status and translational opportunity. *RSC Advances*, *10*(46), 27835–27855. https://doi.org/10.1039/D0RA04971A.

Gulsen, D., & Chauhan, A. (2004). Ophthalmic drug delivery through contact lenses. *Investigative Ophthalmology & Visual Science*, *45*(7), 2342–2347. https://doi.org/10.1167/IOVS.03-0959.

Hamcerencu, M., Desbrieres, J., Popa, M., & Riess, G. (2020). Thermo-sensitive gellan maleate/N-isopropylacrylamide hydrogels: initial "in vitro" and "in vivo" evaluation as ocular inserts. *Polymer Bulletin*, *77*(2), 741–755. https://doi.org/10.1007/S00289-019-02772-5.

Hori, J. (2008). Mechanisms of immune privilege in the anterior segment of the eye: What we learn from corneal transplantation. *Journal of Ocular Biology, Diseases, and Informatics*, *1*(2–4), 94. https://doi.org/10.1007/S12177-008-9010-6.

Hoy, S. M. (2017). Ciclosporin Ophthalmic Emulsion 0.1%: A Review in Severe Dry Eye Disease. *Drugs*, *77*(17), 1909–1916. https://doi.org/10.1007/S40265-017-0834-X.

Jeong, J. H., Nguyen, H. K., Lee, J. E., & Suh, W. (2016). Therapeutic effect of apatinib-loaded nanoparticles on diabetes-induced retinal vascular leakage. *International Journal of Nanomedicine*, *11*, 3101–3109. https://doi.org/10.2147/IJN.S108452.

Kassem, A. A., Salama, A., & Mohsen, A. M. (2022). Formulation and optimization of cationic nanoemulsions for enhanced ocular delivery of dorzolamide hydrochloride using Box-Behnken design: In vitro and in vivo assessments. *Journal of Drug Delivery Science and Technology*, *68*, 103047. https://doi.org/10.1016/J.JDDST.2021.103047.

Keino, H., Horie, S., & Sugita, S. (2018). Immune Privilege and Eye-Derived T-Regulatory Cells. *Journal of Immunology Research*, *2018*(1), 1679197.

Le Bourlais, C., Acar, L., Zia, H., Sado, P. A., Needham, T., & Leverge, R. (1998). Ophthalmic drug delivery systems—Recent advances. *Progress in Retinal and Eye Research*, *17*(1), 33–58. https://doi.org/10.1016/S1350-9462(97)00002-5.

Li, S., Chen, L., & Fu, Y. (2023). Nanotechnology-based ocular drug delivery systems: recent advances and future prospects. *Journal of Nanobiotechnology*, *21*(1), 1–39. https://doi.org/10.1186/S12951-023-01992-2.

Lin, T. T., & He, Y. J. (2008). Ophthalmic Applications of Nanotechnology. *Chinese Journal of Ophthalmology*, *43*(11), 415–435. https://doi.org/10.1007/978-1-59745-375-2_22.

Liu, L. C., Chen, Y. H., & Lu, D. W. (2023). Overview of Recent Advances in Nano-Based Ocular Drug Delivery. *International Journal of Molecular Sciences*, *24*(20), 15352. https://doi.org/10.3390/IJMS242015352.

Madaan, K., Kumar, S., Poonia, N., Lather, V., & Pandita, D. (2014). Dendrimers in drug delivery and targeting: Drug-dendrimer interactions and toxicity issues. *Journal of Pharmacy & Bioallied Sciences*, *6*(3), 139. https://doi.org/10.4103/0975-7406.130965.

Mahaling, B., Srinivasarao, D. A., Raghu, G., Kasam, R. K., Bhanuprakash Reddy, G., & Katti, D. S. (2018). A non-invasive nanoparticle mediated delivery of triamcinolone acetonide ameliorates diabetic retinopathy in rats. *Nanoscale*, *10*(35), 16485–16498. https://doi.org/10.1039/C8NR00058A.

Mahboobian, M. M., Seyfoddin, A., Rupenthal, I. D., Aboofazeli, R., & Foroutan, S. M. (2017). Formulation development and evaluation of the therapeutic efficacy of Brinzolamide containing nanoemulsions. *Iranian Journal of Pharmaceutical Research*, *16*(3), 847. https://pmc.ncbi.nlm.nih.gov/articles/PMC5610741/.

Martell, A. E., & Chelates. (2017). Methods and compositions for daily ophthalmic administration of phentolamine to improve visual performance. *National Institutes of Health*, *13*(3), 633–638. http://califasainc.com/pdf/online://medlineplus.gov/ency/article/003396.htm.

Mohan, S., & Ratra, D. (2023). Intravitreal Implants. *Ocular Pharmacotherapy*, 231–231. https://doi.org/10.5005/jp/books/12924_17.

Momekova, D. B., Gugleva, V. E., & Petrov, P. D. (2021). Nanoarchitectonics of multifunctional niosomes for advanced drug delivery. *ACS Omega*, *6*(49), 33265–33273. https://doi.org/10.1021/ACSOMEGA.1C05083/ASSET/IMAGES/LARGE/AO1C05083_0007.JPEG.

Naylor, A., Hopkins, A., Hudson, N., & Campbell, M. (2019). Tight Junctions of the Outer Blood Retina Barrier. *International Journal of Molecular Sciences*, *21*(1), 211. https://doi.org/10.3390/IJMS21010211.

Pardeshi, S. R., Gholap, A. D., Hatvate, N. T., Gharat, K. D., Naik, J. B., & Omri, A. (2024). Advances in Dorzolamide hydrochloride delivery: Harnessing nanotechnology for enhanced ocular drug delivery in glaucoma management. *Discover Nano*, *19*(1), 1–27. https://doi.org/10.1186/S11671-024-04154-X/TABLES/3.

Patel, A., Cholkar, K., Agrahari, V., & Mitra, A. K. (2013). Ocular drug delivery systems: An overview. *World Journal of Pharmacology*, *2*(2), 47. https://doi.org/10.5497/WJP.V2.I2.47.

Poornachandra, B., Kumar, V. M. B., Jayadev, C., Dorelli, S. H., Yadav, N. K., & Shetty, R. (2017). Immortal Ozurdex: A 10-month follow-up of an intralenticular implant. *Indian Journal of Ophthalmology*, *65*(3), 255. https://doi.org/10.4103/IJO.IJO_634_16.

Preeti, Sambhakar, S., Malik, R., Bhatia, S., Al Harrasi, A., Rani, C., Saharan, R., Kumar, S., Geeta, & Sehrawat, R. (2023). Nanoemulsion: An emerging novel technology for improving the bioavailability of drugs. *Scientifica*, *2023*, 6640103.

Runkle, E. A., & Antonetti, D. A. (2011). The blood-retinal barrier: structure and functional significance. *Methods in Molecular Biology*, *686*, 133–148. https://doi.org/10.1007/978-1-60761-938-3_5.

Sebbag, L., Kirner, N. S., Wulf, L. W., & Mochel, J. P. (2020). Tear film pharmacokinetics and systemic absorption following topical administration of 1% Prednisolone acetate ophthalmic suspension in dogs. *Frontiers in Veterinary Science*, *7*, 571350. https://doi.org/10.3389/FVETS.2020.571350/BIBTEX.

Sebbag, L., Moody, L. M., & Mochel, J. P. (2020). Albumin levels in tear film modulate the bioavailability of medically-relevant topical drugs. *Frontiers in Pharmacology*, *10*, 485655. https://doi.org/10.3389/FPHAR.2019.01560/BIBTEX.

Sen, H. N., Vitale, S., Gangaputra, S. S., Nussenblatt, R. B., Liesegang, T. L., Levy-Clarke, G. A., Rosenbaum, J. T., Suhler, E. B., Thorne, J. E., Foster, C. S., Jabs, D. A., & Kempen, J. H. (2014). Periocular corticosteroid injections in uveitis: Effects and complications. *Ophthalmology*, *121*(11), 2275–2286. https://doi.org/10.1016/J.OPHTHA.2014.05.021.

Silva, B., Gonçalves, L. M., Sao Braz, B., & Delgado, E. (2023). Topical ocular delivery of nanoparticles with epoetin beta in Wistar Hannover rats. *Scientific Reports*, *13*(1), 1–12. https://doi.org/10.1038/s41598-023-28845-0.

Strategic Market Research. (2025). *Global Ocular Drug Delivery Market Report* (Report ID: 02071908). https://www.strategicmarketresearch.com/market-report/ocular-drug-delivery-system-market.

Tai, L., Liu, C., Jiang, K., Chen, X., Wei, G., Lu, W., & Pan, W. (2018). Polypeptide eye absorption enhancer and use thereof. *Nanomedicine: Nanotechnology, Biology, and Medicine*, *13*(6), 2091–2100.

Tang, Z., Fan, X., Chen, Y., & Gu, P. (2022). Ocular Nanomedicine. *Advanced Science*, *9*(15), 2003699.

Truong, D., Wu, K. Y., Nguyen, L., & Tran, S. D. (2024). Advancements in hydrogel technology for ocular drug delivery. *Open Exploration*, *1*(6), 331–352.

Vadlapudi, A. D., & Mitra, A. K. (2013). Nanomicelles: an emerging platform for drug delivery to the eye. *Therapeutic Delivery*, *4*(1), 1–3.

Varela-Fernandez, R., Díaz-Tome, V., Luaces-Rodríguez, A., Conde-Penedo, A., García-Otero, X., Luzardo-alvarez, A., Fernandez-Ferreiro, A., & Otero-Espinar, F. J. (2020). Drug delivery to the posterior segment of the eye: Biopharmaceutic and pharmacokinetic considerations. *Pharmaceutics 12*(3), 269.

Varner, P., Pershing Veterans, J. J., & Pershing, J. J. (2008). Potential bias in ophthalmic pharmaceutical clinical trials. *Clinical Ophthalmology*, *2*(2), 401–411.

Wang, X., Wang, S., & Zhang, Y. (2016). Advance of the application of nano-controlled release system in ophthalmic drug delivery. *Drug Delivery*, *23*(8), 2897–2901.

Wen, J., Galloni, M. Al, Yin, N., & Rashidinejad, A. (2018). Liposomes and Niosomes. In: *Emulsion-Based Systems for Delivery of Food Active Compounds: Formation, Application, Health and Safety*, pp. 263–292.

Youssef, A. A. A., Thakkar, R., Senapati, S., Joshi, P. H., Dudhipala, N., & Majumdar, S. (2022). Design of topical moxifloxacin mucoadhesive nanoemulsion for the management of ocular bacterial infections. *Pharmaceutics*, *14*(6), 1246.

Zarbin, M. A., Montemagno, C., Leary, J. F., & Ritch, R. (2010). Nanotechnology in ophthalmology. *Canadian Journal of Ophthalmology*, *45*(5), 457–476.

Zhang, Y., Shang, Z., Gao, C., Du, M., Xu, S., Song, H., & Liu, T. (2014). Nanoemulsion for Solubilization, Stabilization, and In Vitro Release of Pterostilbene for Oral Delivery. *AAPS PharmSciTech*, *15*(4), 1000.

Zhou, H. Y., Hao, J. L., Wang, S., Zheng, Y., & Zhang, W. S. (2013). Nanoparticles in the ocular drug delivery. *International Journal of Ophthalmology*, *6*(3), 390.

Chapter 21

Challenges in Formulating Ocular Drug Delivery

Himanshu Sharma[1]
Balraj Saini[1]
Pooja Mittal[2,*]
Vishnu Mittal[3]
and Bhawna Garg[3]
[1]Chitkara College of Pharmacy, Chitkara University, Punjab, India
[2]GITAM School of Pharmacy, GITAM (Deemed to be) University, Rudraram, Patancheru, Sangareddy District, Hyderabad, Telangana, India
[3]Adesh Institute of Pharmacy and Biomedical Sciences, Adesh University, Bathinda, Punjab, India

Abstract

Delivering drugs to the eye presents considerable challenges due to complex physiological and anatomical barriers that limit therapeutic effectiveness. Conventional ocular therapies, such as topical eye drops and intravitreal injections, often face issues such as poor drug absorption, low bioavailability, and an inability to provide sustained drug release, particularly for conditions affecting the posterior segment of the eye. This chapter presents a comprehensive overview of advanced ocular drug delivery systems developed to address these limitations. It focuses on emerging technologies such as nanocarriers, sustained-release formulations, in situ gels, and microneedle systems, which enhance drug

* Corresponding Author's Email: poojamittal2009@gmail.com

In: Advances in Nanotechnology for Ocular Drug Delivery
Editors: Anjali Sharma, Ajmer Singh Grewal, Geeta Deswal et al.
ISBN: 979-8-89530-814-1
© 2025 Nova Science Publishers, Inc.

penetration, extend precorneal retention time, and improve patient compliance. The chapter also examines implantable and injectable delivery methods, evaluating their potential to overcome drug stability challenges and improve therapeutic outcomes. Despite encouraging results in preclinical and clinical studies, several hurdles remain in translating these technologies into clinical practice. These include concerns related to biocompatibility, large-scale manufacturing, and regulatory approval. Continued innovation and technological progress are essential to developing more effective and patient-friendly ocular treatments that can successfully overcome existing barriers. With ongoing research and development, these novel delivery systems hold significant promise for enhancing the management of both anterior and posterior segment ocular diseases.

Keywords: ocular drug delivery, nanotechnologies, topical ophthalmic formulation, conventional ocular therapies, drug release, bioavailability

Abbreviations

AMD: Age-Related Macular Degeneration
CLs: Contact Lenses
CMV: Cytomegalovirus
CNTF: Ciliary Neurotrophic Factor
CPP: Cell Penetrating Peptide
CRVO: Central Retinal Vein Occlusion
DME: Diabetic Macular Edema
DR: Diabetic Retinopathy
EMA: European Medicines Agency
IOP: Intraocular Pressure
IV: Intravenous
NPs: Nanoparticles
PCL: Polycaprolactone
PEG: Polyethylene Glycol
PLGA: Poly(lactic-co-glycolic acid)
PVA: Polyvinyl Alcohol
RPE: Retinal Pigment Epithelium
VEGF: Vascular Endothelial Growth Factor

Introduction

The anatomy and physiology of the eye make it a highly complex organ (Stjernschantz and Astin, 2019). The anterior segment constitutes approximately one third of the eye's total structure and includes components such as the ciliary body, lens, iris, cornea, conjunctiva, and aqueous humor. The posterior segment comprises the choroid, optic nerve, neural retina, and sclera. Various disorders can affect both the anterior and posterior segments, potentially leading to visual impairment. Eye drops are among the most commonly used ocular treatments due to their ease of administration and high patient compliance. However, ocular absorption through topical instillation is minimal because of several anatomical and physiological barriers. These include static and dynamic barriers, along with reflex blinking, all of which limit drug penetration. Intravitreal injection is currently the most widely used and recommended method for treating conditions that affect the posterior segment of the eye. Despite its effectiveness, repeated intravitreal injections and ocular punctures can result in serious complications such as endophthalmitis, hemorrhage, retinal detachment, and reduced patient compliance (Cholkar et al., 2013). To address these challenges and enhance ocular bioavailability, a wide range of conventional and advanced drug delivery systems have been developed. These include aqueous gels, liposomes, dendrimers, ocular implants, contact lenses, emulsions, ointments, suspensions, nano-micelles, nanoparticles, microneedles, and in situ thermosensitive gels. The eye, as one of the body's most delicate and immune-privileged organs, presents significant challenges for drug delivery due to its complex network of glands, muscles, nerves, vascular structures, and specialized tissues (Ramos et al., 2021). These anatomical and physiological characteristics, while essential for maintaining ocular health, serve as significant barriers to effective drug administration (Gower et al., 2016). Understanding the two main categories of ocular barriers, static and dynamic, is crucial for addressing these challenges, as both play a major role in influencing drug stability, distribution, and therapeutic efficacy (Akhter et al., 2022). Static barriers are structural in nature and consist of various hydrophilic and hydrophobic layers that hinder drug absorption (Qi et al., 2023). For example, Bowman's layer, although acellular and limited in its regenerative capacity, plays a minor role in drug transport but should be considered due to its progressive thinning with age. The stroma serves as a major barrier to non-polar drugs, while the corneal endothelium, although more permeable than the epithelium, remains restrictive due to its leaky yet regulated transport properties (Koehl, 2018). These structural features contribute to the low bioavailability of systemically administered ocular drugs. In contrast, dynamic barriers involve physiological processes such as tear production, blinking, and nasolacrimal drainage, all of which actively clear drugs from the ocular surface (Raut et al., 2024; Giri et al., 2024). The tear film, composed of mucus, aqueous, and lipid layers, is approximately 6 micrometers thick and has a pH range of 6.5 to 7.6. This variability necessitates the use of pH-sensitive formulations to prevent irritation. The tear film also contains defensive components such as lysozyme, lactoferrin, and immunoglobulin A, which can influence drug efficacy. Additionally, the nasolacrimal drainage system rapidly clears up to 80 percent of a topically applied eye drop within 15 to 30 seconds, significantly limiting drug retention time (Tunic, 2011; Desai et al., 2019). Conjunctival blood flow further complicates drug delivery, requiring advanced formulations to achieve therapeutic concentrations in targeted ocular tissues (Giri et al., 2024; Molokhia et al., 2013). Despite these challenges, conventional drug delivery methods such as topical eye drops remain widely used due to their simplicity and suitability across all age groups (Dorcas et al., 2019). However, the majority of the drug is either drained through the nasolacrimal duct or absorbed systemically, resulting in limited corneal penetration. Topical applications primarily deliver drugs to the anterior segment, which restricts access to the posterior segment. To address these limitations, drug delivery systems must be designed to be safe, effective, and user friendly while ensuring accurate targeting of the therapeutic agents. These constraints diminish the effectiveness of standard dosage forms, underscoring the need for innovative strategies in ocular drug delivery (Desai et al., 2013; Wadhwa et al., 2009).

Novel Ocular Drug Delivery Technologies

Many ocular diseases are chronic and require continuous long-term drug therapy. However, when drugs are administered topically, ocular barriers significantly reduce absorption, often to less than five percent (Barar et al., 2016). Recent advancements in nanotechnology offer promising solutions to the challenges associated with conventional ocular drug delivery methods, particularly those related to low bioavailability and limited drug penetration across ocular barriers. Nanocarriers have the potential to enhance drug transport through these barriers, protect active compounds from degradation by ocular enzymes, and enable sustained and controlled drug release. This approach can increase drug concentrations at the target tissues, reduce dosing frequency, and improve therapeutic outcomes. Various colloidal nanocarriers, including nanoparticles, nano-micelles, nano-wafers, and microneedles, have been investigated for ocular drug delivery. However, nanoparticles larger than ten micrometers may cause ocular irritation and negatively impact the appearance and acceptability of formulations (Rojekar et al., 2024). Despite these limitations, nanocarriers are capable of improving drug penetration into deeper ocular tissues while potentially reducing drug-related toxicity. This makes them a valuable tool in the development of safer and more effective ocular therapeutics.

Nanomicelles

Nanomicelles are colloidal drug delivery systems capable of self-assembling in solution and encapsulating therapeutic agents within their hydrophobic core. Typically ranging in size from 10 to 200 nanometers, they are composed of amphiphilic surfactants or block copolymers (Tawfik et al., 2020). The hydrophobic core serves to sequester lipophilic drugs, while the hydrophilic corona interacts with the surrounding aqueous environment, thereby enhancing the solubility of moderately lipophilic compounds. When applied topically as eye drops, nanomicelles can form clear aqueous solutions, making them particularly well suited for ocular administration. The amphiphilic nature of block copolymers, which contain both hydrophilic and hydrophobic segments, supports the formation of stable nanomicelles. By improving the solubility of poorly water-soluble drugs, nanomicelles can enhance the delivery of hydrophobic therapeutic agents to ocular tissues, offering a promising strategy for the effective treatment of various eye disorders (Dubey et al., 2021).

Nanoparticles

Drug-loaded nanoparticles typically range in size from 50 to 500 nanometers, which allows them to effectively penetrate ocular barriers and deliver medication to ocular tissues through passive or active transport mechanisms. However, nanoparticle solutions may not always be suitable for achieving sustained or controlled drug release over extended periods (Rahić et al., 2020). The surfaces of corneal and conjunctival tissues carry a negative charge, making cationic nanoparticles especially advantageous due to their stronger electrostatic interactions and prolonged retention on the ocular surface compared to anionic nanoparticles. Colloidal nanoparticles can significantly enhance the solubility and transcorneal permeability of poorly water-soluble drugs, thereby improving their therapeutic potential. Both biodegradable and nonbiodegradable nanoparticles have been developed for treating ocular disorders that affect the anterior and posterior segments of the eye. Among the polymers used in nanoparticle formulations, chitosan, a biodegradable and biocompatible anionic polysaccharide, is widely employed for its ability to increase precorneal residence time and improve drug absorption and therapeutic efficacy.

Liposomes and Dendrimers

Liposomes are considered highly effective carriers for ocular drug delivery because of their ability to encapsulate both hydrophobic and hydrophilic drugs. These vesicular systems are composed of lipid bilayers that enclose an internal aqueous core, forming a barrier between the encapsulated drug and the surrounding aqueous environment (López-Cano et al., 2021). Although liposomes are generally biocompatible, their short half-life and inherent instability have raised concerns about their long-term effectiveness, particularly in the treatment of anterior segment ocular conditions. Nonetheless, in vivo studies have demonstrated their therapeutic potential. For example, liposomal formulations have been shown to effectively reduce intraocular pressure in rabbit models of glaucoma. In addition to liposomes, dendrimers represent another class of promising nanocarriers for ocular drug delivery. These are highly branched, star-shaped polymers that allow a high degree of control over size, shape, and surface functionality. Through precise synthesis techniques, dendrimers can be designed with specific architectures and functional groups that enhance drug loading, targeting, and release properties (Malkoch and Gallego, 2020). Their structural versatility and ability to interact with biological membranes make them valuable tools in the development of advanced ocular therapeutics.

Microneedles and Nanowafers

Microneedle drug delivery technology was initially developed for transdermal applications by penetrating the stratum corneum to administer medications through the skin. Building on their success in transdermal systems, researchers have investigated the potential of microneedles for treating both anterior and posterior ocular disorders (Kaushik et al., 2022). This minimally invasive technique allows for the delivery of both hydrophilic and hydrophobic drugs directly to ocular tissues. Notably, in vivo studies have demonstrated the feasibility of using microneedles to deliver medications to the posterior segment of the eye, including the injection of microparticle and nanoparticle suspensions into the suprachoroidal space. In addition, nano wafers, which are small, transparent, disc- or rectangular-shaped membranes, can be applied to the ocular surface using a fingertip (Marcano et al., 2016). These nano wafers enable controlled drug release, increasing contact time and drug residence in the cornea and conjunctiva (Mansuri et al., 2016). Consequently, they may enhance drug absorption into anterior ocular tissues. Furthermore, the biocompatible and biodegradable polymers used to fabricate these nanocarriers are designed to degrade naturally over time, minimizing the need for removal and reducing the risk of ocular irritation.

Non-Invasive Ocular Drug Delivery Methods

All commercially available ocular drugs used to treat retinal disorders are inherently invasive, with intravitreal injection being the most common method for delivering biopharmaceuticals directly to the retina (Barar et al., 2016). However, recent scientific research has increasingly focused on developing non-invasive topical therapies capable of effectively targeting posterior segment diseases. As discussed in the following section, these emerging approaches demonstrate the potential of topically applied formulations to treat retinal conditions, offering a more patient-friendly alternative to current invasive procedures. Topical administration of eye drops remains the most patient-friendly and minimally invasive method for ocular drug delivery. While this approach has shown significant success in treating anterior segment disorders, delivering therapeutically effective drug concentrations to the posterior segment through topical routes remains a major challenge (Sarkar et al., 2025). Several anatomical and physiological barriers, including the tear film, blood-retinal barriers, and the clearance mechanisms of the vitreous and aqueous humor, impede drug transport from the ocular surface to the back of the eye. Despite these challenges, recent advancements suggest that nanoformulations can be designed for administration via topical eye drops or intravitreal injections to effectively reach the posterior segment (Table 1).

Table 1. Therapeutic medications used topically in several preclinical models for disorders affecting the back of the eye

Compound	Formulation	Preclinical data	Reference
TG100801	Solution	Murine CNV model & edema in rat	(Rodrigues et al., 2018)
Pazopanib	Solution	Rat CNV model	(Yafai et al., 2011)
Acrizanib	Suspension	Murine CNV	(Espinosa-Heidmann et al., 2003)
Memantine	Solution	Drug levels in rabbit retina	(Prieto et al., 2012)
Dexamethasone	Iontophoresis	Drug levels in retina & vitreous of rabbit	(Perez et al., 2020)
28-kD single-chain antibody fragment	Sodium caprate	Drug levels in vitreous in rabbit	(Thiel et al., 2002)
Bevacizumab	Annexin A5-based liposomes	Drug levels in retina of rat & rabbit	(Urquhart & Eriksen, 2019)
Transforming growth factor beta 1	Annexin A5-based liposomes	Drug levels in vitreous in rabbit	(Urquhart & Eriksen, 2019)
Acidic fibroblast growth factor	CPP (TAT)	Ischemia- reperfusion model in rat	(Rodrigues et al., 2018)
Calpain inhibitory peptide	CPP (TAT)	Drug levels in rabbit retina	(Loescher et al., 2022)
Bevacizumab	CPP (R6)	Drug levels in vitreous and retina in rat and murine CNV model	(Wong & Wong, 2019)

Challenges in Translating Novel Ocular Drug Delivery Systems

Ocular drug delivery presents numerous challenges that hinder the effective treatment of eye diseases (Cabrera et al., 2019). Anatomical and physiological barriers, such as the blood-retinal and blood-aqueous barriers, significantly restrict drug access to intraocular tissues, especially for larger therapeutic molecules. These unique structural complexities of the eye require the development of innovative strategies to enhance drug bioavailability and therapeutic efficacy. Conventional methods, including eye drops and intravitreal injections, often suffer from poor patient compliance due to their invasive nature and the frequent dosing they require (Torkashvand et al., 2024). Although eye drops are easy to use, they are rapidly cleared from the ocular surface and have limited absorption, which reduces their effectiveness. To overcome these limitations, emerging technologies such as nanoparticles, microneedles, and sustained-release implants are being explored to improve drug retention, penetration, and overall delivery efficiency (Oh and Jung, 2022). However, despite promising preclinical results, the clinical translation of these advanced systems faces significant challenges, including regulatory complexities and the need for extensive safety and efficacy testing. Successfully addressing these issues is essential for advancing ocular therapeutics and improving patient outcomes in the treatment of both anterior and posterior eye diseases.

Novel Approaches to Ocular Drug Delivery

The Anterior Segment of Eye Drug Delivery Technologies

Punctum Plugs and Subconjunctival/Episcleral Implants

Punctum plugs are biocompatible ocular implants inserted into the tear ducts to block tear drainage, thereby increasing tear retention and enabling localized drug delivery. Also referred to as lacrimal plugs or occluders, these implants typically range in size from 2 to 5 millimeters (Renukuntla et al., 2013). As a noninvasive method, punctum plugs provide controlled drug delivery to the anterior segment of the eye. They can be made from biodegradable or non-biodegradable materials, with non-biodegradable types requiring removal once the treatment period ends. A significant advancement in this field is the development of the SmartPlug® punctal plug drug delivery system, which uses a thermosensitive hydrophobic acrylic polymer to treat dry eye syndrome. After insertion, this material changes from a hard, solid state to a soft, gel-like form, ensuring improved comfort and fit. Besides punctum plugs, ocular implants designed for regulated and sustained drug release can also be surgically placed in areas such as the aqueous chamber, episcleral space, or sub-conjunctival region.

Cul-De-Sac Implants and Drug Eluting Contact Lenses

A cul-de-sac refers to a pocket-like depression formed where the palpebral conjunctiva meets the bulbar conjunctiva in the upper or lower eyelid. This anatomical feature serves as the site for inserting cul-de-sac implants, such as Lacrisert® and Ocusert®, which are designed to deliver medications to the anterior segment of the eye (Gote et al., 2019). These implants are commonly used in patients with moderate to severe dry eye and provide sustained release of therapeutic agents. In addition to implants, light-transparent corneal dressings and drug-eluting contact lenses offer advanced solutions for treating anterior ocular disorders. These systems act as drug reservoirs, releasing medications into the post-lens tear film over extended periods. Unlike conventional eye drops, drug-loaded soft contact lenses improve drug penetration through the corneal epithelium and maintain prolonged drug release, thereby enhancing therapeutic effectiveness. Because contact lenses remain in close contact with the cornea, they significantly increase drug bioavailability compared to other noninvasive delivery methods (Franco and De Marco, 2021). Certain soft contact lenses have been shown to extend the delivery of antifungal drugs for up to 21 days. Clinical trials have also demonstrated the safety and efficacy of drug-eluting contact lenses in treating glaucoma, highlighting their superiority over standard eye drops for sustained ocular drug delivery.

Ocular Iontophoresis

Ocular iontophoresis is an active and non-invasive drug delivery technique that uses mild electric currents to enhance drug penetration across ocular barriers. This method facilitates drug delivery through a combination of electrophoresis, which is the movement of charged drug molecules; electroporation, which involves temporary structural changes and pore formation in ocular tissues induced by electric fields; and electro-osmosis, the movement of fluid and neutral molecules (Myles et al., 2005). Unlike invasive surgical techniques, iontophoresis provides a safer and more comfortable option for patients. Drugs can be administered via either the transcorneal or transscleral routes, depending on the targeted area of the eye. The transscleral pathway is especially beneficial for posterior segment diseases because it bypasses the anatomical barriers of the anterior segment, allowing therapeutic agents to reach the back of the eye. Table 2 presents an overview of ocular drug delivery systems currently under clinical investigation for treating anterior segment disorders.

Posterior Segment Ocular Drug Delivery Technologies

To manage disorders affecting the posterior segment of the eye, such as posterior uveitis, central retinal vein occlusion (CRVO), age-related macular degeneration (AMD), diabetic macular edema (DME), and diabetic retinopathy (DR), clinicians are increasingly turning to advanced drug delivery technologies, including ocular implants, to sustain and prolong drug release (Shah et al., 2010). These innovative systems are designed to overcome the limitations of conventional therapies by delivering precise, long-term medication directly to the retina and surrounding tissues. This section provides an overview of intravitreal ocular implants that are currently under clinical evaluation, as well as those already approved and available on the market for therapeutic use.

Durasert™ and Novadur™ Drug Delivery Technology System

The Durasert™ technology system, developed by pSivida Corp. (MA, USA), is designed to deliver drugs at predetermined rates depending on the type of implant used. Drug release can range from several days to several years, providing tailored therapeutic durations. This system consists of a polymeric outer layer and a drug-containing core, with the rate of drug release controlled by the permeability of the polymer coating (Ullrich et al., 2014). One of the earliest applications of this technology was Vitrasert®, the first intravitreal drug delivery system approved for treating cytomegalovirus (CMV) retinitis, which contained the antiviral agent ganciclovir. The Durasert™ technology also includes a small diffusion hole in the implant that enables consistent drug release over extended periods, such as six to eight months (Port et al., 2017). Another application of this platform is Retisert®, a surgically implanted intravitreal device that delivers corticosteroids to treat chronic non-infectious posterior uveitis. This condition, characterized by inflammation of the choroidal capillaries (choroiditis), can lead to severe complications, including optic nerve damage and irreversible vision loss if left untreated (Port et al., 2017).

Novadur™ and I-vation™ TA: Drug Delivery Technologies

The Novadur™ drug delivery system incorporates therapeutic agents within a PLGA (poly lactic-co-glycolic acid) polymer matrix. Upon contact with the vitreous humor, this biodegradable and biocompatible polymer gradually breaks down into lactic and glycolic acid, allowing for controlled and sustained drug release over time (Gote & Pal, 2019). In contrast, the I-vation™ TA system consists of a titanium helical coil implant coated with a non-biodegradable polymer, designed to provide extended drug delivery without degradation of the implant itself (Mariz, 2018). In a Phase I clinical trial evaluating the safety and preliminary efficacy of I-vation™ TA, 31 patients with diabetic macular edema (DME) participated. The study observed a slight increase in intraocular pressure, indicating that the implant was generally well tolerated. Additionally, treatment with the I-vation™ TA implant resulted in a reduction in macular thickness from baseline, suggesting potential therapeutic benefits in managing DME.

Using Encapsulated Cell Technology with Hollow Microneedles and Microsurgical Cannulas to Administer Suprachoroidal Drugs

Renexus® (NT-501) is an ocular implant containing genetically modified human retinal pigment epithelial (RPE) cells transfected with a plasmid to produce ciliary neurotrophic factor (CNTF) using encapsulated cell technology (Ibeanu et al., 2020). This innovative system is currently undergoing Phase III clinical trials for the treatment of retinitis pigmentosa, glaucoma, and dry age-related macular degeneration (AMD) (ClinicalTrials.gov Identifier: NCT03316300). The implant features a hollow capsule composed of a polymeric matrix that houses the CNTF-secreting cells. The capsule is semipermeable, allowing therapeutic proteins to diffuse through while preventing immune cell infiltration, thereby minimizing immune response. Once implanted in the pars plana, the device is anchored to the sclera, ensuring localized and sustained drug delivery. This approach provides continuous release of therapeutic agents directly to the posterior segment, improving drug concentrations at the target tissue over extended periods. Furthermore, the suprachoroidal space has shown considerable promise for targeted drug delivery, as anatomical studies reveal that therapeutic agents can effectively disperse across this region following administration (Gabai et al., 2023). A variety of advanced ocular drug delivery platforms, including Renexus® and others, are currently being evaluated in clinical trials to address diseases of the posterior segment, as outlined in Table 2.

Challenges and Opportunities for Ocular Medication Administration Systems

Developing new technologies that provide safer and more effective treatments for ocular disorders remains a significant challenge due to the limitations of current ocular drug delivery systems (Giri et al., 2024).

Table 2. Currently available ocular drug delivery systems in clinical trials for treatment of posterior segment disorders

Drug (brand name)	Mode of administration	Excipient controlling release characteristic of drug	Target indication	Developmental stage	Clinical trial No.
Dexamethas one (Ozudex®)	Intravitreal implant	PLGA (Novadur®)	Macular edema posterior uveitis	Launched	NCT01427751
Ganciclovir (Vitrasert®)	Intravitreal implant	PVA/EVA	CMV retinitis	Launched	NCT00000135
Fluocinolone acetonide (Retisert®)	Intravitreal implant	PVA	Posterior uveitis	Launched	NCT00570830
Verteporfin (Visudyne®)	IV injection	Liposome	Wet AMD	Launched	NCT00242580
Dexamethasone (Dexycu®)	Intravitreal implant	Acetyl triethyl citrate	Postoperative inflammation	Launched	NCT02547623
Difluprednate (Durezol®)	Eye drops	Emulsion	DME	Off-label	NCT00429923

A major drawback of conventional eye drops is their poor pharmacological bioavailability, especially in reaching targeted intraocular tissues. For instance, nanoformulations such as liposomes, dendrimers, nanomicelles, nanoparticles, nanowafers, and microneedles can enhance drug delivery to the conjunctiva and cornea by enabling controlled and sustained release, thereby improving drug absorption in anterior segment tissues (Soni et al., 2019).

However, for diseases affecting the posterior segment of the eye, all currently available treatment options are invasive. Intravitreal injections, though commonly used, carry significant risks including pain, hemorrhage, and retinal detachment (Aiello et al., 2004). Developing a noninvasive, sustained drug delivery system for the posterior eye remains a key challenge for ocular drug delivery experts, requiring innovations that can overcome anatomical barriers while maintaining therapeutic efficacy and patient comfort (Dosmar et al., 2022).

Conclusion

There are substantial challenges to the large-scale production and dissemination of knowledge regarding novel ocular drug delivery systems. Despite these obstacles, nanotechnology holds tremendous therapeutic potential for treating a wide range of eye disorders and may eventually replace conventional ocular medications. The advancement and modernization of both new and existing formulations increasingly rely on integrated and concurrent processes. Researchers aiming to target specific ocular tissues must navigate numerous anatomical and physiological barriers that hinder effective drug delivery. Although significant progress has been made, the quest for safe, effective, and biocompatible drug delivery platforms continues. Current efforts focus on enhancing the in vivo performance of traditional formulations, particularly through nanotechnology-driven innovations. Large-scale formulation development is actively underway for various nanocarrier systems, including dendrimers, solid lipid nanoparticles, micelles, lipid-based vesicles, micro- and nano-emulsions, nanosuspensions, microneedles, and polymeric nanoparticles. Compared to conventional methods, these nanotechnology-based systems offer several advantages such as reduced dosing frequency, minimized dose-dependent side effects, improved drug retention, and controlled release profiles. They also help mitigate fluctuations in plasma drug concentrations. Recent advancements have demonstrated the potential of nanotechnology to reduce toxicity while enhancing targeted drug accumulation in ocular tissues. As scientific innovation continues, the development of nanomedicine in clinical settings appears increasingly feasible, promising extended precorneal residence time, improved bioavailability, and less frequent injections, marking a significant step forward in ocular therapeutics.

References

Aiello, L. P., Brucker, A. J., Chang, S., Emmett T Cunningham, J., D'amico, D. J., Harry W Flynn, J., Grillone, L. R., Hutcherson, S., Liebmann, J. M., & O'brien, T. P. (2004). Evolving guidelines for intravitreous injections. *Retina*, *24*(5), S3-S19.

Akhter, M. H., Ahmad, I., Alshahrani, M. Y., Al-Harbi, A. I., Khalilullah, H., Afzal, O., Altamimi, A. S., Najib Ullah, S. N. M., Ojha, A., & Karim, S. (2022). Drug delivery challenges and current progress in nanocarrier-based ocular therapeutic system. *Gels*, *8*(2), 82.

Barar, J., Aghanejad, A., Fathi, M., & Omidi, Y. (2016). Advanced drug delivery and targeting technologies for the ocular diseases. *BioImpacts: BI*, *6*(1), 49.

Cabrera, F. J., Wang, D. C., Reddy, K., Acharya, G., & Shin, C. S. (2019). Challenges and opportunities for drug delivery to the posterior of the eye. *Drug discovery today*, *24*(8), 1679-1684.

Cholkar, K., Dasari, S. R., Pal, D., & Mitra, A. K. (2013). Eye: Anatomy, physiology and barriers to drug delivery. In *Ocular transporters and receptors* (pp. 1-36). Elsevier.

Desai, A., Shukla, M., Maulvi, F., & Ranch, K. (2019). Ophthalmic and otic drug administration: novel approaches and challenges. *Novel Drug Delivery Technologies: Innovative Strategies for Drug Re-positioning*, 335-381.

Desai, D., Wang, J., Wen, H., Li, X., & Timmins, P. (2013). Formulation design, challenges, and development considerations for fixed dose combination (FDC) of oral solid dosage forms. *Pharmaceutical development and technology*, *18*(6), 1265-1276.

Dorcas, W., Emilliene, E., Estella, T. F., Joseph, E. N., Kovin, N., Jaggernath, J., Timothy, W., Ngo, N. V., Therese, A. O. M., & Ntungwen, F. C. (2019). An overview of herbal traditional eye care practices and the development of eye health promotion strategies in Cameroon. *Journal of Advances in Medical and Pharmaceutical Sciences,*, *20*(4), 1-16.

Dosmar, E., Walsh, J., Doyel, M., Bussett, K., Oladipupo, A., Amer, S., & Goebel, K. (2022). Targeting ocular drug delivery: an examination of local anatomy and current approaches. *Bioengineering*, *9*(1), 41.

Dubey, R. K., Kumawat, B. K., Patel, V. K., Prajapati, V. K., & Lal, G. (2021). A systemic review on: ophthalmic drug delivery system and its future perspective challenges. *World Journal of Pharmaceutical Research*, 10(13), 480-497.

Espinosa-Heidmann, D. G., Suner, I. J., Hernandez, E. P., Monroy, D., Csaky, K. G., & Cousins, S. W. (2003). Macrophage depletion diminishes lesion size and severity in experimental choroidal neovascularization. *Investigative ophthalmology & visual science*, *44*(8), 3586-3592.

Franco, P., & De Marco, I. (2021). Contact lenses as ophthalmic drug delivery systems: A review. *Polymers*, *13*(7), 1102.

Gabai, A., Zeppieri, M., Finocchio, L., & Salati, C. (2023). Innovative strategies for drug delivery to the ocular posterior segment. *Pharmaceutics*, *15*(7), 1862.

Giri, B. R., Jakka, D., Sandoval, M. A., Kulkarni, V. R., & Bao, Q. (2024). Advancements in Ocular Therapy: A Review of Emerging Drug Delivery Approaches and Pharmaceutical Technologies. *Pharmaceutics*, *16*(10), 1325.

Gote, V., & Pal, D. (2019). Ocular implants in the clinic and under clinical investigation for ocular disorders. *EC Ophthalmology, 10*(8), 660-666.

Gote, V., Sikder, S., Sicotte, J., & Pal, D. (2019). Ocular drug delivery: present innovations and future challenges. *Journal of Pharmacology and Experimental Therapeutics*, *370*(3), 602-624.

Gower, N. J., Barry, R. J., Edmunds, M. R., Titcomb, L. C., & Denniston, A. K. (2016). Drug discovery in ophthalmology: past success, present challenges, and future opportunities. *BMC Ophthalmology*, *16*, 11.

Ibeanu, N., Egbu, R., Onyekuru, L., Javaheri, H., Tee Khaw, P., R. Williams, G., Brocchini, S., & Awwad, S. (2020). Injectables and depots to prolong drug action of proteins and peptides. *Pharmaceutics*, *12*(10), 999.

Kaushik, A., Mazumder, R., Padhi, S., Mazumder, A., Budhori, R., & Manormaa, S. (2022). Novel Approaches in Ocular Drug Delivery-A Revolution. *International Journal of Applied Pharmaceutics*, *14*(3), 1-11.

Koehl, K. L. (2018). *Optimizing Ocular Gene Therapy for Long Term Intraocular Pressure Control in ADAMTS10-Open Angle Glaucoma*. Michigan State University.

Loescher, M., Seiz, C., Hurst, J., & Schnichels, S. (2022). Topical drug delivery to the posterior segment of the eye. *Pharmaceutics*, *14*(1), 134.

López-Cano, J. J., González-Cela-Casamayor, M. A., Andrés-Guerrero, V., Herrero-Vanrell, R., & Molina-Martínez, I. T. (2021). Liposomes as vehicles for topical ophthalmic drug delivery and ocular surface protection. *Expert opinion on drug delivery*, *18*(7), 819-847.

Malkoch, M., & Gallego, S. G. (2020). *Dendrimer Chemistry: Synthetic approaches towards complex architectures*. Royal Society of Chemistry.

Mansuri, S., Kesharwani, P., Jain, K., Tekade, R. K., & Jain, N. (2016). Mucoadhesion: A promising approach in drug delivery system. *Reactive and functional polymers*, *100*, 151-172.

Marcano, D. C., Shin, C. S., Lee, B., Isenhart, L. C., Liu, X., Li, F., Jester, J. V., Pflugfelder, S. C., Simpson, J., & Acharya, G. (2016). Synergistic cysteamine delivery nanowafer as an efficacious treatment modality for corneal cystinosis. *Molecular Pharmaceutics*, *13*(10), 3468-3477.

Mariz, M. J. M. L. M. (2018). *Ocular drug delivery insert technology: from polymer synthesis to in vitro testing* 00500:: Universidade de Coimbra].

Molokhia, S. A., Thomas, S. C., Garff, K. J., Mandell, K. J., & Wirostko, B. M. (2013). Anterior eye segment drug delivery systems: current treatments and future challenges. *Journal of Ocular Pharmacology and Therapeutics*, *29*(2), 92-105.

Myles, M. E., Neumann, D. M., & Hill, J. M. (2005). Recent progress in ocular drug delivery for posterior segment disease: emphasis on transscleral iontophoresis. *Advanced drug delivery reviews*, *57*(14), 2063-2079.

Oh, S. J., & Jung, J. H. (2022). Sustainable drug release using nanoparticle encapsulated microneedles. *Chemistry–An Asian Journal*, *17*(16), e202200333.

Perez, V. L., Wirostko, B., Korenfeld, M., From, S., & Raizman, M. (2020). Ophthalmic drug delivery using iontophoresis: recent clinical applications. *Journal of Ocular Pharmacology and Therapeutics*, *36*(2), 75-87.

Port, A. D., Orlin, A., Kiss, S., Patel, S., D'Amico, D. J., & Gupta, M. P. (2017). Cytomegalovirus retinitis: a review. *Journal of Ocular Pharmacology and Therapeutics*, *33*(4), 224-234.

Prieto, E., Puente, B., Uixera, A., Garcia de Jalon, J., Perez, S., Pablo, L., Irache, J., Garcia, M., & Bregante, M. (2012). Gantrez AN nanoparticles for ocular delivery of memantine: in vitro release evaluation in albino rabbits. *Ophthalmic research*, *48*(3), 109-117.

Qi, Q., Wei, Y., Zhang, X., Guan, J., & Mao, S. (2023). Challenges and strategies for ocular posterior diseases therapy via non-invasive advanced drug delivery. *Journal of Controlled Release*, *361*, 191-211.

Rahić, O., Tucak, A., Omerović, N., Sirbubalo, M., Hindija, L., Hadžiabdić, J., & Vranić, E. (2020). Novel drug delivery systems fighting glaucoma: Formulation obstacles and solutions. *Pharmaceutics*, *13*(1), 28.

Ramos, M. F., Brassard, J., & Masli, S. (2021). Immunology and pathology in ocular drug development. *Toxicologic Pathology*, *49*(3), 483-504.

Raut, A. K., Mohapatra, S., SiddiquI, G., Rajak, S. K., Sonar, R., Basu, S., Joshi, V., & Singh, V. (2024). The Human Cornea: Unraveling Its Structural, Chemical, and Biochemical Complexities. *Chemistry & Biodiversity*, 2024, e202402224.

Renukuntla, J., Shah, S. J., Patel, M., Vadlapudi, A. D., & Mitra, A. K. (2013). Drug Delivery to Anterior Segment of the Eye. *Treatise on Ocular Drug Delivery*, *1*, 42.

Rodrigues, G. A., Lutz, D., Shen, J., Yuan, X., Shen, H., Cunningham, J., & Rivers, H. M. (2018). Topical drug delivery to the posterior segment of the eye: addressing the challenge of preclinical to clinical translation. *Pharmaceutical research*, *35*(12), 245.

Rojekar, S., Parit, S., Gholap, A. D., Manchare, A., Nangare, S. N., Hatvate, N., Sugandhi, V. V., Paudel, K. R., & Ingle, R. G. (2024). Revolutionizing Eye Care: Exploring the Potential of Microneedle Drug Delivery. *Pharmaceutics*, *16*(11), 1398.

Sarkar, T., Gogoi, N. R., Jana, B. K., & Mazumder, B. (2025). Formulation Advances in Posterior Segment Ocular Drug Delivery. *Journal of Ocular Pharmacology and Therapeutics*, 41(3), 101-130.

Shah, S. S., Denham, L. V., Elison, J. R., Bhattacharjee, P. S., Clement, C., Huq, T., & Hill, J. M. (2010). Drug delivery to the posterior segment of the eye for pharmacologic therapy. *Expert review of ophthalmology*, *5*(1), 75-93.

Soni, V., Pandey, V., Tiwari, R., Asati, S., & Tekade, R. K. (2019). Design and evaluation of ophthalmic delivery formulations. In *Basic fundamentals of drug delivery* (pp. 473-538). Elsevier.

Stjernschantz, J., & Astin, M. (2019). Anatomy and physiology of the eye. Physiological aspects of ocular drug therapy. In *Biopharmaceutics of ocular drug delivery* (pp. 1-25). CRC Press.

Tawfik, S. M., Azizov, S., Elmasry, M. R., Sharipov, M., & Lee, Y.-I. (2020). Recent advances in nanomicelles delivery systems. *Nanomaterials*, *11*(1), 70.

Thiel, M., Coster, D., Standfield, S., Brereton, H., Mavrangelos, C., Zola, H., Taylor, S., Yusim, A., & Williams, K. (2002). Penetration of engineered antibody fragments into the eye. *Clinical & Experimental Immunology*, *128*(1), 67-74.

Torkashvand, A., Izadian, A., & Hajrasouliha, A. (2024). Advances in ophthalmic therapeutic delivery: A comprehensive overview of present and future directions. *Survey of ophthalmology*, 69(6), 967-983.

Tunic, F. (2011). Lacrimal and Nasolacrimal Systems. *Sheep & Goat Medicine-E-Book: Sheep & Goat Medicine-E-Book*, 406.

Ullrich, F., Fusco, S., Chatzipirpiridis, G., Pané, S., & Nelson, B. J. (2014). Recent progress in magnetically actuated microrobotics for ophthalmic therapies. *European Ophthalmic Review*, 8(2), 120.

Urquhart, A. J., & Eriksen, A. Z. (2019). Recent developments in liposomal drug delivery systems for the treatment of retinal diseases. *Drug Discovery Today*, *24*(8), 1660-1668.

Wadhwa, S., Paliwal, R., Paliwal, S. R., & Vyas, S. (2009). Nanocarriers in ocular drug delivery: an update review. *Current Pharmaceutical Design*, *15*(23), 2724-2750.

Wong, C. W., & Wong, T. T. (2019). Posterior segment drug delivery for the treatment of exudative age-related macular degeneration and diabetic macular oedema. *British Journal of Ophthalmology*, *103*(10), 1356-1360.

Yafai, Y., Yang, X. M., Niemeyer, M., Nishiwaki, A., Lange, J., Wiedemann, P., King, A. G., Yasukawa, T., & Eichler, W. (2011). Anti-angiogenic effects of the receptor tyrosine kinase inhibitor, pazopanib, on choroidal neovascularization in rats. *European Journal of Pharmacology*, *666*(1-3), 12-18.

Chapter 22

Future Trends in Nanotechnology-Based Ocular Drug Delivery

Abhishek Chandola[1]
Deepika Rani[1]
Ujjwal Nautiyal[3]
and Ranjit Singh[1,*]
[1]AVIPS, Shobhit University, Gangoh, Saharanpur, Uttar Pradesh, India
[3]Himalayan School of Pharmaceutical Sciences, Swami Rama Himalayan University, Swami Rama Nagar, Jollygrant, Uttarakhand, India

Abstract

Ocular drug delivery plays a vital role in managing a variety of eye diseases, including glaucoma, macular degeneration, and diabetic retinopathy. Despite its importance, achieving effective drug delivery to ocular tissues remains challenging due to the eye's complex anatomy, low drug bioavailability, rapid precorneal clearance, and risks of systemic or local side effects. Conventional delivery methods such as eye drops and intravitreal injections often lack the ability to provide sustained and targeted drug release, which limits their therapeutic effectiveness. Recent advancements in ophthalmic nanotechnology-based drug delivery systems offer promising solutions to overcome these obstacles. The design of nanoparticles, nanocarriers, and other nanoscale systems aims to improve drug penetration, enhance bioavailability, enable controlled and prolonged drug release, and reduce adverse effects. Additionally, innovative technologies including smart nanoparticles, gene delivery platforms, and multifunctional nanomaterials are emerging

* Corresponding Author's Email: ranjitsps@gmail.com

In: Advances in Nanotechnology for Ocular Drug Delivery
Editors: Anjali Sharma, Ajmer Singh Grewal, Geeta Deswal et al.
ISBN: 979-8-89530-814-1
© 2025 Nova Science Publishers, Inc.

as powerful tools to enable more effective, non-invasive, and personalized ocular therapies. This chapter provides a comprehensive review of current challenges in ocular drug delivery and highlights the transformative potential of nanotechnology in this field.

Keywords: nanotechnology, ocular drug delivery, nanocarriers, dendrimers, ophthalmic nanomedicine, smart nanoparticles

Abbreviations

AAVs: Adeno-Associated Viruses
AMD: Age Related Macular Degradation
CNV: Corneal Neovascularization
EMA: European Medical Agency
FDA: Food and Drug Administration
G/W: Gel in Water
HA: Hyaluronic-Acid
NE: Nanoemulsion
PAMAM: Poly(amidoamines)
PEG: Polyethylene Glycol
PVA: Polyvinyl Alcohol
RPE: Retinal Pigment Epithelium

Introduction

Ocular drug delivery involves administering medications to treat various eye diseases, yet it faces significant challenges due to natural protective mechanisms such as tear turnover, drainage pathways, and anatomical barriers including the blood-retina barrier. An ideal ocular drug delivery system must overcome these obstacles to deliver drugs precisely to targeted sites within the eye. To this end, novel strategies have been developed to enhance drug absorption, prolong release, and enable targeted delivery (Ghate & Edelhauser, 2006). One major challenge is effectively treating conditions like age-related macular degeneration. While eye drops provide a convenient means of delivering medication to the anterior segment, reaching deeper structures such as the retina is more complex and often requires advanced delivery systems like intravitreal implants to ensure adequate drug concentration and sustained therapeutic effects (Patel, 2013). Conventional delivery methods, including topical applications (eye drops and ointments), systemic administration, and periocular injections, are primarily effective for anterior segment disorders but suffer from poor bioavailability and rapid clearance due to continuous tear production and drainage. Moreover, the blood-retina and blood-aqueous barriers severely limit drug penetration into the posterior segment, further reducing treatment efficacy (Maharjan et al., 2019; Liu et al., 2023). These challenges have prompted the development of innovative drug delivery systems aimed at improving ocular bioavailability and overcoming physiological and biological barriers.

Contribution and Benefits of Nanotechnology in Overcoming Conventional Limitations

Ocular drug delivery using nanotechnology involves the precise administration of therapeutic agents to the eye through nanoscale carriers or systems (Liu et al., 2023). As previously discussed, conventional ocular drug delivery methods face numerous limitations. Nanotechnology offers several advantages that help overcome these challenges:

Enhanced Bioavailability

Eye drops are the most commonly used dosage form in ocular therapy. However, they generally exhibit poor bioavailability, often less than five percent, due to limited pre-corneal retention and insufficient penetration into ocular tissues. Consequently, frequent dosing is necessary to maintain therapeutic drug levels on the ocular surface (Xu et al., 2013). Research by Ye et al. (2013) demonstrated that the use of prodrugs combined with nanotechnology can significantly enhance both drug retention time and bioavailability.

Prolonged-Release

Conventional ocular therapies often require frequent application of eye drops or ointments. To overcome this limitation, researchers have developed nanotechnology-based delivery systems using materials such as hydrogels, in situ-gelling systems, liposomes, niosomes, polymeric inserts, and implants. These advanced formulations improve bioavailability and enable sustained drug release, resulting in enhanced therapeutic outcomes (Gaudana et al., 2009).

Targeted Delivery

Retinal angiogenesis and degenerative diseases are among the leading causes of vision impairment worldwide. Delivering drugs to the posterior segment of the eye is particularly challenging due to anatomical barriers. Tight junctions in ocular tissues such as the retina and cornea restrict drug diffusion, and systemic delivery is further hindered by the blood-retinal barrier. Nanotechnology enables targeted drug delivery to specific ocular tissues, thereby minimizing systemic exposure and enhancing efficacy (Suri et al., 2020).

Versatility

Nanocarriers can deliver a wide range of therapeutic agents, including hydrophobic and hydrophilic drugs, proteins, and nucleic acids. Their size, surface charge, and composition make nanoparticles and liposomes highly versatile drug delivery systems. These carriers enhance drug solubility and stability, increase bioavailability, and protect drugs from degradation (Lu et al., 2021). Hydrophilic drugs can be efficiently encapsulated within nanocarriers to improve delivery (Kumari et al., 2014). Lipid-based nanoparticles, such as liposomes and solid lipid nanoparticles, have shown efficacy in delivering RNA-based therapeutics, as evidenced by mRNA COVID-19 vaccines (Cullis & Hope, 2017). Furthermore, polymeric nanocarriers like chitosan and polyethylene glycol enhance protein and peptide stability, facilitate cellular uptake, and enable targeted delivery by shielding biomolecules from enzymatic degradation (Danhier et al., 2012).

Advanced Nanocarriers

Nanomicelles

Nanomicelles are spherical nanoparticles formed by the spontaneous assembly of amphiphilic molecules in aqueous environments. These molecules contain both hydrophilic and hydrophobic segments, enabling the encapsulation of lipophilic drugs within their hydrophobic core. Typically ranging from 10 to 100 nanometers in size, nanomicelles can be designed for prolonged drug delivery to the ocular surface. Their structure protects drugs from degradation and extends circulation time. Additionally, their unique configuration improves adhesion to the eye's surface, thereby increasing residence time and enhancing bioavailability (Maharjan et al., 2019).

Nanosuspension

A nanosuspension is a colloidal dispersion where drug particles are reduced to nanoscale size and stabilized using polymers or surfactants. These can be prepared by aqueous or non-aqueous methods and are mainly employed to improve the solubility of hydrophobic drugs. In ocular drug delivery, nanosuspensions enhance drug solubility and increase residence time on the eye's surface. They offer promising solutions for treating conditions such as glaucoma, macular degeneration, diabetic retinopathy, and uveitis by improving both delivery efficiency and patient comfort (Liu et al., 2023).

Nanoemulsions and Microemulsions

Nanoemulsions (NEs) consist of two immiscible liquids, typically oil and water, stabilized by amphiphilic surfactants. Their droplet sizes generally range from 20 to 500 nanometers, with an average size around 100 nanometers. These formulations are transparent and stable. Microemulsion-based systems have demonstrated clinical potential by reducing the dosing frequency of drugs such as pilocarpine from four times to twice daily. Due to their excellent permeability across corneal and conjunctival barriers, nanoemulsions are especially effective for treating anterior segment disorders (Liu et al., 2023).

Dendrimers

Posterior segment eye diseases are important targets for nanotechnology-based therapies; however, delivering drugs topically to these regions remains a significant challenge due to the long distances and multiple physiological barriers drugs must overcome to reach the retina and choroid (Ghate & Edelhauser, 2006; Kambhampati & Kannan, 2013). Poly(amidoamine) dendrimers (PAMAMs) have emerged as promising nanocarriers due to their highly branched, three-dimensional structure, which provides a high density of functional groups for enhanced drug loading, targeted delivery, and improved biocompatibility. Their amphiphilic properties enable effective encapsulation and transport of a variety of ocular therapeutics, making them suitable candidates for treating posterior segment diseases (Liu et al., 2012).

Liposomes and Niosomes

Liposomes are bilayer lipid vesicles with amphiphilic properties, allowing encapsulation of both hydrophilic and hydrophobic drugs. When modified with polyethylene glycol (PEG), their circulation time increases and their ability to cross the blood-retina barrier improves. This makes them promising candidates for subretinal drug delivery in ocular therapy (Rusu et al., 2022).

Nanowafers

Nanowafers are small, transparent discs embedded with drug-loaded nanoreservoirs designed for direct application to the ocular surface using a fingertip. These innovative devices enable controlled and sustained drug release, enhancing contact time with the eye and improving drug penetration, particularly in the anterior segment. Moreover, nanowafers provide a protective polymer membrane that supports corneal healing, making them beneficial in treating conditions such as corneal neovascularization (CNV) and dry eye disease. Typically fabricated from biodegradable and biocompatible polymers, nanowafers gradually degrade, ensuring sustained therapeutic effects while minimizing irritation (Gote et al., 2019).

Hydrogels

Hydrogels are biocompatible, three-dimensional polymeric networks that enable the sustained and controlled release of drugs, helping to maintain stable therapeutic concentrations while reducing the need for frequent dosing. This property makes them particularly suitable for managing chronic ocular conditions such as age-related macular degeneration and diabetic retinopathy. Incorporating nanoparticles, nanomicelles, or microneedles within hydrogels can further refine drug release kinetics and prolong residence time on the ocular surface or within targeted tissues. Hyaluronic acid (HA)-based hydrogels have demonstrated significant potential in regenerative ophthalmology, especially for supporting retinal progenitor cell transplantation in the subretinal space. To enhance hydrogel stability, mechanical properties, and shelf-life, synthetic polymers such as PEG and polyvinyl alcohol (PVA) are commonly integrated into the formulations (Liu et al., 2023; Rusu et al., 2022).

Stimuli-Responsive and Smart Drug Delivery Systems

Emerging nanocarriers are increasingly designed with stimuli-responsive capabilities to enhance precision and control in ocular drug delivery (Guidi et al., 2024). These smart systems can respond to external triggers such as pH changes, temperature fluctuations, light exposure, and ultrasound waves. For instance, light-sensitive nanoparticles release their drug payload when activated by specific wavelengths, allowing on-demand and localized therapy. Ultrasound-triggered liposomes utilize ultrasonic energy to prompt targeted drug release, thereby improving bioavailability while minimizing systemic side effects and enhancing treatment specificity (Long et al., 2021).

Gene and RNA-Based Therapies Using Nanotechnology

Gene therapy for inherited retinal diseases has advanced significantly with the use of viral vectors such as adeno-associated viruses (AAVs). These vectors enable targeted delivery to photoreceptor and retinal pigment epithelial (RPE) cells, minimizing immune responses and reducing the required dosages. Animal studies have demonstrated long-term gene expression with positive clinical outcomes (Drag et al., 2023). Nanotechnology offers non-viral alternatives—custom-engineered nanoparticles protect genetic material, enhance targeting, and reduce immunogenicity. Lipid-based nanoparticles have been used to deliver genes for conditions such as Leber congenital amaurosis and retinitis pigmentosa. Surface modifications further improve targeting and uptake by retinal cells. These platforms hold great potential to provide safer and more effective gene therapy solutions (Patel et al., 2019; Wang et al., 2018).

Integration of Nanotechnology with Digital Health

Smart devices such as wearable sensors and nanotechnology-integrated contact lenses are opening new avenues for ocular drug delivery. These devices can monitor intraocular pressure, detect disease biomarkers, and release drugs in response to real-time physiological signals. For example, contact lenses embedded with glucose-responsive nanocarriers have been developed to release insulin for diabetic patients, representing a significant step toward personalized ocular therapy (Kim et al., 2022).

Long-Acting and Sustained Drug Delivery

Biodegradable implants embedded with nanocarriers are being investigated for long-acting ocular drug delivery. These systems can release drugs over several months, making them well-suited for chronic conditions such as glaucoma and diabetic retinopathy. Hydrogel-based nanocarrier platforms also demonstrate potential in maintaining sustained therapeutic levels, thereby reducing the need for frequent administration (Han et al., 2023).

Challenges Faced While Using Nanotechnology Based Ocular Drug Delivery

Manufacturing Challenges

Scaling up nanocarrier production poses significant challenges, particularly in maintaining consistency, stability, and efficacy. The fabrication process requires precise control over parameters such as particle size, surface charge, and drug-loading capacity, often necessitating advanced technologies like microfluidics and 3D printing. Ensuring batch-to-batch reproducibility remains difficult, especially when balancing cost-efficiency. Additionally, there is an urgent need for innovative sterilization methods that preserve the physicochemical properties of nanocarriers. For instance, manufacturing liposomal nanocarriers for ocular drug delivery must be carefully optimized to prevent aggregation and ensure uniform drug loading. Techniques such as high-pressure homogenization have been explored to enhance formulation consistency (Gaudana et al., 2009).

Regulatory Challenges

Nanotechnology-based ocular drugs encounter complex regulatory challenges due to their novel mechanisms of action and the absence of standardized evaluation protocols. Regulatory agencies such as the FDA and EMA demand comprehensive data on safety and efficacy, which significantly extends development time and increases costs. Furthermore, the lack of universally accepted guidelines for nanomedicines often results in delays in approval and market entry (Ciolino et al., 2009). For instance, the ophthalmic nanodrug Restasis® (cyclosporine-loaded nanoemulsion) underwent extensive regulatory scrutiny regarding its stability, ocular bioavailability, and long-term safety before receiving approval for commercialization.

Commercial Challenges

The commercialization of nanotechnology-based ocular therapies faces multiple obstacles, including high development costs, intellectual property complexities, and market uncertainty. Pharmaceutical companies often hesitate to invest due to uncertain returns on investment. Additionally, both clinicians and patients must be convinced of these therapies' superior efficacy, safety, and cost-effectiveness compared to conventional treatments. Long-acting intravitreal nanocarriers for age-related macular degeneration (AMD) have experienced commercial setbacks, largely because of the high costs associated with clinical trials and limited funding opportunities. Companies developing biodegradable nanocarriers for AMD have struggled to secure substantial financial support and strategic partnerships necessary to advance their products (Yasukawa et al., 2004).

Biological and Toxicological Challenges

Nanocarriers carry potential biological risks such as cytotoxicity, immunogenicity, and unintended biodistribution. While their ability to penetrate biological barriers is advantageous, their small size may lead to accumulation in non-target tissues, raising concerns about long-term safety. A thorough understanding of interactions between nanocarriers and ocular tissues, the tear film, and the immune system is essential for developing safe ocular nanomedicines. For example, polymeric nanoparticles used in ocular drug delivery have been shown to cause transient inflammation due to immune activation. To minimize these adverse effects, surface modifications including PEGylation are being investigated to enhance biocompatibility and reduce immune responses (del Amo and Urtti, 2008).

Patient Compliance

Patient compliance remains a challenge even though nanotechnology-based ocular drug delivery systems provide improved therapeutic outcomes. Certain formulations, such as ocular implants or injectable nanosuspensions, require specialized administration techniques that may be inconvenient or intimidating for patients. Public acceptance of these advanced systems depends on awareness and perceptions of nanotechnology in healthcare. While nanotechnology-integrated contact lenses show great promise, their adoption is limited by concerns related to comfort, cost, and ease of use (Hughes et al., 2005).

Future Trends of Nanotechnology Based Ocular Drug Delivery

Future advancements in nanotechnology are poised to revolutionize ocular drug delivery by improving therapeutic efficacy, enhancing patient compliance, and overcoming existing limitations. Key trends shaping the future of this field include:

Bio-Inspired and Biomimetic Nanocarriers

Researchers are developing bio-inspired nanocarriers that mimic biological systems to enhance biocompatibility and targeting efficiency. For example, exosome-like nanoparticles derived from natural cellular membranes can evade immune detection and provide prolonged circulation time, thereby improving drug delivery to retinal tissues. Additionally, peptide-modified liposomes are being engineered to increase targeting specificity toward ocular tissues (Danhier et al., 2012).

Artificial Intelligence (AI)-Driven Nanocarrier Design

The integration of artificial intelligence and machine learning with nanotechnology is enabling smarter design and optimization of nanocarriers. AI algorithms can analyze vast datasets to predict optimal formulations that enhance drug-loading efficiency, reduce toxicity, and streamline development. This approach accelerates innovation while minimizing costs and the number of experimental iterations (Wang et al., 2023).

3D Printing and Nanofabrication Technologies

Advancements in 3D printing have facilitated the development of customized ocular implants and drug delivery devices. For example, researchers have successfully created 3D-printed biodegradable ocular inserts capable of providing sustained drug release over extended periods. Furthermore, nanofabrication techniques are being refined to engineer highly tunable nanocarriers for targeted treatment of a wide range of ocular diseases (Xu et al., 2021).

Theranostic Nanocarriers

Theranostics, which combines therapeutic and diagnostic capabilities within a single nanocarrier, is an emerging innovation in ocular drug delivery. By incorporating imaging agents, these systems enable real-time monitoring and precise targeting. For example, gold nanoparticles functionalized with fluorescent markers have demonstrated effectiveness in both drug delivery and retinal imaging. Silica-coated gold nanoparticles have been used to treat choroidal neovascularization (CNV), leading to significant reductions in lesion size and improvements in visual acuity (Kim et al., 2022).

Regulatory and Commercial Advancements

As nanotechnology-based ocular therapies evolve, regulatory bodies such as the FDA and EMA are developing new frameworks to ensure safety and efficacy. Efforts are underway to streamline approval pathways for nanomedicines, facilitating faster clinical translation. Additionally, collaborations between academia, pharmaceutical companies, and regulatory agencies are fostering commercialization and wider adoption of these advanced therapies.

Conclusion

Nanotechnology-based ocular drug delivery represents a transformative advancement in ophthalmic therapeutics. By overcoming the limitations of conventional delivery methods, nanotechnology provides targeted, sustained, and personalized treatment options that are reshaping the field of ocular care. Continuous research and innovation are driving the development of sophisticated nanocarrier systems, enabling more effective and efficient drug delivery strategies. The integration of nanotechnology with emerging technologies such as artificial intelligence, biomimetic nanocarriers, and theranostic platforms holds great potential to enhance treatment precision and improve patient outcomes. Future advancements, especially in the design of biodegradable and smart nanocarriers, aim to increase patient convenience and compliance. With sustained research efforts and strategic investment, nanotechnology is positioned to deliver safer, more effective, and patient-friendly solutions for a broad spectrum of ocular diseases in the near future.

References

Ciolino, J. B., Dohlman, C. H., & Kohane, D. S. (2009). Contact lenses for drug delivery. *Seminars in Ophthalmology*, 24(3), 156–160.

Cullis, P. R., & Hope, M. J. (2017). Lipid nanoparticle systems for enabling gene therapies. *Molecular Therapy*, 25(7), 1467–1475.

Danhier, F., Ansorena, E., Silva, J. M., Coco, R., Le Breton, A., & Préat, V. (2012). PLGA-based nanoparticles: An overview of biomedical applications. *Journal of Controlled Release*, 161(2), 505–522.

del Amo, E. M., & Urtti, A. (2008). Current and future ophthalmic drug delivery systems: A shift to the posterior segment. *Drug Discovery Today*, 13(3–4), 135–143.

Drag, S., Dotiwala, F., & Upadhyay, A. K. (2023). Gene therapy for retinal degenerative diseases: Progress, challenges, and future directions. *Investigative Ophthalmology & Visual Science*, 64(7), 39.

Gaudana, R., Jwala, J., Boddu, S. H. S., & Mitra, A. K. (2009). Recent perspectives in ocular drug delivery. *Pharmaceutical Research*, 26(5), 1197–1216.

Ghate, D., & Edelhauser, H. F. (2006). Ocular drug delivery. *Expert Opinion on Drug Delivery*, 3(2), 275–287.

Gote, V., Sikder, S., Sicotte, J., & Pal, D. (2019). Ocular drug delivery: Present innovations and future challenges. *Journal of Pharmacology and Experimental Therapeutics*, 370(3), 602–624.

Guidi, L., Cascone, M. G., & Rosellini, E. (2024). Light-responsive polymeric nanoparticles for retinal drug delivery: Design cues, challenges and future perspectives. *Heliyon*, 10(5), e26616.

Han, H., Li, S., Xu, M., Zhong, Y., Fan, W., Xu, J., Zhou, T., Ji, J., Ye, J., & Yao, K. (2023). Polymer- and lipid-based nanocarriers for ocular drug delivery: Current status and future perspectives. *Advanced Drug Delivery Reviews*, 196, 114770.

Kambhampati, S. P., & Kannan, R. M. (2013). Dendrimer nanoparticles for ocular drug delivery. *Journal of Ocular Pharmacology and Therapeutics*, 29(2), 151–165.

Kim, T. Y., Mok, J. W., Hong, S. H., Jeong, S. H., Choi, H., Shin, S., Joo, C. K., & Hahn, S. K. (2022). Wireless theranostic smart contact lens for monitoring and control of intraocular pressure in glaucoma. *Nature Communications*, 13(1), Article 6855.

Kumari, A., Singla, R., Guliani, A., & Yadav, S. K. (2014). Nanoencapsulation for drug delivery. *EXCLI Journal*, 13, 265–286.

Liu, L. C., Chen, Y. H., & Lu, D. W. (2023). Overview of recent advances in nano-based ocular drug delivery. *International Journal of Molecular Sciences*, 24(20), Article 15352.

Liu, S., Jones, L., & Gu, F. X. (2012). Nanomaterials for ocular drug delivery. *Macromolecular Bioscience*, 12(5), 608–620.

Long, K., Yang, Y., Lv, W., Jiang, K., Li, Y., Lo, A. C. Y., Lam, W. C., Zhan, C., & Wang, W. (2021). Green light-triggered intraocular drug release for intravenous chemotherapy of retinoblastoma. *Advanced Science*, 8(20), Article 2101754.

Lu, H., Zhang, S., Wang, J., & Chen, Q. (2021). A review on polymer and lipid-based nanocarriers and its application to nano-pharmaceutical and food-based systems. *Frontiers in Nutrition*, 8, Article 783831.

Maharjan, P., Cho, K. H., Maharjan, A., Shin, M. C., Moon, C., & Min, K. A. (2019). Pharmaceutical challenges and perspectives in developing ophthalmic drug formulations. *Journal of Pharmaceutical Investigation*, 49(2), 215–228.

Patel, A. (2013). Ocular drug delivery systems: An overview. *World Journal of Pharmacology*, 2(2), 47–64.

Patel, S., Ryals, R. C., Weller, K. K., Pennesi, M. E., & Sahay, G. (2019). Lipid nanoparticles for delivery of messenger RNA to the back of the eye. *Journal of Controlled Release*, 303, 91–100.

Rusu, C., Dosmar, E., Walsh, J., Doyel, M., Bussett, K., Oladipupo, A., Amer, S., & Goebel, K. (2022). Targeting ocular drug delivery: An examination of local anatomy and current approaches. *Bioengineering*, 9(10), Article 504.

Suri, R., Beg, S., & Kohli, K. (2020). Target strategies for drug delivery bypassing ocular barriers. *Journal of Drug Delivery Science and Technology*, 55, 101389.

Wang, N., Zhang, Y., Wang, W., Ye, Z., Chen, H., Hu, G., & Ouyang, D. (2023). How can machine learning and multiscale modeling benefit ocular drug development? *Advanced Drug Delivery Reviews*, 196, 114772.

Wang, Y., Rajala, A., & Rajala, R. V. S. (2018). Nanoparticles as delivery vehicles for the treatment of retinal degenerative diseases. In *Advances in Experimental Medicine and Biology* (Vol. 1074, pp. 117–123). Springer New York LLC.

Xu, Q., Kambhampati, S. P., & Kannan, R. M. (2013). Nanotechnology approaches for ocular drug delivery. *Middle East African Journal of Ophthalmology*, 20(1), 26–37.

Xu, X., Awwad, S., Diaz-Gomez, L., Alvarez-Lorenzo, C., Brocchini, S., Gaisford, S., Goyanes, A., & Basit, A. W. (2021). 3D printed punctal plugs for controlled ocular drug delivery. *Pharmaceutics*, 13(9), Article 1421.

Yasukawa, T., Ogura, Y., Tabata, Y., Kimura, H., Wiedemann, P., & Honda, Y. (2004). Drug delivery systems for vitreoretinal diseases. *Progress in Retinal and Eye Research*, 23(3), 253–281.

Ye, T., Yuan, K., Zhang, W., Song, S., Chen, F., Yang, X., Wang, S., Bi, J., & Pan, W. (2013). Prodrugs incorporated into nanotechnology-based drug delivery systems for possible improvement in bioavailability of ocular drugs delivery. *Asian Journal of Pharmaceutical Sciences*, 8(4), 207–217.

Index

C

L

M

N

O

P

Q

R

S

T

U

V

W

X